Lecture Notes in Electrical Enginee

Volume 1099

The book series *Lecture Notes in Electrical Engineering* (LNEE) publishes the latest developments in Electrical Engineering—quickly, informally and in high quality. While original research reported in proceedings and monographs has traditionally formed the core of LNEE, we also encourage authors to submit books devoted to supporting student education and professional training in the various fields and applications areas of electrical engineering. The series cover classical and emerging topics concerning:

- Communication Engineering, Information Theory and Networks
- Electronics Engineering and Microelectronics
- Signal, Image and Speech Processing
- Wireless and Mobile Communication
- Circuits and Systems
- Energy Systems, Power Electronics and Electrical Machines
- Electro-optical Engineering
- Instrumentation Engineering
- Avionics Engineering
- Control Systems
- Internet-of-Things and Cybersecurity
- Biomedical Devices, MEMS and NEMS

For general information about this book series, comments or suggestions, please contact leontina.dicecco@springer.com.

To submit a proposal or request further information, please contact the Publishing Editor in your country:

China

Jasmine Dou, Editor (jasmine.dou@springer.com)

India, Japan, Rest of Asia

Swati Meherishi, Editorial Director (Swati.Meherishi@springer.com)

Southeast Asia, Australia, New Zealand

Ramesh Nath Premnath, Editor (ramesh.premnath@springernature.com)

USA, Canada

Michael Luby, Senior Editor (michael.luby@springer.com)

All other Countries

Leontina Di Cecco, Senior Editor (leontina.dicecco@springer.com)

**** This series is indexed by EI Compendex and Scopus databases. ****

Ashwani Kumar · S. N. Singh · Pradeep Kumar
Editors

Decarbonisation and Digitization of the Energy System

Proceedings of the 2nd International Conference on Smart Grid Energy Systems and Control, SGESC 2023

Springer

Editors
Ashwani Kumar
Department of Electrical Engineering
National Institute of Technology
Kurukshetra
Kurukshetra, Haryana, India

S. N. Singh
Department of Electrical Engineering
Indian Institute of Technology Kanpur
Kanpur, Uttar Pradesh, India

Pradeep Kumar
Department of Electrical Engineering
National Institute of Technology
Kurukshetra
Kurukshetra, Haryana, India

ISSN 1876-1100 ISSN 1876-1119 (electronic)
Lecture Notes in Electrical Engineering
ISBN 978-981-99-7629-4 ISBN 978-981-99-7630-0 (eBook)
https://doi.org/10.1007/978-981-99-7630-0

This Springer imprint is published by the registered company Springer Nature Singapore Pte Ltd.
The registered company address is: 152 Beach Road, #21-01/04 Gateway East, Singapore 189721, Singapore

Paper in this product is recyclable.

Preface

The traditional electricity generation system deals with the generation, transmission and utilization of power from large centralized generating stations through grids. The electricity is distributed at different voltage levels to the end customers. With the implications of the fossil fuel-based generation, there is increasing penetration of distributed energy resources (DERs) into the system. With the main share of solar PV and wind power, the requirement of fast energy storage systems has become essential with technological disruptions in the energy sector. It may pose many challenges and have created opportunities for both existing and new stakeholders in the emerging electricity markets. The hierarchy of electrical energy distribution systems has changed to give support to the prosumers and offered adoption of potential role to distributed ledger technologies (DLTs). This has necessitated the new digital technology and distributed generation and renewable energy sources with the influx of storage devices to take care of the demand flexibility in operation and management of local grids with the locally supported ancillary services. The new technology with DERS and support services has given incentives to the innovation in electricity market design, its operation and dispatch management, with the real-time operation of micro or mini power resources. The new technologies have given birth to new platform models and are changing the landscape of current facilities and electricity business.

Thus, the power grid has become very vast and complex in nature and operation of the power system requires proper security, interoperability, and cost-effectiveness. The new technology and innovation with the integration of RES, smart devices, fast and secure communication for perfect coordination among devices and integration of hardware and software demand quick decisions with advanced sensing technologies and control techniques, capturing and analyzing data regarding power usage, delivery and generation in practically real-time. The information technologies with bi-directional communication and electricity flow have enabled both utilities and customers to monitor, predict and manage energy usage and environmental sustainability.

With increased interconnection of variable renewable sources generation, mainly solar photovoltaic and wind that may be on-land and offshore wind and is replacing conventional generation of power to hybrid in nature. Thus, there is a need to address the issues related to the impact of RES in electrical energy sector that is transitioning toward a decentralized and digital framework-based operating model. Thus, the decentralization in distributed energy services and need of energy decarbonization has promoted the electricity sector due to the continuous influx of DER and emergence of many players in the electricity market operation. These electric transportation networks are responsible for the generation-transmission-distribution power chain and may face the challenging tasks of digitizing and integrating unmanageable and large-scale giga-watt generation with wind and photovoltaic, as well as integrating distributed resources and electric vehicles and charging stations.

Therefore, control mechanisms will have to be modernized and in addition, electricity systems in the future will have to consider other actors, such as smart transport systems and services, which will be important and integral part of the quick real-time boosters of these systems. This transition is part of the 5 Ds: decentralization, digitalization, decarbonization, democratization, and decreasing consumption which are the drivers of the current electrical energy transition. This energy transition is affecting services related to distribution of energy and also helping the transition toward a competitive and demand-side driven based model that may open up many opportunities for service innovation for better operation and management. Another energy revolution is the "decentralization of energy sources," i.e., as wind, solar and storage technologies have more competitive investment costs and these will become more affordable, and therefore, there will be greater penetration of distributed generation in the coming future and the transition to new operation of the power sector may have many implications and multiple challenges at different phases of operation, management including the technological drivers and regulatory framework.

Data-Driven Analytics in Energy Systems

Artificial Intelligence and Machine Learning, IoT and ICT for Smart Grids, Data Analytics, Energy Efficiency, Big Data, Digital Twin, Energy Management, Demand Response/Demand Side Management, State Estimation Tools, Forecasting in Energy Systems, System Modelling and Simulation, Co-Simulation and Real-Time Simulation, Cloud Computation/Edge Computation, Data Acquisition and Monitoring, Data Management, Distributed Optimization, Modern Heuristic Optimization, High-Performance Computing for Grid Analysis.

Kurukshetra, India — Ashwani Kumar
Kanpur, India — S. N. Singh
Kurukshetra, India — Pradeep Kumar

Contents

Editors and Contributors

About the Editors

Dr. Ashwani Kumar did his B.Tech. in Electrical Engineering from Govind Vallabh Pant University of Agriculture and Technology, Pant Nagar, Uttarakhand in 1988 with honors, did his M.Tech. in Power Systems from Punjab University Chandigarh in 1993 with honors, and his Ph.D. degree from IIT Kanpur in 2003 in the area of power systems. He holds a post-doctoral position in 2008 at Tennessee Tech. University, USA. He is currently Professor in the Department of Electrical Engineering at NIT Kurukshetra and coordinator in School of Renewable Energy and Efficiency, SREE. He has an interest in the areas of power system operation, Power Systems Restructuring, Distributed Generation and Renewable Energy Systems, FACTS applications to Power Systems, Demand Side Management, and Smart grid/microgrid technology. He has completed sponsored projects awarded by DST and a project during his Post-doc. Program at Tennessee Technological University USA. He has guided 14 Ph.D. and 55 M.Tech. scholars. He has 120 publications to his credit in archival journals of repute and 165 publications in IEEE Confs and Scopus-indexed confs. He has received POSOCO Power System awards in a joint venture of Power Grid and FITT IIT Delhi for guiding the best M.Tech. thesis among the top 25 in India consecutively in the years 2013–2018. He is a recipient of best paper awards from the National System Conference NSC NIT Surathkal, Literati award by the *International Journal of Emerald*

IJESM. He has organized IEEE conferences and International conferences on Smart Grid, Energy Systems and Control, SGESC-2021, and 2023. He visited the USA, Italy, Australia, Netherlands, University of Malaysia for academic activities. 5 Ph.D. scholars and 3 M.Tech. scholars are doing a thesis under his guidance currently. He is involved with various administrative activities in the Institute and is presently coordinator of the School of Renewable Energy and Efficiency, SREE. He is a fellow member of the Institution of Engineers, a Fellow IETE, India, a Senior member of IEEE, life member of ISTE.

Prof. Sri Niwas Singh, FIEEE (USA), FNAE, FIET (UK), IE(I), IETE(I), FAvH Director, IIITM Gwalior

Honour: Ph.D. Electrical (Power Systems), Indian Institute of Technology, Kanpur

Experience: (At IIT Kanpur, 2002 onward) Chairman, JEE (Advance) 2016 Head, Electrical Engineering Department (Feb 2017 to April 2017) Member, Green Cell, IIT Kanpur (2014 to Dec 2015) Member, Environmental Advisory Committee, IIT Kanpur (2014 to Dec 2015) Professor

Area of Interest: Electricity Market & Economics HVDC Transmission & FACTS Technology Wind & Solar Forecasting Power Power System Operation & Control Active Distributed System & Smart Grid AI Applications in Power Systems. Power Quality

Office Phone: 0751- 2449801

Email: snsingh@iiitm.ac.in

Website: https://home.iitk.ac.in/~snsingh/

Prof. S. N. Singh obtained his M. Tech. and Ph. D. in Electrical Engineering from the Indian Institute of Technology Kanpur in 1989 and 1995. Presently, Prof. Singh is Director, Atal Bihari Bajpayee Indian Institute of Information Technology and Management Gwalior (MP), India (on leave from Professor (HAG), Department of Electrical Engineering, Indian Institute of Technology Kanpur, India). Before joining IIT Kanpur as an Associate Professor, Dr. Singh had worked with UP State Electricity Board as an Assistant Engineer from 1988 to 1996, with Roorkee University (now IIT Roorkee) as an Assistant Professor from 1996 to 2000, and with the Asian Institute of Technology, Bangkok, Thailand as an Assistant Professor from 2001 to 2002. He was the Vice Chancellor of Madan Mohan

Malviya University of Technology Gorakhpur from April 2017 to July 2020. Dr. Singh received several awards including the Young Engineer Award 2000 of the Indian National Academy of Engineering (INAE), the Khosla Research Award of IIT Roorkee, and the Young Engineer Award of CBIP New Delhi (India), 1996. Prof. Singh is the recipient of the Humboldt Fellowship of Germany (2005, 2007) and the Ottomonsted Fellowship of Denmark (2009–10). Prof. Singh became the first Asian to receive the 2013 IEEE Educational Activity Board Meritorious Achievement Award in Continuing Education. He is also the recipient of the INAE Outstanding Teacher Award 2016 and IEEE R10 region (Asia-Pacific) Outstanding Volunteer Award 2016. Dr. Singh is appointed as IEEE Distinguished Lecturer of the Power & Energy Society in 2019 and Industry Application Society from 2019–2021. He is also the recipient of the NPSC 2020 Academic Excellence Award and the 2021 IEEE Industry Application Society (IAS) Outstanding Educator/Mentor Award.

His research interests include power system restructuring, FACTS, power system optimization & control, security analysis, wind power, etc. Prof. Singh has published more than 500 papers (h-index=57, Citation=12k+) in International/National journals/ conferences and supervised 40 Ph.D.s (8 Ph.D.s under progress). He has also written 30 book chapters, 8 Edited books, and 2 textbooks; one on Electric Power Generation, Transmission and Distribution, and the second is Basic Electrical Engineering, published by PHI, India. Prof. Singh has completed three dozen technical projects in India and abroad. His two NPTEL (YouTube) video lectures on HVDC Transmission and Power System Operation & Control are very popular. Prof. Singh was Chairman, IEEE UP Section for 2013 & 2014, IEEE R10 (Asia-Pacific) Conference & Technical Seminar Coordinator 2015–18 and R10 Vice-Chair, Technical Activities (2019–2020). Presently Prof. Singh is Immediate Past Chairman of IEEE, India Council. Dr Singh is Fellows of IEEE (USA), IET (UK), INAE, IE(I), IETE, AvH.

Dr. Pradeep Kumar is Assistant Professor in the Department of Electrical Engineering at the National Institute of Technology (NIT) in Kurukshetra, India. His areas of research are artificial intelligence Applications in Power Systems, Smart Grid, Substation Design especially, Earthing and Bonding based on IEEE-80 2000 (with ETAP), Direct Stroke Lightning Protection using IEEE 998, and Insulation coordination, Transmission Line Design and Planning, Power Quality.

Contributors

S. Anand Electrical and Electronics Engineering, Nitte Meenakshi Institute of Technology Yelahanka, Bangalore, India

Baidyanath Bag Department of Electrical Engineering, National Institute of Technology, Raipur, Chhattisgarh, India

Rajnish Bhasker Department of Electrical Engineering, VBSPU, Jaunpur, India

Ambati Bhimaraju Sardar Vallabhbhai National Institute of Technology Surat, Surat, Gujarat, India

Deepak Bhonsle C. K. Pithawla College of Engineering and Technology, Surat, Gujarat, India

Bharat Bhushan Department of Electrical Engineering, Delhi Technological University, New Delhi, India

Saurabh Chanana National Institute of Technology, Kurukshetra, India

Tamal Chatterjee Department of Electrical Engineering, National Institute of Technology, Raipur, Chhattisgarh, India

Dharminder Chaudhary Department of Computer Science and Engineering, Amrita School of Computing, Amrita Vishwa Vidyapeetham, Chennai, India

B. Datta Department of Electrical Engineering, National Institute of Technology, Arunachal Pradesh, India

Mrinal Kanti Dey National Institute of Technology, Kurukshetra, India

Ashutosh Dixit Department of Electrical Engineering, Graphic Era Deemed to be University, Dehradun, Uttarakhand, India

M. S. P. Durgarao Department of Computer Science and Engineering, Amrita School of Computing, Amrita Vishwa Vidyapeetham, Chennai, India

Abhishek Gautam National Institute of Technology Hamirpur, H.P, India

Sourabh Ghosh Motilal Nehru National Institute of Technology Allahabad, Prayagraj, India

Srikanth Gollapudi NIT, Jamshedpur, India

S. K. Gupta Department of Electrical Engineering, Chandigarh University, Mohali, India

Sandeep Gupta Department of Electrical Engineering, Graphic Era (Deemed to be University), Dehradun, Uttarakhand, India

Surisetty Mahesh Chandra Gupta Department of Computer Science and Engineering, Amrita School of Computing, Chennai, India

Degala Isaac National Institute of Technology Kurukshetra, Thanesar, India

G. Jahnavi Osmania University, Hyderabad, India

Trapti Jain IIT Indore, Indore, India

Sandeep Jangir NRSC, Indian Space Research Organisation, Hyderabad, India

Sandeep Kakran NIT Kurukshetra, Haryana, India

Amit Kumar National Institute of Technology Kurukshetra, Thanesar, India

Ashwani Kumar Department of Electrical Engineering, National Institute of Technology Kurukshetra, Kurukshetra, Haryana, India;
National Institute of Technology Hamirpur, H.P, India

Gautam Kumar Department of Electrical Engineering, National Institute of Technology, Jamshedpur, Jharkhand, India

Narendra Kumar Department of Electrical Engineering, Delhi Technological University, New Delhi, India

Pradeep Kumar National Institute of Technology Kurukshetra, Kurukshetra, India

Rajesh Kumar National Institute of Technology, Kurukshetra, Haryana, India

Sanjay Kumar Department of Electrical Engineering, National Institute of Technology, Jamshedpur, Jharkhand, India

Sumeet Kumar National Institute of Technology, Kurukshetra, Haryana, India

T. Kranthi Kumar Department of Science and Humanities, Sreenidhi Institute of Science and Technology, Hyderabad, India

Meet Kumari Department of Electronics and Communication Engineering, Chandigarh University, Mohali, India

Nibha Kumari Motilal Nehru National Institute of Technology Allahabad, Prayagraj, India

Pranjali Kumari Department of Electrical Engineering, National Institute of Technology, Jamshedpur, Jharkhand, India

Shomi Kumari Sardar Vallabhbhai National Institute of Technology Surat, Surat, Gujarat, India

Ashokkumar Lakum Lukhdhirji Engineering College, Morbi, Gujarat, India

Aeidapu Mahesh Sardar Vallabhbhai National Institute of Technology Surat, Surat, Gujarat, India

Anchal Maurya Department of Electrical Engineering, NIT Kurukshetra, Kurukshetra, Haryana, India

Siddhant Singh Maurya IIT Indore, Indore, India

Srilatha Namilakonda Department of Electrical Engineering, Osmania University, Hyderabad, India

Joshi Sukhdev Nirbheram Sardar Vallabhbhai National Institute of Technology Surat, Surat, Gujarat, India

Mahesh Pandya Lukhdhirji Engineering College, Morbi, Gujarat, India

Pasupuleti Gnaneshwar Parikshith Department of Computer Science and Engineering, Amrita School of Computing, Amrita Vishwa Vidyapeetham, Chennai, India

Ajay Kumar Prajapat NIT Kurukshetra, Haryana, India

Dhruv Rajput KIET Group of Institutions, Delhi-NCR, Ghaziabad, India

Kartik Saini Department of Electrical Engineering, Delhi Technological University, New Delhi, India

Parvesh Saini Department of Electrical Engineering, Graphic Era Deemed to be University, Dehradun, Uttarakhand, India

Nitin Kumar Saxena KIET Group of Institutions, Delhi-NCR, Ghaziabad, India; Electrical and Computer Engineering Department, University of Denver, Denver, United States

M. Shobha Department of Electrical Engineering, National Institute of Technology, Arunachal Pradesh, India

Jaya Shukla Department of Electrical Engineering, VBSPU, Jaunpur, India

Asheesh K. Singh Motilal Nehru National Institute of Technology Allahabad, Prayagraj, India

Ashutosh Singh IIT (ISM), Dhanbad, India;
SOET, IGNOU, New Delhi, India

Bharat Singh PhD Scholar, Electrical Engineering Department, NIT Kurukshetra, Kurukshetra, Haryana, India

Navneet K. Singh Motilal Nehru National Institute of Technology Allahabad, Prayagraj, India

S. N. Singh Atal Bihari Vajpayee-Indian Institute of Information Technology and Management, Gwalior, India

Soumyendra Singh Department of Computer Science and Engineering, Amrita School of Computing, Chennai, India

Tanmay Soni Department of Computer Science and Engineering, Amrita School of Computing, Chennai, India

A. Sriker Osmania University, Hyderabad, India

Abhinav Srivastava KIET Group of Institutions, Delhi-NCR, Ghaziabad, India

R. K. Srivastava Department of Electrical Engineering, IIT BHU, Varanasi, India

Padmanabh Thakur Department of Electrical Engineering, Graphic Era Deemed to be University, Dehradun, Uttarakhand, India

Yarragoti Ravi Theja Department of Computer Science and Engineering, Amrita School of Computing, Amrita Vishwa Vidyapeetham, Chennai, India

Sukriti Tiwari National Institute of Technology Hamirpur, H.P, India

Praveen Kumar Tyagi KIET Group of Institutions, Delhi-NCR, Ghaziabad, India

Amod C. Umarikar IIT Indore, Indore, India

Tarun Varshney Electrical Electronics and Communication Engineering, SET, Sharda University, Greater Noida, India

Dev Verma KIET Group of Institutions, Delhi-NCR, Ghaziabad, India

Priyanka Yadav Department of Electrical Engineering, Graphic Era Deemed to be University, Dehradun, Uttarakhand, India

Subhash Yadav JSS Academy of Technical Education Noida, Noida, India

G. Yesuratnam Osmania University, Hyderabad, India

Optimal Energy Scheduling and Feasibility Analysis in Microgrid Considering the Hospital Load Model with Isolated Grid

Bharat Singh and Ashwani Kumar

Abstract In the modern power system, the intermittent nature of renewable energy sources plays a vital role in meeting the load demand. In this context, the contribution of energy storage devices has also been significant. The optimal combination of hybrid renewable energy sources has become a fundamental analysis for distribution network operators (DNO). The main contribution of the research work is: (i) obtaining the optimal generation scheduling of the micro-combined heat and power (CHP), Solar photovoltaic (PV), wind turbine (WT) and battery energy storage (BESS); (ii) economic dispatch analysis of Microgrid; (iii) techno-economic analysis of heat units; (iv) the net present cost (NPC) has been minimized; (v) the feasibility analysis has been determined to combine the energy dispatch and techno-economic analysis. The concerned work has been implemented for obtaining the feasibility analysis of the realistic hospital load model of Vayusenabad, New Delhi, India, using the Hybrid Optimization of GAMS and Multiple Energy Resources (HOMER) software with MATLAB interfacing.

Keywords Battery energy storage · HOMER · Renewable energy sources · Tecno-economic analysis

1 Introduction

At the small-scale Microgrid level, it is critical to determine the optimal energy storage scheduling incorporated with hybrid renewable energy sources. The critical load, like viz hospital load, military base camps, information, security and educational research and development sector, etc., have needed the uninterrupted power supply. Therefore, in this research paper, the hospital load of Delhi, NCR region of India, has been investigated for feasibility analysis on the techno-economic ground.

B. Singh (✉) · A. Kumar
PhD Scholar, Electrical Engineering Department, NIT Kurukshetra, Kurukshetra 136119, Haryana, India
e-mail: bharat_6180045@nitkkr.ac.in

A. Kumar et al. (eds.), *Decarbonisation and Digitization of the Energy System*, Lecture Notes in Electrical Engineering 1099, https://doi.org/10.1007/978-981-99-7630-0_1

As per the Microgrid analysis's techno-economic analysis, much literature is available. The solar PV with an optimal size of battery energy has been determined in literature [1] to solve the techno-economic study for small load [2] analysis. Moreover, the remote area-wise analysis has been considered in the literature [3] as the Islanding mode of operation. The combined heat and power (CHP), micro-turbine (MT), photovoltaic (PV), fuel cell (FC), small wind turbine (WT), and battery energy storage system (BESS)-based hybrid renewable energy sources (HRES) were taken into consideration for the unit commitment problem along with the network constants [4]. The solar PV output has been determined based on the techno-economic analysis [5]. In this context, the network-based constraints have also been considered in [6] for the size determination of energy storage.

Apart from the unit commitment problem, the optimal dispatch analysis with the feasibility analysis plays a crucial role. In this context, many issues have come into the picture, viz. optimal scheduling of HRES (PV/WT/MT-CHP/FC) along with BESS [7], optimal dispatch of BESS [8], the optimal size of BESS [9], network constraints [10], power loss minimization [11], voltage control with reactive power dispatch [12], etc.

The demand side management consists of a demand response program (DRP) with the hospital industrial and domestic load with electric vehicle (EVs) load [13] for the impact of **Covid'19**. The author has considered the feasibility analysis in the literature [14] using HRES, Electric vehicles (EVs), and demand side management. In this context, the EV load with the distribution network configuration has been considered for loss minimization [15]. Moreover, the impact of BESS has been considered for the distribution system in literature considering the DRP [16]. The benefit maximization-based approach was introduced in the literature [17] for optimal energy scheduling with the EVs and BESS. The energy-saving-based analysis was taken into account for the Microgrid feasibility analysis [18]. The voltage-based analysis is also a critical issue in the Microgrid feasibility analysis for power quality and reactive power balance issues [19]. Therefore the volt/VAr analysis has been considered in the literature [20] for voltage control and power loss management.

As per the above literature survey, most authors have considered the socioeconomic-based analysis, but very few considered the feasibility analysis. Others did not consider hybrid renewable energy management based on techno-economic and feasibility-based analyses. Therefore, the main contribution of this research paper is to minimize the net present cost (NPC). In this scenario, the feasibility-based analysis of a realistic system load has been carried out to combine energy management and techno-economic analysis.

Moreover, the realistic hospital load of the Delhi NCR region of India has been taken for the optimal cost–benefit-based analysis. The rest of the paper consists of as follows: (i) the problem formulation has been represented in Sect. 2; (ii) the Microgrid architecture model with hybrid energy sources has been represented in Sect. 3; (iii) the results have been discussed in Sect. 4; (iv) finally, in Sect. 5, the research work is concluded.

2 Problem Formulation

In this section, the mathematical formulation has been done for the analysis. The mathematical formulation for the various energy sources has also been taken for the analysis as follows.

2.1 Objective Function Formulation

In this paper, the single objective function has been considered to minimize the net present cost (NPC) as following Eq. (1).

$$NPC = \text{Cost}_{\text{annual}}^{\text{Total}} \frac{\{rt(1+rt)^{yr} - 1\}}{rt(1+rt)^{yr}} \tag{1}$$

where, $\text{Cost}_{\text{annual}}^{\text{Total}}$ is the total annual cost of the MG, which consists of

$$rt = \frac{i - f}{1 + f}. \tag{2}$$

The rt is the rate of interest with ith period along with affiliation of parameter f.

$$\text{Cost}_{\text{annual}}^{\text{Total}} = \text{Cost}_{PV} + \text{Cost}_{WT} + \text{Cost}_{CHP} + \text{Cost}_{Conv} + \text{Cost}_{BESS} \tag{3}$$

where the cost values are represented for PV, WT, CHP, and BESS.

2.2 Power Balance

The active power generation of the system is represented in Eq. (4).

$$P_{\text{gen}}^{k} + Pdis_{\text{BESS}}^{k} = Pd_{\text{Load}}^{k} + Pch_{\text{BESS}}^{k} + P_{\text{Loss}}^{k} \tag{4}$$

where Pd_{Load}^{k} is the load demand for the kth hour, $Pdis_{\text{BESS}}^{k}$ and Pch_{BESS}^{k} are discharging and charging power of BESS, whereas, P_{Loss}^{k} is the power loss and the total generated power (P_{gen}^{k}) represented as,

$$P_{\text{gen}}^{k} = Pg_{PV}^{k} + P_{WT}^{k} + P_{CHP}^{k} \tag{5}$$

where the generated power is given for PV, CHP and WT at kth time duration.

2.3 Modelling of the Energy Sources

In this section, various energy sources have been modelled for the analysis. The solar PV, wind turbine (WT), CHP-MT units, and BESS cost-based model are represented.

Solar PV model

The mathematical model of the solar PV is represented in Eq. (6)

$$P_{\text{solar}}(I_\beta) = N_{PV} \cdot P_{\text{rated}}^{PV} \frac{G}{G0} \cdot \{1 - T_c(T_A - 25)\} \cdot \eta_{\text{inv}} \eta_{rl} \tag{6}$$

In Eq. (6), the solar PV output is represented by the ambient temperature (T_A) and efficiency of solar PV [12]. The PV cell temperature is modelled in Eq. (7).

$$T_C = T_A\{1 - \frac{\eta_{mppt}}{0.9}\} \frac{G\_NOCT}{G_0\_NOCT} \cdot \{T_{C\_NOCT} - T_{A\_NOCT}\} \tag{7}$$

where, G_NOCT and G_0_NOCT are solar irradiation at standard test and at nominal operated PV cell temperature, respectively. T_{C_NOCT} and T_{A_NOCT} are the nominal temperature for solar and ambient, respectively, as represented in Fig. 1.

Wind turbine

In this paper, the quadratic wind turbine model has been taken for analysis as given in Eq. (8).

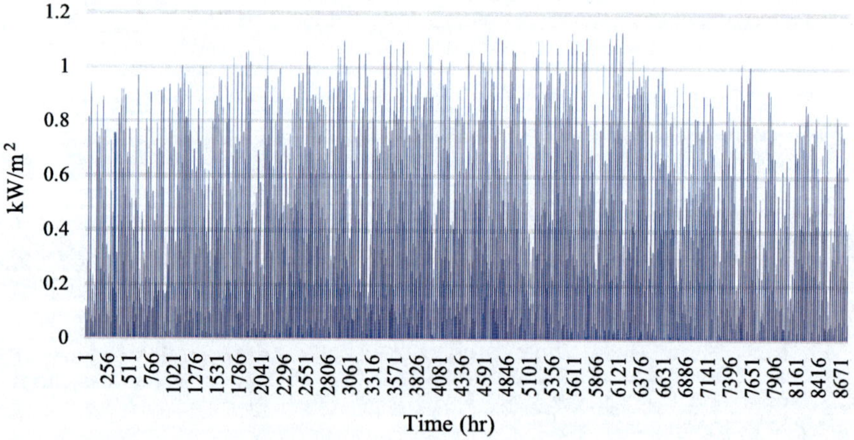

Fig. 1 Annual global solar irradiation for Vayusenabad, New Delhi, India (28°30.8′ N, 77°14.7′ E)

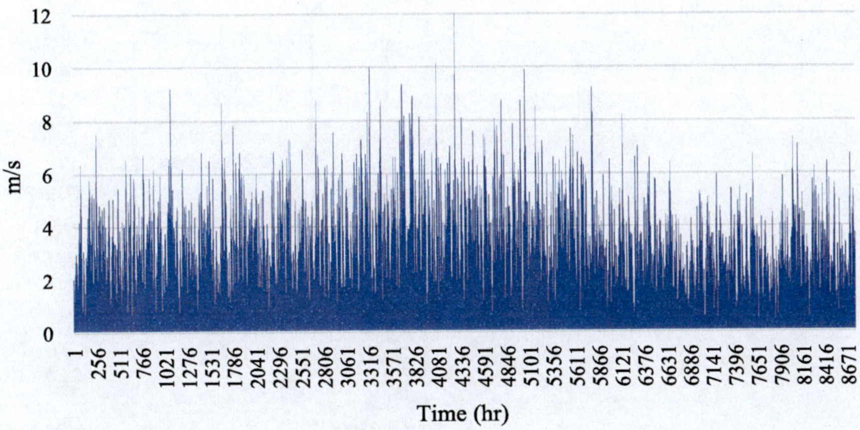

Fig. 2 Annual wind speed of Delhi N.C.R., India

$$P_{wind} = \begin{cases} \mathrm{P_{rated}} \cdot \left((v - v_{in})^3 / (v_r - v_{in})^3\right); & v_{\mathrm{cutin}} \le v \le v_r \\ \mathrm{P_{rated}}; & v_r \le v \le v_{\mathrm{out}} \\ 0; & v > v_{\mathrm{out}} \text{ and } \mathrm{v} < \mathrm{v_{cutin}} \end{cases} \quad (8)$$

The annual wind speed curve is depicted in Fig. 2.

CHP cost model

In this paper, the CHP and MT-based model has been carried out for analysis [13]. The cost equation of CHP and MT-based generation has been carried out for the analysis of Eq. (9).

$$\mathrm{Cost}_{CHP} = \left(\frac{P_{\mathrm{gen}}^{CHP}}{\eta_{CHP}}\right) Csu_{\mathrm{fuel\_}CHP} \quad (9)$$

Battery Energy Storage System (BESS)

The mathematical modelling of the energy storage system has been considered in this paper for analysis [14]. The BESS's charging and discharging power have been determined to power the energy requirements in the Microgrid.

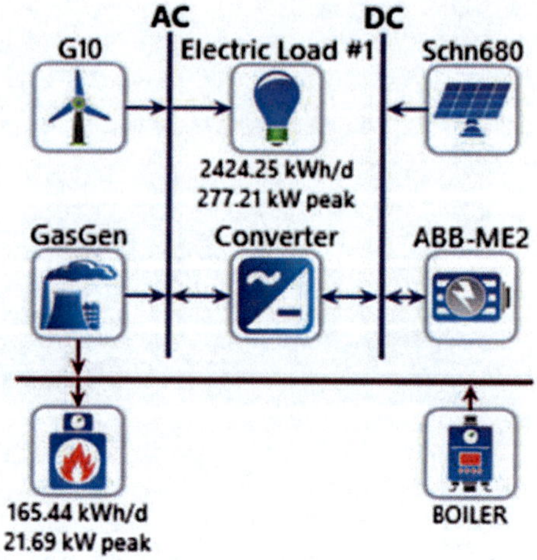

Fig. 3 Hybrid model with load using the HOMER

3 Islanding Mode for Microgrid Architecture

In this mode of operation, the Microgrid has been disconnected from the upcoming grid. Therefore, the grid is disconnected from the main distribution substation, and distributed generation units maintain the supply. The thermal CHP units with boiler, WT, PV, and BESS with converter have been taken for analysis, as shown in Fig. 3.

This paper has analysed the hospital load with the heat load model. Therefore, the CHP unit has been used to meet the electric and thermal loads simultaneously. In Fig. 3, the hybrid energy sources with BESS consider the hospital load having the thermal demand. The total daily load demand obtained is 2424.25 kWh/day and 277.21 kW per week. The heat load demand obtained is 165.44 kWh/day, having a 21.69 kW peak demand.

4 Results and Discussion

In this section, the results obtained have been discussed. Furthermore, the hybrid energy sources of CHP, MT, PV, and WT, along with BESS in the Islanding mode of operation, have been considered for the analysis as follows.

Table 1 The cost-based analysis of various energy source combinations

Source	Parameters	Configurations				
		PV + CHP – MT + BESS	PV + WT + CHP – MT + BESS	PV + BESS	PV + WT + BESS	WT + CHP MT + BESS
Cost	NPC ($)	1,438,786	1,543,121	1,700,645	1,737,624	1,040,000
	COE ($)	0.12105	0.130729	0.148	0.150686	9.918389
	Operating cost ($/yr)	30,540.55	45,923.64	25,854.15	26,011.23	1,202,962
	Initial capital ($)	1,043,972	949,442.3	1,366,415	1,401,364	88,300,000
System	Renewable fraction (%)	73.87	33.66	93.32	93.32	54.52

4.1 *Techno-Economic Analysis*

In this section, the techno-economic-based analysis has been determined for the HREs. The total cost of the system ($1,438,786.00), and the Levelized Cost of Energy (0.12105 $/kWh), are obtained for the combination of PV, CHP, and BESS.

Moreover, the renewable penetration of solar PV obtained is 72.87%, and the remaining is 29.23% by CHP and BESS. The total capital cost ($1,040,000), the operating cost ($168,289), the replacement cost ($58,481), and the salvage cost (–$7,088) are obtained for the optimal system configuration of PV + CHP + BESS. It is mentioned in Table 1 that the optimal combination of PV + CHP + BESS is accepted among the other five combinations of energy sources. Therefore, the PV + MT-CHP + BESS is the optimal solution for the Islanding mode of operation.

4.2 *Outputs of Energy Sources*

In this subsection, the output of energy sources has been discussed.

Solar PV

In this scenario, the PV model of Schneider Conext-CoreXC, having a power rating of 680kW, has been used. The output for the solar PV system has been obtained for the annual demand, as shown in Fig. 4.

The maximum rating of solar PV is 680 kW, 12 V, with each solar panel parallelly connected with four numbers of columns and ten rows.

CHP unit

In Fig. 5, the CHP-MT units' yearly power output is shown.

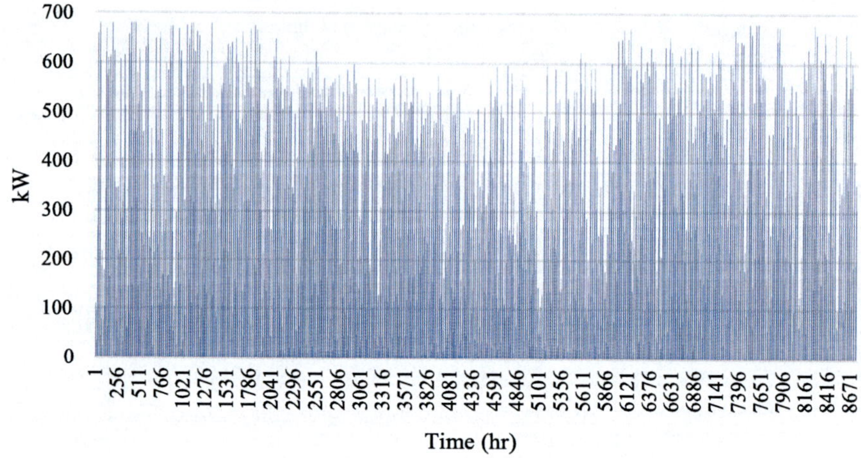

Fig. 4 Annual power output for the Schneider Conext-CoreXC 680 kW with Generic PV Power Output

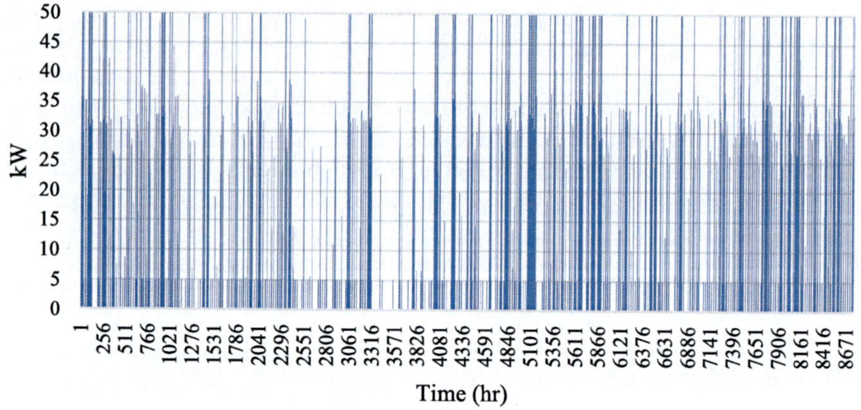

Fig. 5 CHP with MT annual power output

The MT with CHP of rated (100 kW), mean thermal output (83 kW), minimum thermal output (20.4 kW), and maximum thermal output (70.7 kW) is taken for the analysis. The annual production of MTCHP units obtained is 72660 kWh, which is 5.54% of the total output.

Battery energy storage system

The state of charge (SOC) percentage of BESS is depicted in Fig. 6. The BESS (ABB PS- BatME 2) is configurated as the nominal voltage is 3.68 V, nominal capacity is 0.346 KWh, capacity is 94 Ah, efficiency is 95%, maximum charging current is 47A, and discharging current is 47A, is considered for the analysis.

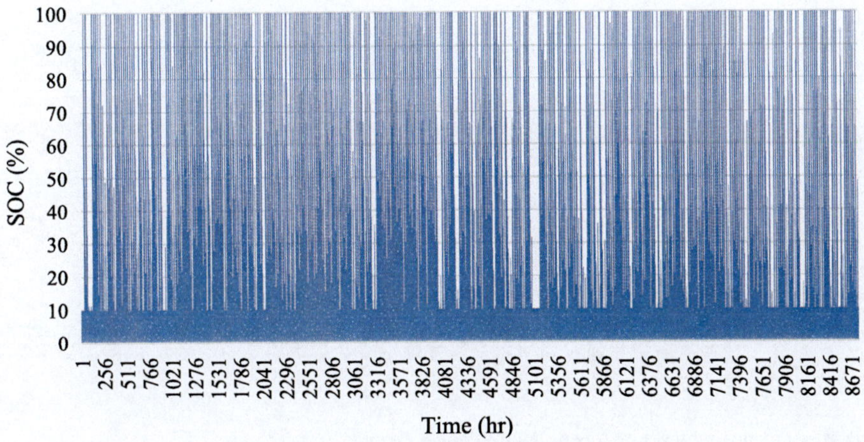

Fig. 6 State of Charge (SOC) for BESS ABB PS-BatME2 State of Charge

The annual discharging and charging power output of BESS are shown in Fig. 7.

Wind turbine power output

In Fig. 8, the annual wind power generation is shown.

The annual power output from the small WT-based unit is represents.

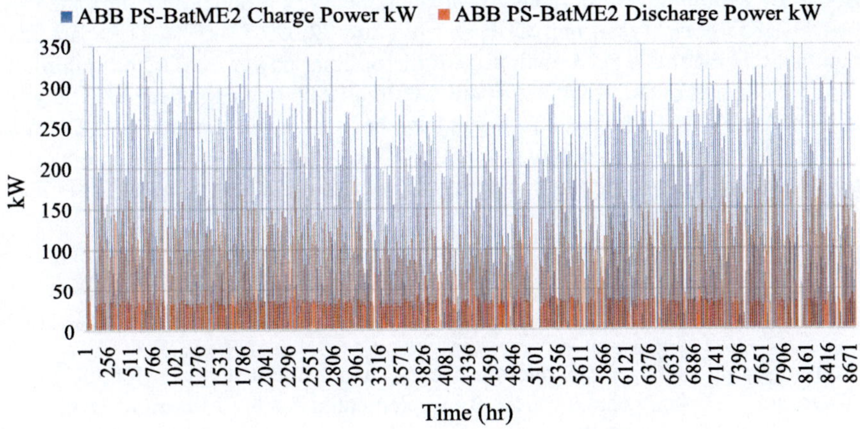

Fig. 7 Hourly charging and discharging power output profile of BESS

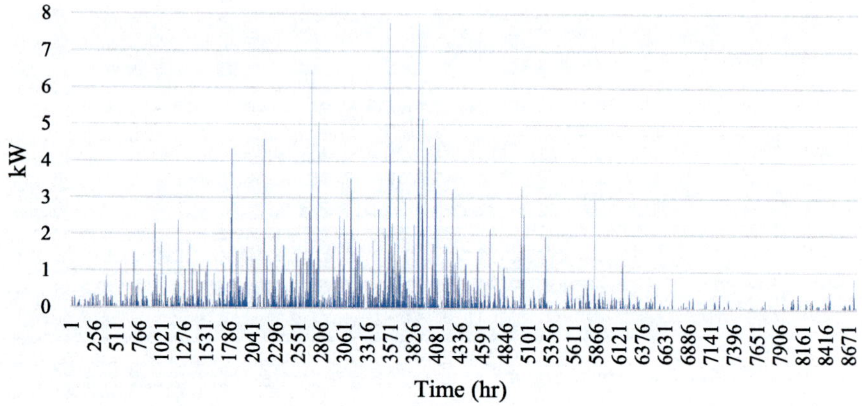

Fig. 8 Annual wind turbine Generic 10 kW Power Output

4.3 *Power Output Summary*

This section shows power generation and various combinations in Table 2. The optimal schedule has been obtained for the combination of PV + CHP-MT + BESS, having the minimum NPC (Sect. 4.1). The load following (LF) dispatch strategy (Singh and Sharma 2022a) has been implemented in this combination. The power output of PV is 680 kW, MT and CHP is 50 kW, and the maximum number of BESS (1kWh) of 1980 is given in Table 1. The annual energy production contributed by PV is 1238589 kWh/year, and CHP and MT are 72,659.73 kWh. There are five combinations that have been analyzed for the optimal power output and minimum cost of the system. The solar PV, small CHP and MT units for heat load supply, BESS of 1 kWh of each, and converters have been taken for the analysis. The fuel consumption of the boiler obtained is 4939.007 L. The capital cost, the production cost of energy, and the operation and maintenance cost are also given in Table 2.

5 Conclusions

In this research paper, the economic energy management analysis for the minimization of NPC has been successfully implemented for the Microgrid level. The minimum NPC has been obtained for the best combination of CHP − MT + PV + BESS in the off-grid mode of operation. The annual demand for hospital load has been analyzed with zero unmet power. The size of various generating units has been achieved successfully. Moreover, the results are highlighted in Tables 1 and 2, considering the optimal combination of energy sources and minimum NPC. Therefore, this research article represents the feasible and optimal analysis of various energy source

Table 2 Power output for the hybrid energy source generation

Sources	Parameters	Configurations				
		PV + CHPMT + BESS	PV + WT + CHPMT + BESS	PV + BESS	PV + WT + BESS	WT + CHPMT + BESS
Dispatch	Dispatch strategy	LF	CC	CC	CC	CC
System	Renewable fraction (%)	73.87	33.66	93.32	93.32	54.52
MT CHP	Architecture (kW)	50	50			50
	Hours	3049	6345			3551
	Production (kWh)	72,659.73	234,835.1			144,788.7
	Fuel (mÂ^3)	28,693.82	83,964.66			50,866.22
	O&M cost ($/yr)	1524.5	3172.5			1775.5
	Fuel cost ($/yr)	8608.147	25,189.4			15,259.87
PV Schn 680	Architecture (kW)	680.08	680.08	1000	1000	
	MPPT (kW)	680.08	680.08	680.08	680.08	
	Capital cost ($)	612,072	612,072	900,000	900,000	
	Production (kWh/yr)	1,238,589	1,238,589	1,700,909	1,700,909	
WT G10	Architecture		1		1	1170
	G10/Capital cost ($)		35,000		35,000	4.10E + 07
	G10/ Production (kWh/yr)		1241.89		1241.89	1,453,011
	G10/O&M Cost ($)		120		120	140,400
BESS ABB-ME2	Architecture	1980	792	2178	2178	300,960
	Autonomy (hr)	6.102627	2.441051	6.71289	6.71289	927.5994
	Annual throughput (kWh/yr)	190,126.3	105,725.9	210,198.5	209,901.5	417,776.8
	Operating hours (hours)	0	0	0	0	0

(continued)

Table 2 (continued)

Sources	Parameters	Configurations				
		PV + CHPMT + BESS	PV + WT + CHPMT + BESS	PV + BESS	PV + WT + BESS	WT + CHPMT + BESS
	Nominal capacity (kWh)	684.9216	273.9686	753.4138	753.4138	104,108.1
	Usable nominal capacity (kWh)	616.4294	246.5718	678.0724	678.0724	93,697.27
Converter	Architecture (kW)	245.6181	430.2813	424.5965	424.4248	1569.186
	Rectifier mean output (kW)	0	7.882064	0	0	49.0762
	Inverter mean output (kW)	89.85439	79.27724	96.3371	96.21582	44.15965
Boiler	Fuel consumption (L)	4939.007	2113.598	7219.751	7219.751	4328.417

combinations for the small hospital load model (Vayusenabad, New Delhi, India (28°30.8′ N, 77°14.7′ E) in the off-grid mode.

This research analysis has also become a vital support for the DNO to model energy source combinations for the hospital load to supply the electric and heat demand. The HOMER software has been used for the analysis, with an execution time of 9.87 s. The simulation model has been implemented on Windows 10, intel i-7 processor with 8 GB RAM processing speed of 3.7 GHz.

References

1. Bandyopadhyay S, Mouli GRC, Qin Z, Elizondo LR, Bauer P (2020) Techno-economical model based optimal sizing of PV-battery systems for microgrids. IEEE Trans Sustain Energy 11(3):1657–1668. https://doi.org/10.1109/TSTE.2019.2936129
2. Krishan O, Suhag S (2019) Techno-economic analysis of a hybrid renewable energy system for an energy poor rural community. J Energy Storage 23(April):305–319. https://doi.org/10.1016/j.est.2019.04.002
3. Akinyele D (2018) Analysis of photovoltaic mini-grid systems for remote locations: a techno-economic approach. Int J Energy Res 42(3):1363–1380. https://doi.org/10.1002/er.3886
4. Singh B, Sharma AK (2022) Network constraints economic dispatch of renewable energy sources with impact of energy storage. Int J Comput Digit Syst 11(1):423–440. https://doi.org/10.12785/ijcds/110135

5. Mahmud MAP, Huda N, Farjana SH, Lang C (2019) Techno-economic operation and environmental life-cycle assessment of a solar PV-driven islanded microgrid. IEEE Access 7:111828–111839. https://doi.org/10.1109/ACCESS.2019.2927653
6. Singh B, Sharma AK (2022) Impact of battery storage with DG integration in distribution network using combined dispatch strategy for loss minimisation. Int J Comput Digit Syst 11(1):289–302. https://doi.org/10.12785/ijcds/110124
7. Singh B, Kumar A (2020) Sizing of energy storage with network constraints economic dispatch of fuel cell micro-turbine and renewable energy sources. (5):881–891. https://doi.org/10.35940/ijeat.E9871.069520
8. Singh B, Kumar Sharma A (2021) Optimal placement of DG with battery energy storage using CPLS and combined dispatch strategy. Lect Notes Electr Eng 667:317–325. https://doi.org/10.1007/978-981-15-5313-4_30
9. Moradi H, Esfahanian M, Abtahi A, Zilouchian A (2018) Optimization and energy management of a standalone hybrid microgrid in the presence of battery storage system. Energy 147:226–238. https://doi.org/10.1016/j.energy.2018.01.016
10. Singh B, Rawat SS (2020) Optimal placement of DG with battery energy storage in distribution network for power loss minimization using combined dispatch and combined PLS strategy. Int J Eng Adv Technol 9(5):1073–1081. https://doi.org/10.35940/ijeat.e1007.069520
11. Singh B, Rawat SS (2021) Impact of storage devices with renewable integrated distribution network for power loss minimization. Int J Eng Adv Technol 10(3):180–192. https://doi.org/10.35940/ijeat.C2265.0210321
12. Singh B, Singh Rawat S, Negi J (2021) Reactive power dispatch in distribution system for power loss minimization and voltage control. Int J Adv Eng Manag 3:895. https://doi.org/10.35629/5252-0302895900
13. Chawla D, Wagh P, Ali S, Jadhav U, Ghewade B (2020) Impact of COVID 19 pandemic on tuerculosis. Indian J Forensic Med Toxicol 14(4):6807–6810. https://doi.org/10.37506/ijfmt.v14i4.12689
14. Li C et al (2013) Techno-economic feasibility study of autonomous hybrid wind/PV/battery power system for a household in Urumqi, China. Energy 55:263–272. https://doi.org/10.1016/j.energy.2013.03.084
15. Singh B, Sharma AK (2022) Impact of electric vehicle load demand and energy storage device in integrated renewable energy sources. Smart Sci. 1–25. https://doi.org/10.1080/23080477.2022.2074658
16. Vallem VVSNM, Kumar A (2020) Optimal energy dispatch in microgrids with renewable energy sources and demand response. Int Trans Electr Energy Syst 30(5):1–27. https://doi.org/10.1002/2050-7038.12328
17. Singh B, Sharma AK (2022) Benefit maximization and optimal scheduling of renewable energy sources integrated system considering the impact of energy storage device and Plug-in Electric vehicle load demand. J Energy Storage 54:105245. https://doi.org/10.1016/j.est.2022.105245
18. Singh B, Negi J, Rawat SS (2021) Energy savings in renewable integrated distribution network and impact of storage devices 9(3):63–71
19. Nejabatkhah F, Li YW, Tian H (2019) Power quality control of smart hybrid AC/DC microgrids: an overview. IEEE Access 7:52295–52318. https://doi.org/10.1109/ACCESS.2019.2912376
20. Singh B, Kumar Sharma A (2022) Impact of D-STATCOM and OLTC with integrated volt/var control in distribution system for power loss minimization and voltage control. Smart Sci 1–21. https://doi.org/10.1080/23080477.2022.2074657

Optimal Allocation and Sizing of Distributed Generation in IEEE-85 BUS System Considering Various Load Models Using Multi-objective Metaheuristic Algorithms

Sumeet Kumar and Ashwani Kumar

Abstract Present-day distribution systems are bidirectional due to the integration of renewable energy sources and distributed generation. For efficient operation of the distribution systems, the loss reduction and voltage management are of the essential operational requirements. In India, the overall transmission and distribution losses for the 2018–2019 financial year were 20.66% and that needs major concerns for better efficiency of the system. The distribution system experiences voltage deviation and stability issues due to the improper management of the reactive power management and load growth. In order to limit these losses and better voltage profile, the system shall be planned with better control of reactive power and reduced losses. The losses and voltage can be better managed with the distribution generation integration into the existing system with optimal location and size. In this paper, DGs (distributed generation) in the radial distribution network have been optimally located and sized to minimize the line loss and enhance the voltage profile. This study aims to develop a multi-objective optimization model that considers various distribution load models while maximizing both technical and financial advantages. The appropriate positioning and sizing of DG resources in distribution networks are significantly influenced by load models and therefore, different load models have been incorporated in this study. The multi-objective function (MOF) contains the cost of active power loss reduction, voltage deviation enhancement, and the cost of installing DGs. The effects of four different load models are investigated using metaheuristic techniques. The study is carried out on an IEEE 85-bus radial distribution network as a test system. A comparison of results using Whale Optimization (WO), Grey Wolf Optimization (GWO), and Firefly Algorithm (FA) is analyzed to verify the effectiveness of the applied techniques.

S. Kumar (✉) · A. Kumar
National Institute of Technology, Kurukshetra, Haryana, India
e-mail: sumeet_32114301@nitkkr.ac.in

A. Kumar
e-mail: ashwani.k.sharma@nitkkr.ac.in

A. Kumar et al. (eds.), *Decarbonisation and Digitization of the Energy System*, Lecture Notes in Electrical Engineering 1099, https://doi.org/10.1007/978-981-99-7630-0_2

Keywords Active and reactive power · Distributed generation · Distribution system · An Optimization technique

1 Introduction

The complicated structure of the demand–supply, rapidly increasing power consumption, and contemporary electrical gadgets make it difficult to maintain economical and reliable distribution systems. In the competitive power market, the sustainability of the distributed generation (DG) resources is playing key role for the better operation and management. The main objective of a modern power distribution system is to provide quality and uninterrupted power supply to each consumer. Therefore, an effective distribution control system should increase system efficiency overall through loss reduction and management of power quality. A passive distribution method was used in the past, however present distribution n system is having bidirectional power flows due to the integration of renewable energy sources and distributed generation meeting the increasing load demand of various types. However, there are numerous challenges in the competitive regime of distribution systems in terms of the losses, voltage and reactive power requirements and compensation; DGs cost, distributed energy resources and their location, and energy storage facilities, which require fundamental change in the operation and management of the network. Distributed energy resources location and sizing, dispatching of the units, practical load consideration, load growth are key issues that require attention for the distribution system planner.

There are a number of cost-effective solutions available to enhance the operation and performance of distribution networks. These methods include strengthening feeders, rearranging networks, installing dispersed generation, placing reactive power sources, and integration of distributed generation with renewable sources. The planning and operation of distribution system require its efficient operation and management with reduced losses and higher efficiency. With the introduction of the competitive power system structure, it has added the new sources of power generation along with conventional sources in the system for sustainable green energy. The distribution network thus has grown with bidirectional power flow acting as active distribution system. The additional components, increasing load growth, the customers, and electricity providers are affected by distributed generation in terms of voltage profile, power flow, continuity, stability, power supply quality, and short circuit level [1]. To overcome these effects there is a need to address the issues of operation and management with distributed generation having proper location and sizes. Various optimization techniques are being used for optimal placement and sizing of the distributed generation. In certain ways, the distribution network operator has almost no effect on the placement and sizing of DG because the choice to locate it depends on stockholders, the availability of the fuel source, land rights of way, and climatic circumstances [2]. However, these DGs require the proper integration in the network which has impact on the loss's reduction and voltage profile improvement with optimal sizes. The load pattern and its nature are essential for better management as they have impact on the losses and voltage profile.

Small sources with modular power technologies, ranging from 1 kW to 50 MW, are known as distributed generation. They produce electricity closer to the points of consumption. Conventional and non-conventional both types of distributed generation are available. Solar power, wind power, small hydro, natural gas generators, biogas, biomass, and geothermal power are the non-conventional types, and in conventional types of fuel cells, diesel generators, sterling engines, gas turbines, and internal combustion reciprocating engines are there [3].

Various techniques and approaches have been proposed by different authors to address the issues of optimal sizing and placement of distributed generation considering different load models and different types of DGs. The number, size, and position of multi-distributed generation (multi-DG) units in distribution networks with different load models have been chosen by authors in [4] using a multi-objective index-based Particle Swarm Optimization technique. In [5] voltage-dependent load model of the distribution system has been taken into consideration for the appropriate location and size of distributed generators. A cuckoo search algorithm-based multi-objective index-based technique was implemented in [6] to optimize the DG's size and site under various load scenarios. A soft computing technique has been proposed in [7] for the Optimal sizing and sitting of DG with considering different distribution load models. In [8] Various load models' effects on distributed generation planning are proposed by considering single and multiple DG units. The optimal positioning and size of DG units considering different load models have been addressed in [9] using a unique multi-objective quasi-oppositional grey wolf optimizer approach. The impacts of load models and load demand in the distribution system when distributed generators are present have been taken into consideration in [10]. Performance of Voltage Step Constraints and Load Models in the Optimal Location and Size of Distributed Generation, using Incremental Power Flow and the Exhaustive Search Approach, [11].

All loads that are taken into account are of constant P, Q loads during the conventional power flow analysis. This presumption is unworkable in the actual operation of a dynamic, complicated power system. Such load modeling may produce contradictory findings and erroneous conclusions, which might result in inaccurate assessments of power loss, cost, deferral values, and other system indices [12]. Loads are often modeled as voltage or frequency dependent in actual power system static or dynamic studies. Loads come in a variety of forms and classifications. Voltage-dependent loads are classified based on the ZIP model as polynomial load model and exponential load model. In the previous proposed works in literature, load models are formulated based on their types and categories as constant, residential, commercial, and industrial loads.

In this paper, the exponential load model (ZIP model) is considered and problem is formulated for the optimal allocation and sizing of DG units. Different nature-based optimization techniques have been utilized to find the optimal DG placement and sizing considering multi-objectives in the objective function. The multi-objective function contains minimization of cost of line losses, real power generated by DG, and voltage deviation. The main contribution of the papers is:

- Improving the technical, economic, and environmental benefits by integrating different types of DGs into the distribution network.
- Three different nature-based evolutionary algorithms have been implemented and their result has been compared to get the best optimization among them.
- The lowering of DG unit generation costs, improvement of bus voltage variation, and reduction of power loss.
- To research various load models used in a real-world distribution system operation situation.
- The IEEE 85-BUS test radial distribution system was used to compare the solutions of various load models and types of DGs.

This paper is organized as follows: Sect. 2 introduces the formulation of distribution system load models. The multi-objective function formulation is done in Sect. 3. In a discussion in Sect. 4, the simulation findings for the test system are provided. Finally, the conclusions and references of the suggested work are provided.

2 Formulation of Distribution System Load Model

2.1 *Exponential Load Model*

The power-voltage relationship at the load bus is represented by exponential equations in the exponential model. These equations are essentially described in the ZIP model, except they contain fewer coefficients. To explain the algebraic connection between active & reactive power with applied voltage V, the exponential load model comprises two coefficients (exponents) termed n_p and n_q.

$$\mathrm{P} = \mathrm{P}_0\left(\frac{v}{v_0}\right)^{n_p} \tag{1}$$

$$\mathrm{Q} = \mathrm{Q}_0\left(\frac{v}{v_0}\right)^{n_q} \tag{2}$$

where P and Q are the real and reactive power of the actual load, V is the applied voltage of the load bus, V_0 is nominal voltage, P_0, Q_0 are the nominal active and nominal reactive power of the load, and n_p, n_q are exponential model coefficients.

The load behavior is expressed by the exponential models using the exponent's n_p and n_q as follows:

- Constant Impedance load (CI) when n_p=n_q=2
- Constant Current load (CC) when n_p=n_q=1
- Constant Power load (CP) when n_p=n_q=0

2.2 Polynomial Load Model (ZIP)

In practical distribution systems, the actual load is the combination of different nature of loads like constant power, constant current, and constant impedance. Such types of loads can be represented by the ZIP model, also known as the polynomial load model. In order to calculate the actual power (active power), the ZIP model adds Constant Impedance (CI), Constant Current (CC), and Constant Power (CP) to create a polynomial equation that describes the connection between the load power's characteristics and the applied voltage. Equations (3)–(4) mentioned here reflect the ZIP model algebraically:

$$P = P_0\left[a_p\left(\frac{V}{V_0}\right)^2 + b_p\left(\frac{V}{V_0}\right) + c_p\right] \tag{3}$$

$$Q = Q_0\left[a_q\left(\frac{V}{V_0}\right)^2 + b_q\left(\frac{V}{V_0}\right) + c_q\right] \tag{4}$$

where P denotes the load's real active power demand, Q denotes its actual reactive power demand, V denotes the load's actual voltage at the load bus, V_0 denotes the load's nominal voltage, P_0 denotes the load's nominal active power, and Q_0 denotes the load's nominal reactive power. The ZIP load model parameters for active power are a_p, b_p, and c_p, and a_q, b_q, and c_q are the ZIP load model coefficients for reactive power; where the values of $a_p = 0.1$, $b_p = 0.1$, $c_p = 0.8$, $a_q = 0.1$, $b_q = 0.1$, and $c_q = 0.8$.

3 Mathematical Formulation

Finding the optimal location and size of DG units considering the impact of different distribution load models to reduce the overall cost of the system while taking equality and inequality limitations into consideration is the main goal of the multi-objective problem formulated here.

3.1 Energy Loss Costs (CL)

The annual cost of power loss is given below:

$$\mathrm{CL} = (\mathrm{TPRL}) * (\mathrm{Kp} + \mathrm{Ke} * \mathrm{Lsf} * 8760)\$ \tag{5}$$

The loss factor is presented below as a function of the load factor (Lf).

$$\mathrm{Lsf} = \mathrm{k} * \mathrm{Lf} + (1 - \mathrm{k}) * \mathrm{Lf}^2 \tag{6}$$

where CL = annual cost of energy losses, TPRL = total real power loss of the system, LSF = loss factor, k = 0.2, Lf = 0.47, Kp = 57.6923 \$/kW, and Ke = 0.00961538 \$/kWh.

3.2 Cost of the DG for Reactive and Actual Power

$$\mathrm{C}(\mathrm{P_{DG}}) = \alpha * \mathrm{P_{DG}^2} + \mathrm{b} * \mathrm{P_{DG}} + \mathrm{c}\$/\mathrm{MWh} \tag{7}$$

The following cost coefficients are used: $\alpha = 0$, b = 20, c = 0.25

Based on the most complex power that DG can deliver, the cost of reactive power is determined as follows:

$$\mathrm{C}(\mathrm{Q_{DG}}) = \left[\mathrm{C}(\mathrm{S_{Dg}}) - \mathrm{C}\left(\sqrt{\mathrm{S_{DgMAX}^2} - \mathrm{Q_{Dg}^2}}\right)\right] * \mathrm{k} \tag{8}$$

$$\mathrm{S_{DgMAX}} = \frac{\mathrm{P_{DgMAX}}}{\cos\varnothing} \tag{9}$$

To conduct the analysis, the power factor has been taken as 0.9 lagging and unity with $\mathrm{P_{DgMAX}} = 1.1 * \mathrm{P_{DG}}$. k is between 0.05 and 0.1, 0.1 is used in this study.

3.3 Reflection of Voltage Deviation on Cost

The distribution system voltage may change as a result of the penetration of DG units. The voltage violation should thus be kept to a minimum. The voltage deviation is described as follows.

$$V_D = \sum_{i=1}^{N_l} |V_i - V_M| \tag{10}$$

The following is the formulation of how the voltage variation impacts the cost:

$$C_{VD} = V_D W_{VD} \tag{11}$$

where V_D is the voltage deviation at load buses, $V_M = 1.0p.u$ is the maximum allowable voltage, C_{VD} is the cost due to voltage deviation, W_{VD} is the economic operator of voltage deviation.

3.4 Multi-objective Function

$$\min F = \alpha_1 CL + \alpha_2 \mathrm{C(P_{DG})} + \alpha_3 C_{VD} \tag{12}$$

where F is the multi-objective function, CL is the annual cost of line loss, $\mathrm{C(P_{DG})}$ is the total annual DG generation cost, C_{VD} is the annual voltage deviation cost, α_1, α_2, and α_3 are the weight factor of the objective function and the values are 0.786, 0.0265, and 0.1875 respectively.

The constraints are:

Voltage limits:

$$0.95 \le V_i \le 1.06 \tag{13}$$

Power balance limitation:

without considering DG units:

$$P_{Grid} = \sum\nolimits_{j=1}^{N_l} P_d(j) + P_{loss} \tag{14}$$

$$Q_{Grid} = \sum\nolimits_{j=1}^{N_l} Q_d(j) + Q_{loss} \tag{15}$$

with considering DG units

$$P_{Grid} + \sum\nolimits_{i=1}^{N_{DG}} P_{DGi} = \sum\nolimits_{j=1}^{N_l} P_d(j) + P_{loss} \tag{16}$$

$$Q_{Grid} + \sum\nolimits_{i=1}^{N_{DG}} Q_{DGi} = \sum\nolimits_{j=1}^{N_l} Q_d(j) + Q_{loss} \tag{17}$$

Upper and lower limits of DG

$$60\ \text{KW} \le P_{DG} \le 3500\ \text{KW} \tag{18}$$

4 Simulation Results and Discussion

Figure 1 depicts the IEEE-85 Bus standard 12.66 kV radial distribution system, which contains 85 buses and 84 branches. The one-line diagram of the IEEE test system is shown in Fig. 1. The analysis is carried out for a multi-objective function using metaheuristic technique and the results are obtained and compared with three algorithms. The total load on the system is 2.3788 MVAR and 2.661 MW, respectively are the total load on this test system bus and line data were extracted from [13]. This simulation takes into account the backward-forward sweep approach to carry out the distribution system load flow analyses. The DG's size and location are obtained for a test system at the given load scenario. The maximum iteration size taken is 100 and the population size is taken as 50. The simulation study of the test system has been carried out using MATLAB 2021a software and an Intel Core I3 10th Gen CPU with 8.0 GB of RAM. Three different optimization techniques, whale optimization from [14], grey wolf optimization from [15], and firefly algorithm from [16], are implemented for the analysis of optimal DG allocation and the results are compared to validate the above multi-objective problem formulation.

The losses, DG size, minimum voltage, maximum voltage bus, cost of emery loss are obtained for different types of loads at unity power factor is given in Table 1. Comparing the results with different techniques, the firefly algorithm has performed well in three cases and losses obtained are minimal. The voltage profile curve with different load models and different techniques at unity power factor is shown in Fig. 2. The blue line in the graph represents the base case without DG placement and the other three colors represent results with the three DG placement.

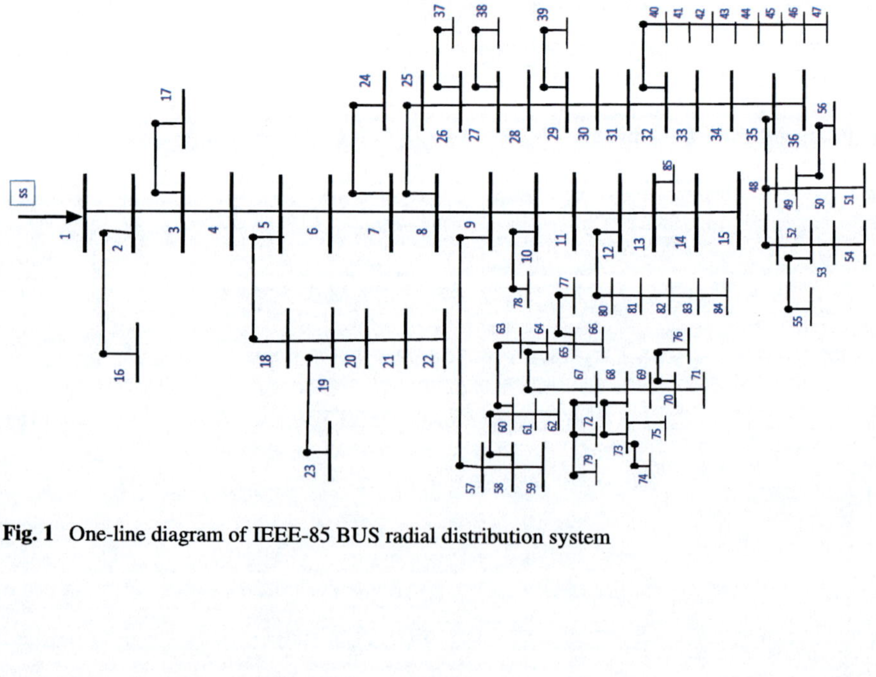

Fig. 1 One-line diagram of IEEE-85 BUS radial distribution system

Table 1 Comparison of different parameters with different algorithms at unity power factor

	Constant Power (CP)			
Parameters	Base	Whale	Gwo	Firefly
Active power loss	212.8156	142.1956	145.7648	116.6554
Reactive power loss	133.6105	81.2419	82.5248	70.1582
Number of DG	Nil	3	3	3
Optimal DG location	Nil	68/34/53	47/9/40	60/63/36
Optimal DG size P KW	Nil	794/1087/633	791/344/980	796/626/878
Optimal DG size Q Kvar	Nil	0/0/0	0/0/0	0/0/0
Minimum voltage	0.90752	0.97209	0.95566	0.96967
Minimum voltage bus number	54	84	76	84
Maximum voltage	0.9969	1.006	1.0026	0.99854
Maximum voltage bus number	2	53	47	2
Cost of line/Energy losses in $	17,130.642	11,446.0725	11,733.3756	9390.2087
Cost of Pdg ($/MWH)	Nil	50.53	42.55	46.25
Cost of Q ($/MVARH)	Nil	4.3181	3.6364	3.9524
Execution time	0.093163	165.2648	163.3334	163.6892
	Constant Current (CC)			
Parameters	Base	Whale	Gwo	Firefly
Active power loss	179.9943	120.8651	100.3181	110.6617
Reactive power loss	113.1624	69.4993	61.8868	65.1737
Number of DG	Nil	3	3	3
Optimal DG location	Nil	49/65/40	2/72/48	85/69/56
Optimal DG size P KW	Nil	1073/749/542	60/748/713	540/876/878
Optimal DG size Q Kvar	Nil	0/0/0	0/0/0	0/0/0
Minimum voltage	0.91536	0.97449	0.96078	0.97833
Minimum voltage bus number	54	84	84	43
Maximum voltage	0.99711	1.004	0.9982	0.99871
Maximum voltage bus number	2	49	2	2
Cost of line/Energy losses in $	14,488.688	9729.0668	8075.1324	8907.7369
Cost of Pdg ($/MWH)	Nil	47.53	30.67	46.13
Cost of Q ($/MVARH)	Nil	4.0618	2.6213	3.9422
Execution time	0.060565	167.5929	163.1372	164.3701
	Constant Impedance (CI)			
Active power loss	156.8548	113.2255	117.9073	107.1056
Reactive power loss	98.7214	63.9807	66.2219	61.468
Number of DG	Nil	3	3	3
Optimal DG location	Nil	84/69/48	41/31/40	31/41/76
Optimal DG size P KW	Nil	213/635/1401	997/432/268	583/755/827

(continued)

Table 1 (continued)

	Constant Power (CP)			
Optimal DG size Q Kvar	Nil	0/0/0	0/0/0	0/0/0
Minimum voltage	0.9213	0.97956	0.95872	0.9752
Minimum voltage bus number	54	15	76	84
Maximum voltage	0.99728	1.0063	0.99854	0.99877
Maximum voltage bus number	2	48	41	2
Cost of line/Energy losses in $	12,626.0652	9114.113	9490.9752	8621.4891
Cost of Pdg ($/MWH)	Nil	45.23	34.19	43.55
Cost of Q ($/MVARH)	Nil	3.8653	2.9221	3.7218
Execution time	0.066352	168.8701	163.5817	163.9136
	ZIP load			
Active power loss	204.2579	132.1819	128.2455	112.2748
Reactive power loss	128.279	76.1429	79.4284	67.3835
Number of DG	Nil	3	3	3
Optimal DG location	Nil	48/60/68	32/50/27	30/35/68
Optimal DG size P KW	Nil	1399/267/802	321/590/165	627/695/898
Optimal DG size Q Kvar	Nil	0/0/0	0/0/0	0/0/0
Minimum voltage	0.9095	0.97311	0.94091	0.96951
Minimum voltage bus number	54	84	76	84
Maximum voltage	0.99695	1.0004	0.99776	0.99855
Maximum voltage bus number	2	48	2	2
Cost of line/energy losses in $	16,441.7856	10,640.0121	10,323.1502	9037.5894
Cost of Pdg ($/MWH)	Nil	49.61	21.77	44.65
Cost of Q($/MVARH)	Nil	4.2395	1.8605	3.8158
Execution time	0.071562	166.0708	163.2554	163.4411

The optimal DG placement and sizing problem may have a solution with higher sizes of DG units, for power loss reduction, but the cost of DG allocation would rise. Therefore, in order to conduct a realistic feasibility analysis of DG installation at a site, it is required to examine various aspects of cost, power loss, and voltage enhancement at the same time taking a multi-objective case. Maximum three DGs are taken into consideration in this study to obtain the best location and size for DGs. It is observed from Table 1, that grey wolf optimization algorithm gives the lowest DG power generation cost.

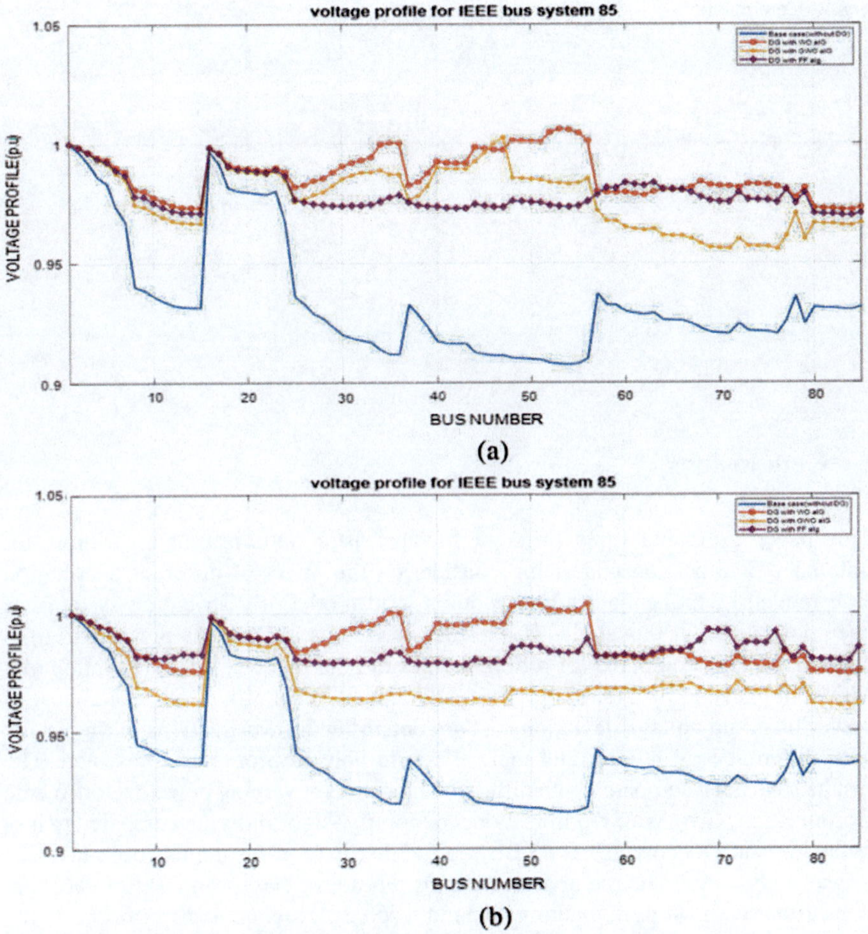

Fig. 2 Voltage profiles for various load models in an 85-BUS system with three DG operating at unity power factor **a** Constant Power (CP) **b** Constant Current (CC) **c** Constant Impedance (CI) **d** ZIP load

In Table 2, the results are obtained with three DG at 0.9 lagging power factor case and are tabulated. The lowest annual line loss cost is obtained with firefly algorithm and the lowest DG generation cost is obtained with Grey Wolf Optimization algorithm. The improved voltage profile curve with 0.9 lagging power factor for different distribution load models taking different heuristic techniques is presented in Fig. 3. The voltage profile is better with DGs for all the cases of loads.

Fig. 2 (continued)

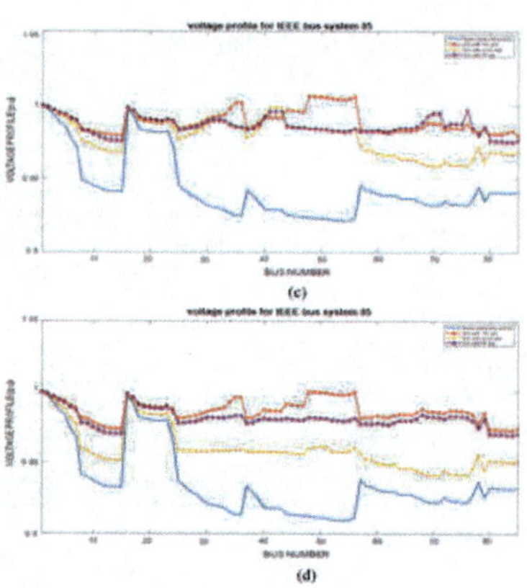

5 Conclusions

This paper presented three different metaheuristic optimization algorithms for optimal DG allocation and sizing considering the effect of different distribution load models. Exponential and polynomial load model of distribution system has been considered in this study. The results of various approaches presented in the paper have been compared for different types of load models. The formulation of a MOF with several different objectives, DG annual generation cost, annual cost of power loss, and voltage deviation cost are determined. Investigations of the results with different load models and their effect on voltage profile are presented. The results have been compared with different algorithms at varying power factor. Whale Optimization, Grey Wolf Optimization, and Firefly Algorithms are compared to loss reduction and DG cost. IEEE-85 BUS radial distribution system has been taken as a case study to validate the above scenarios. Future research can take into account the shifting demand profile taking demand response program and its impact on the operation and management considering the EVs load and storage devices.

Table 2 Comparison of different parameters with different algorithms at 0.9 lag power factor

	Constant Power (CP)			
Parameters	Base	Whale	Gwo	Firefly
Active power loss	212.8156	66.3423	79.6937	65.3305
Reactive power loss	133.6105	32.9205	44.7748	33.1067
Number of DG	Nil	3	3	3
Optimal DG location	Nil	43/68/52	11/12/35	54/72/56
Optimal DG size P KW	Nil	621/809/1082	60/60/1388	887/873/481
Optimal DG size Q Kvar	Nil	300/391/524	29/29/672	429/422/232
Minimum voltage	0.90752	0.9871	0.95542	0.9815
Minimum voltage bus number	54	84	76	84
Maximum voltage	0.9969	1.035	0.99854	1.0304
Maximum voltage bus number	2	52	35	54
Cost of line/Energy losses in $	17,130.642	5340.2395	6414.9643	5258.7928
Cost of Pdg ($/MWH)	Nil	50.49	30.41	45.07
Cost of Q ($/MVARH)	Nil	4.9038	2.9528	4.3772
Execution time	0.14696	167.3439	165.2409	163.4633
	Constant Current (CC)			
Parameters	Base	Whale	Gwo	Firefly
Active power loss	179.9943	64.4497	51.52	44.8128
Reactive power loss	113.1624	31.0324	29.2536	23.3105
Number of DG	Nil	3	3	3
Optimal DG location	Nil	69/85/56	33/24/11	51/33/58
Optimal DG size P KW	Nil	749/189/1414	1196/60/367	725/715/882
Optimal DG size Q Kvar	Nil	362/91/684	579/29/177	351/346/427
Minimum voltage	0.91536	0.99091	0.96321	0.98657
Minimum voltage bus number	54	84	76	76
Maximum voltage	0.99711	1.0344	0.99869	1.0224
Maximum voltage bus number	2	56	2	51
Cost of line/Energy losses in $	14,488.688	5187.8935	4147.114	3607.2193
Cost of Pdg ($/MWH)	Nil	47.29	32.71	46.69
Cost of Q ($/MVARH)	Nil	4.5929	3.1763	4.5346
Execution time	0.083512	166.4707	165.3276	168.0667
	Constant Impedance (CI)			
Parameters	156.8548	53.3583	54.7656	36.5692
Active power loss	98.7214	26.3495	28.2068	19.3764
Reactive power loss	Nil	3	3	3
Number of DG	Nil	34/79/55	30/52/11	63/61/48
Optimal DG location	Nil	1412/608/229	1653/90/60	869/481/864
Optimal DG size P KW	Nil	683/294/110	800/43/29	420/232/418

(continued)

Table 2 (continued)

	Constant Power (CP)			
Optimal DG size Q Kvar	0.9213	0.98948	0.97141	0.98975
Minimum voltage	54	84	76	84
Minimum voltage bus number	0.99728	1.0278	1.0075	1.0083
Maximum voltage	2	55	30	48
Maximum voltage bus number	12,626.0652	4295.0924	4408.3728	2943.6476
Cost of line/Energy losses in $	Nil	45.23	36.31	44.53
Cost of Pdg ($/MWH)	Nil	4.3928	3.5261	4.3248
Cost of Q ($/MVARH)	0.12053	169.7483	167.7229	167.5108
Execution time	0.12053	169.7483	167.7229	167.5108
	ZIP load			
Parameters	204.2579	49.2881	109.1829	62.8579
Active power loss	128.279	25.0413	52.928	31.4008
Reactive power loss	Nil	3	3	3
Number of DG	Nil	36/84/79	49/4/21	54/77/50
Optimal DG location	Nil	1357/375/736	1885/60/80	607/914/821
Optimal DG size P KW	Nil	657/181/356	912/29/38	293/ 442/397
Optimal DG size Q Kvar	0.9095	0.99233	0.96494	0.98505
Minimum voltage	54	22	76	84
Minimum voltage bus number	0.99695	1.021	1.0342	1.032
Maximum voltage	2	36	49	54
Maximum voltage bus number	16,441.7856	3967.4539	8788.7008	5059.7585
Cost of line/Energy losses in $	Nil	49.61	40.75	47.09
Cost of Pdg ($/MWH)	Nil	4.8183	3.9575	4.5735
Cost of Q ($/MVARH)	0.10908	168.2883	168.3482	167.3802
Execution time	0.10908	168.2883	168.3482	167.3802

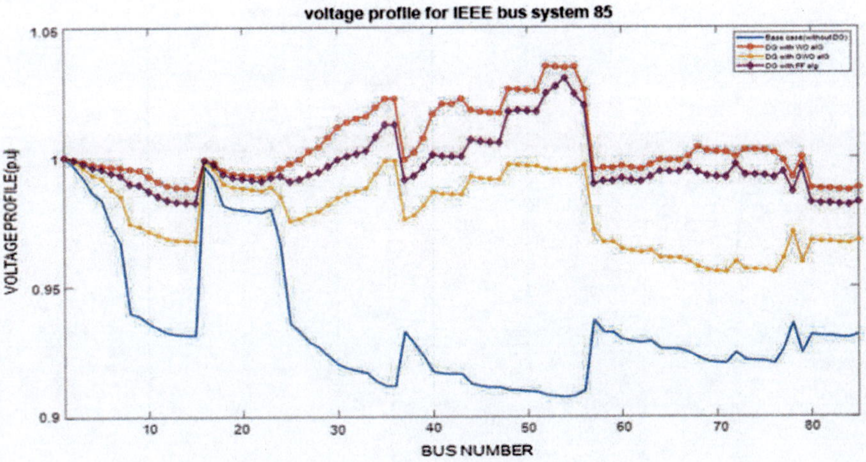

(a)

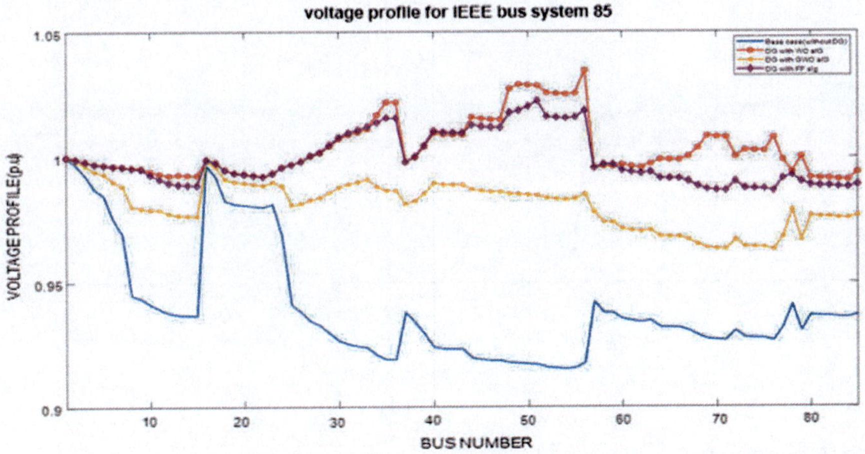

(b)

Fig. 3 Voltage profiles of 85-BUS system with three DG at 0.9 lagging power factor for different types of load models **a** Constant Power (CP) **b** Constant Current (CC) **c** Constant Impedance (CI) **d** ZIP load

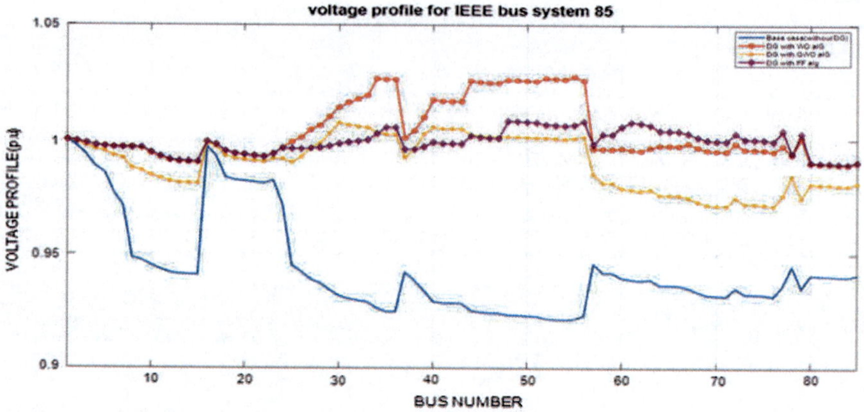

(c)

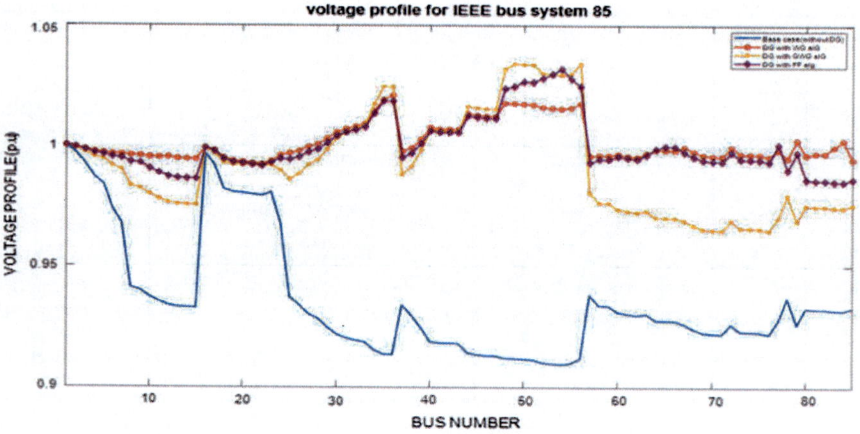

(d)

Fig. 3 (continued)

References

1. Davis MW (2002) Distributed resource electric power systems offer significant advantages over central station generation and t & d power systems. In: Power engineering society summer meeting, vol 1. IEEE, pp 62–69
2. Georgilakis PS, Hatziargyriou ND (2013) Optimal distributed generation placement in power distribution networks: models, methods, and future research. IEEE Trans Power Syst 28(3):3420–3428. https://doi.org/10.1109/TPWRS.2012.2237043
3. Viral R, Khatod D (2012) Optimal planning of distributed generation systems in distribution system: a review. Renew Sustain Energy Rev 16(7):5146–5165
4. El-Zonkoly A (2011) Optimal placement of multi-distributed generation units including different load models using particle swarm optimization. Gener, Transm Distrib IET 5:760–771. https://doi.org/10.1049/iet-gtd.2010.0676
5. Kumar M, Nallagownden P, Elamvazuthi I (2017) Optimal placement and sizing of distributed generators for voltage-dependent load model in radial distribution system. Renew Energy Focus 19–20:23–37. https://doi.org/10.1016/j.ref.2017.05.003
6. Prakash R, Lokeshgupta B, Sivasubramani S (2018) Optimal site and size of DG with different load models using cuckoo search algorithm. In: 2018 IEEE international conference on power electronics, drives and energy systems (PEDES), pp 1–6. https://doi.org/10.1109/PEDES.2018.8707724
7. Agnihotri G, Dubey M, Bohre A (2016) Optimal sizing and sitting of DG with load models using soft computing techniques in practical distribution system. Gener, Transm Distrib. 10. https://doi.org/10.1049/iet-gtd.2015.1034
8. Singh D, Misra RK, Singh D (2007) Effect of load models in distributed generation planning. IEEE Trans Power Syst 22(4):2204–2212. https://doi.org/10.1109/TPWRS.2007.907582
9. Kumar S, Mandal KK, Chakraborty N (2021) Optimal placement of different types of DG units considering various load models using novel multiobjective quasi-oppositional grey wolf optimizer. Soft Comput 25:4845–4864. https://doi.org/10.1007/s00500-020-05494-3
10. Mohanty PK, Lal DK (2021) Effects of load models and load growth in distribution system in presence of distributed generator. In: 2021 1st international conference on power electronics and energy (ICPEE), pp 1–6. https://doi.org/10.1109/ICPEE50452.2021.9358482
11. Payasi RP, Singh AK, Singh D (2013) Effect of voltage step constraint and load models in optimal location and size of distributed generation. In: 2013 international conference on power, energy and control (ICPEC), pp 710–716. https://doi.org/10.1109/ICPEC.2013.6527748
12. Singh D, Misra R (2007) Effect of load models in distributed generation planning. IEEE Trans Power Syst 22(4):2204–2212
13. Sukraj K, Thangaraj Y, Raju H, Thirumalai M (2019) Simultaneous allocation of shunt capacitor and distributed generator in radial distribution network using modified firefly algorithm. 1–5. https://doi.org/10.1109/ICSSS.2019.8882829
14. Mirjalili S, Mirjalili SM, Saremi S, Mirjalili S (2020) Whale optimization algorithm: Theory, literature review, and application in designing photonic crystal filters. In: Studies in computational intelligence, vol 811. Springer, pp 219–238. https://doi.org/10.1007/978-3-030-12127-3_13
15. Joshi H, Arora S (2017) Enhanced grey wolf optimization algorithm for global optimization. Fundamenta Informaticae 153:235–264. https://doi.org/10.3233/FI-2017-1539
16. Yang X-S, Xingshi H (2013) Firefly algorithm: recent advances and applications. Int J Swarm Intell 1. https://doi.org/10.1504/IJSI.2013.055801

A Sustainable Privacy-Preserving Aggregation Authentication Protocol for Smart Grid

Dharminder Chaudhary, Tanmay Soni, Soumyendra Singh, M. S. P. Durgarao, and Surisetty Mahesh Chandra Gupta

Abstract Smart grid (SG) is an advanced framework for controlling computers, automation, new technologies, and smart equipment that is more efficient and dependable. Smart grid possesses advanced self-healing capabilities and enables electricity customers to become active participants. Both security and privacy are two very important attributes of SG communication. The proposed model is efficient for communication between smart grids which uses a privacy-preserving aggregation mechanism. The model structure uses super increasing sequences to organize data and then encrypts it using the homomorphic Paillier approach. Data aggregation is done directly on cipher at local gateways without following any decryption for data transfers from users to smart grid operation center. The aggregate outcome of the source data may be accessed at the operation center that saves time. Finally, our findings demonstrate that the proposed method supports batch verification that makes it cost-effective in terms of computing and communication as compared to existing architectures. The proposed design is fast symmetric encryption, and it is suitable for massive user groups, and it can be adopted to support the flexible and rapid growth of residential scales in smart grids.

Keywords Security · Privacy · Smart grid · Privacy-preserving aggregation · Paillier's technique

1 Introduction

Sustainable development and green energy are now becoming very mainstream around the world and by combining the power grid and communication technologies, it is possible to achieve more precise dispatch and management. The smart grid is an electrical grid that provides information about consumer behavior in a highly

D. Chaudhary (✉) · T. Soni · S. Singh · M. S. P. Durgarao · S. M. C. Gupta
Department of Computer Science and Engineering, Amrita School of Computing, Amrita Vishwa Vidyapeetham, Chennai, India
e-mail: manndharminder999@gmail.com

A. Kumar et al. (eds.), *Decarbonisation and Digitization of the Energy System*, Lecture Notes in Electrical Engineering 1099, https://doi.org/10.1007/978-981-99-7630-0_3

secure, reliable, and efficient manner. Since we can not stock electricity easily, it is very important to match the transmission capacity with the power supply in power grid. The smart grid, designed to replace conventional electric power infrastructure, can improve the grid's efficiency, safety, and dependability through bidirectional power and communication flows. It is attracting the attention of industrialists and researchers due to the diverse involvements of smart grid. It appears that the traditional grid scheme will not meet the requirements and demands of the 21st century.

For the smart grid, Wu et al. [1] developed a key management scheme. But the computation cost of [1] was very high with this it was not secure against reply and man-in-the-middle attack (MITM). Additionally, it didn't offer security and perfect forward secrecy. Xia et al. [2] reviewed [1] and showed that [1] was not protected from MITM attack. Xia et al. [2] then proposed a secure key distribution using a trusted third party. However, it does not give smart meters anonymity and does not provide absolute forward opacity. Then Park et al. [3] pointed out that Xia et al. [2] is vulnerable to impersonation attack. In addition, Park et al. pointed out, on the basis of this weakness, that the claim of Xia et al. for tolerance against the unknown key share attack is wrong. Tsai et al. [4] proposed a new anonymous key distribution strategy for smart grid scenarios using an identity-based encryption technique and an identity-based signing mechanism. But it was not safeguarded from MITM attack. Odelu et al. [5] promised exceptional privacy and presented a new scheme for mutual authentication distribution in smart grids. However, the major drawbacks of Odelu et al's scheme are hefty computation and communication costs. According to Braeken et al. [13], Chen et al.'s [12] scheme fails to deliver security from DoS attack even it uses an anonymous authentication and key exchange technique. Braeken et al. [13] presented a protocol for smart grid communication. Unfortunately, there is no data confidentiality, anonymity, and in registration phase, there is no secure channel. Abbasinezhad-Mood et al. [14], in 2018, proposed a smart grid authentication scheme but it failed to provide data confidentiality, anonymity, and unsecured against impersonation attack, insider attack, and guessing identity attack. Based on hybrid cryptography, Kumar et al. [6] advanced an approach to secure data transfer between a remote smart meter and an energy distribution system. However, it could not offer safeguard from forgery attack. In [7], Garg et al. offered a simple yet reliable authentication method for use in a smart grid context. Integrating the elliptic curve with the Menezes–Qu–Vanstone key exchange algorithm and hash functions to achieve authentication security and anonymity. However, it was vulnerable to a forgery attack. In 2022, Palacios et al. [15] proposed a privacy-preserving aggregate protocol to measure residential loads. Zhai et al. [22] proposed two privacy-preserving outsourcing algorithms for multidimensional data aggregation. Ming et al. [23] proposed a privacy-preserving data aggregation model which can withstand failures for smart meters. Singh et al. [24] proposed a secure privacy-preserving data aggregation and classification (SP-DAC) model. Wu et al. [25] proposed a robust model which supports strong privacy preservation. In the same year, Li et al. [16] used fog computing to design an advanced privacy-preserving aggregate signcryption for Smart Grid. Zhang et al. [17] used fine-grained access to design a distributed privacy-preserving aggregation framework for smart grid. All of these protocols can't work on encrypted data. Moreover, we need an efficient privacy-preserving aggregation framework for smart grid.

2 System Model, Security Specification, and Design Objectives

There is a discussion of outline of the system model (see Fig. 1), security requirements, and design goals in this section.

2.1 System Model

In the proposed scheme we will be discussing how to communicate user's electricity consumption data with the control center (CC), in smart grid communications while maintaining privacy. Here, there is a traditional residential area (RA), that includes a smart grid control center, central gateway (CG) linked to a significant number of residential users $\upsilon = \{U_1, U_2, \ldots U_n\}$. The CG is a versatile workshop that is mostly used for relaying and aggregation. The relaying component assists residential users in transmitting traffic to the control center (CC), while the aggregation component is in charge of compressing residential user's energy consumption data, i.e., a trustworthy control authority (CA) situated at the control center, as well as assisting the CA in relaying replies back to the (residential user) U_i present in the residential area RA.

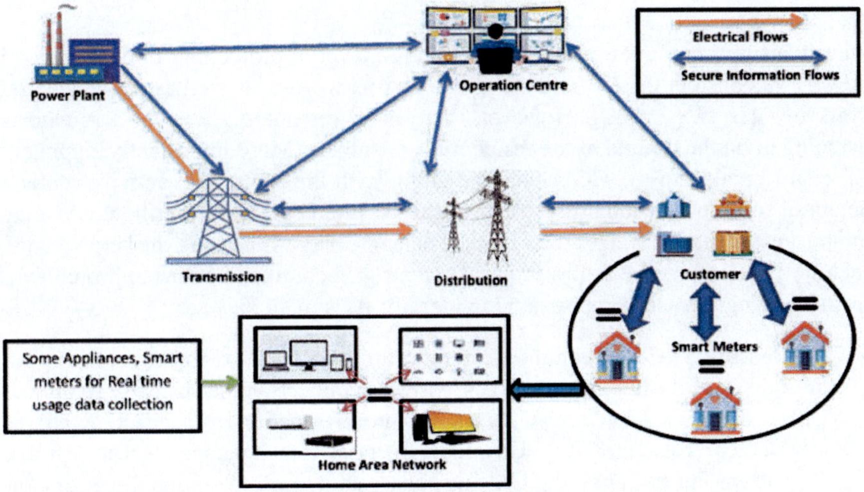

Fig. 1 Illustration of communication architecture of the proposed design

Then the CG will execute certain verification operations during the aggregation and relaying phase to ensure the data's validity and integrity. Each consumer $U_i \in \upsilon$ is outfitted with a number of smart meters that form a home area network (HAN) and have capabilities to track real-time information electronically, related to the electricity uses. Through the relay of the CG, this near-instantaneous information will be transmitted to the CA after a certain set period of time. After collecting information from household consumers, the CA may gain real-time situational information. That helps CA to make energy usage more competitive and shift certain demands to off-peak hours by implementing dynamic pricing.

2.2 Communication Model

In RA, every user $U_i \in \upsilon$ communicates with the nearby CG using comparatively inexpensive Wireless networks. In other words, each $U_i \in \upsilon$ can connect with the CG overtly or covertly within the CG's Wireless coverage. However, the distance between the residential area and the control server is large so the contact between the CG and the CA is done using wired links or other low-delay, high-bandwidth links. However, despite the fact that smart grid connectivity has a low latency and high bandwidth, the communication reliability of the CG-to-CA communication remains a challenge because hundreds of thousands of household users spread around various residential neighborhoods in a region can record their energy consumption data all at the same time.

2.3 Security Specification

Safe smart grid communications require a high level of protection. In the proposed model, we consider the CA and the CG to be trustworthy, as well as the residential users $\upsilon = \{U_1, U_2, \dots U_n\}$. However, there is an opponent A_d in the RA who is listening in on the reports of the residential consumers. More importantly, opponent A_d could break into the CG's database along with the smart grid activity center's database so that individual user reports can be intercepted. Furthermore, A_d may mount any successful attacks to intimidate data integrity. As a result, the below-stated security criteria must be followed in safe smart grid communications to prohibit A_d from knowing the user's reports and to identify A_d's malicious acts.

- **Confidentiality:** Confidentiality is about to protect user's report from the opponent A_d, i.e., if A_d intercept the RA's Wireless connection, he/she can't figure out what's in the reports; and even if A_d extracts the information from the CG databases and/or the control center, it is also unable to classify each person electricity usage information. Furthermore, the CA's responses should fill the requirement of confidentiality that means only legitimate user can read the data in RA.

- **Data Integrity and Authentication:** It refers to verifying an encrypted report which is submitted by a legitimate residential person and is not changed throughout the delivery, i.e., if opponent A_d modifies and/or forges a report, certain deceptive operations can be identified. Only accurate reports will be received by the CA, which will be useful for tracking energy use. Besides this, the CA's replies should be authenticated such that residential consumers can access accurate and trustworthy details.

2.4 Design Objectives

The proposed design aims to provide an effective and privacy-preserving aggregation protocol to establish a safe communication in the smart grid, based on the above concept and security specifications. The following two goals, in particular, should be met.

- **Guarantee assured by the proposed model:** The proposed plan or scheme should satisfy confidentiality, data integrity, and authentication. The privacy of users in RA could be disclosed if the security majors are not taken into consideration by smart grid and someone can alter the electricity usage report of users.
- **Communication-effectiveness achieved by proposed design:** While the correspondence between the CA and the CG is characterized by low latency and high bandwidth, the proposed method should also assume communication effectiveness so that the near genuine user's electricity usage reports can be quickly communicated to the CA.

3 Preliminaries

3.1 Bilinear Pairing

Let q be a large prime number and G_1, G_2 be two cyclic groups of order q. A map $e : G_1 \times G_1 \rightarrow G_2$ is said to be a efficiently computable and non-degenerated bilinear map if $e(aR_1, bR_2) = e(R_1, R_2)^{ab} \in G_2$ for all $a, b \in Z_q^*$ and for any $R_1, R_2 \in G_1$ and $e(P, P) \neq 1_{G_2}$ where P be a generator of G_1. For a more detailed explanation of assumptions of complexity and the pairing technique, we refer to [9].

Definition 1 A probabilistic algorithm Gen is a bilinear parameter generator that output (P, q, e, G_1, G_2) after getting security parameter α as input, where G_1, G_2 are groups of order q, P is a generator of G_1 and e is a efficiently computable and non degenerated bilinear map.

Definition 2 (*CDH: Computational Diffie–Hellman Problem*) For given $(aP,\ bP,\ P) \in G$, compute $abP \in G$ where $a,\ b \in Z_q^*$ are unknown.

Definition 3 (*BDH: Bilinear Diffie–Hellman Problem*) For a given tuple $(aP,\ bP,\ cP,\ P) \in G$, compute $e(P,\ P)^{abc} \in G_2$ where $a,\ b,\ c \in Z_q^*$ are unknown.

Definition 4 (*DBDH: Decisional Bilinear Diffie–Hellman Problem*) For a given tuple $(aP, bP, cP, P) \in G$ and $Y \in G_2$, decide whether Y is a random element in G_2 or $Y = e(P,\ P)^{abc} \in G_2$, where $a,\ b,\ c \in Z_q^*$ are unknown.

3.1.1 Paillier Cryptosystem

Homomorphic properties can be accomplished by the Paillier Cryptosystem, which is commonly desirable in several privacy-preserving applications [10, 11]. Specifically, Encryption, Decryption, and Key generation are used in the Paillier Cryptosystem.

1. Key Generation: Suppose $|p_1| = |p_2| = \alpha_1$ are given where α_1 is the security parameter and $p_1,\ p_2$ are two large prime numbers then compute RSA modulus $n = p_1.p_2$ and $\beta = lcm(p_1 - 1, p_2 - 1)$. Now compute $\gamma = (L(g^{\beta}\ mod\ n^2))^{-1}$ where $g \in Z_{n^2}^*$ be a generator and L be a function defined as $L(t) = \frac{t-1}{n}$. Then the private key $S_k = (\beta, \gamma)$ and the public key is $P_k = (n, g)$.
2. Encryption: For any given message $m \in Z_n$, the cipher can be computed as $c = E(m) = g^m.r^n\ mod\ n^2$ where $r \in Z_{n^2}$ chosen randomly.
3. Decryption: For any given cipher $c \in Z_{n^2}^*$, the plaintext can be computed as $m = D(c) = L(c^{\beta\ mod\ n^2}).\gamma\ mod\ n$.

4 Proposed Smart Grid Communication Architecture

In this section, we provide an efficient and privacy-preserving aggregation approach for secure smart grid communications, which is divided into four sections: system setup, user report creation, report aggregation while protecting privacy, as well as safe report viewing and reaction. We utilize ℓ dimensional electric uses data $(d_1,\ d_2,\ \ldots,\ d_\ell)$ and consider that the number of users in a RA is constant ψ. We also assume that each dimensional item's value is smaller than D. Most of the symbols are available in Table 1.

4.1 System Setup

In this phase, the trusted authority CA firstly generates (q, P, G_1, G_2, e) for given security parameters (α_1, α_2) by running $Gen(\alpha_1)$ then it computes the public key $N = p_1.q_1,\ g$ where $|p_1| = |q_1| = \alpha_2$ are two primes and the corresponding secret

Table 1 Notations

Notation	Description
U_i	ith user
e	Bilinear pairing
G_1	Elliptic curve points group
G_2	Cyclic target group
q	Prime number
A_d	Adversary
λ_d	Advarsery
RA	Residential area
CC	Control center
CG	Central gateway
CA	Control authority
HAN	Home area network

key β, γ of Paillier Cryptosystem. With this, CA also chooses a sequence of super increasing elements $s = (s_1 = 1, s_2, ..., s_\ell)$ which satisfies $|s_i| \geq \alpha_1$, $\sum_{j=1}^{i-1} s_j.\psi.D < s_i$ for $i = 2, ..., \ell$, and $\sum_{j=1}^{\ell} s_j.\psi.D < N$. With the help of s, CA calculates another sequence g_i where $g_i = g^{s_i}$ for $1 \leq i \geq \ell$. In addition CA chooses $R_1, R_2 \in G_1, \delta, x \in Z_q^*$, two hash function h_1, h_2 where $h_1 : \{0, 1\}^* \rightarrow G_1$ and $h_2 : \{0, 1\}^* \rightarrow Z_q^*$ then calculates $y = xP$ and $e(P, P)^\alpha$. At last, CA publishes $(q, P, G_1, G_2, e, N, R_1, R_2, g_1, g_2, ..., g_\ell, e(P, P)^\alpha, y, h_1, h_2)$ as system parameters and keeps the secret keys $(\beta, \gamma, s, \delta, x)$. A user u_i of HAN chooses the secret key $x_i \in Z_q^*$ and computes a public key $y_i = x_i.P$. Further GW chooses x_g as private key and computes $y_g = x_g.P$ as public key. With this, CA computes

$$t_{i1} = h_1(u_i||ID_{ra}||\delta) \quad (1)$$

$$t_{i2} = h_2(u_i||ID_{ra}||x) \quad (2)$$

and generates

$$ak_i = (\delta P + t_{i1}.y, t_{i1}P, t_{i2}P, T_{i1}R_1 + T_{i2}R_2). \quad (3)$$

4.2 *Generation of User Report*

Every user u_i in HAN collect ℓ types of data $(d_{i1}, d_{i2}, ..., d_{i\ell})$ in every τ minutes, where τ represent a fix time, e.g., $\tau = 20$ $minutes$ and $d_{ij} \leq D$. Now user choose $r_i \in Z_{n'}^*$ randomly and computes

$$C_i = g_1^{d_{i1}}.g_2^{d_{i2}}......g_\ell^{d_{i\ell}}.r_i^n \; mod \; n^2 \quad (4)$$

Now user uses his/her private key and creates signature as

$$\varsigma_i = x_i.h_1(C_i||ID_{ra}||u_i||T_1) \tag{5}$$

where ID_{ra} and T_1 denotes identity of residential area and time stamp. Finally user u_i report $(\varsigma_i||C_i||ID_{ra}||u_i||T_1)$ to GW.

4.3 Aggregation of Privacy-Preserving Report

In this phase, GW firstly checks T_1, then it verifies the signature ς_i. After receiving encrypted data from ψ, the user computes $(\varsigma_i||C_i||ID_{ra}||u_i||T_1)$ for $1 \le i \le \psi$. The GW verifies the signature by computing $e(P, \varsigma_i) = e(y_i, h_1(C_i||ID_{ra}||u_i||T_1))$ and $e(P, x_i.h_1(C_i||ID_{ra}||u_i||T_1))$ and equate them. If both are equal then signature is accepted. GW can use batch verification in order to make verification more efficient as

$$e\left(P, \sum_{i=1}^{\psi} \varsigma_i\right) = e\left(P, \sum_{i=1}^{\psi} x_i h_1(C_i||ID_{ra}||u_i||T_1)\right)$$
$$= \prod_{i=1}^{\psi} e(P, x_i h_1(C_i||ID_{ra}||u_i||T_1)) = \prod_{i=1}^{\psi} e(y_i, h_1(C_i||ID_{ra}||u_i||T_1)). \tag{6}$$

Now GW aggregates the privacy-preserving report by computing $C = \prod_{i=1}^{\psi} C_i \ mod \ n^2$, $\varsigma_g = x_g h_1(C||ID_{ra}||ID_{gw}||T_2)$, and GW reports $(C||ID_{ra}||ID_{gw}||T_2||\varsigma_g)$ to CA.

4.4 Secure Report Reading and Response

After getting the message $(C||ID_{ra}||ID_{gw}||T_2||\varsigma_g)$ from GW, first CA verifies $e(P, \varsigma_g) = e(y_g, h_1(C||ID_{ra}||ID_{gw}||T_2))$, then read the encrypted data by performing the following steps:

$$C = \prod_{i=1}^{\psi} C_i \ mod \ N^2 = \prod_{i=1}^{\psi} g_1^{d_{i1}}.g_2^{d_{i2}}......g_\ell^{d_{i\ell}}.r_i^n \ mod \ N^2$$
$$= g_1^{\sum_{i=1}^{\psi} d_{i1}}.g_2^{\sum_{i=1}^{\psi} d_{i2}}....g_\ell^{\sum_{i=1}^{\psi} d_{i\ell}}.\left(\prod_{i=1}^{\psi} r_i\right)^N \ mod \ N^2$$
$$= g^{s_1 \sum_{i=1}^{\psi} d_{i1}}.g^{s_2 \sum_{i=1}^{\psi} d_{i2}}....g^{s_\ell \sum_{i=1}^{\psi} d_{i\ell}}.\left(\prod_{i=1}^{\psi} r_i\right)^N \ mod \ N^2$$

$$= g^{s_1 \sum_{i=1}^{\psi} d_{i1} + s_2 \sum_{i=1}^{\psi} d_{i2} .... + s_\ell \sum_{i=1}^{\psi} d_{i\ell}} . \left(\prod_{i=1}^{\psi} r_i \right)^N mod\ N^2 \tag{7}$$

1. In this step we take $M = s_1 \sum_{i=1}^{\psi} d_{i1} + s_2 \sum_{i=1}^{\psi} d_{i2} .... + s_\ell \sum_{i=1}^{\psi} d_{i\ell}$ and $Q = \prod_{i=1}^{\psi} r_i$ then $C = g^M.Q^N\ mod\ N^2$. CA use its secret key (β, γ) to recover M.

$$M = s_1 \sum_{i=1}^{\psi} d_{i1} + s_2 \sum_{i=1}^{\psi} d_{i2} .... + s_\ell \sum_{i=1}^{\psi} d_{i\ell} mod N^2 \tag{8}$$

2. CA uses algorithm (1) to recover and store the aggregated data $(D_1, D_2, ..., D_\ell)$ where $D_j = \sum_{i=1}^{\psi} d_{ij}$.

Correctness: Let $X_\ell = M$ that means

$$X_\ell = s_1 \sum_{i=1}^{\psi} d_{i1} + s_2 \sum_{i=1}^{\psi} d_{i2} + ...+ s_\ell \sum_{i=1}^{\psi} d_{i\ell}\ mod(N). \tag{9}$$

Since $d_{ij} \leq D$ that implies

$$s_1 \sum_{i=1}^{\psi} d_{i1} + s_2 \sum_{i=1}^{\psi} d_{i2} .... + s_{(\ell-1)} \sum_{i=1}^{\psi} d_{i(\ell-1)} < s_1 \sum_{i=1}^{\psi} D + s_2 \sum_{i=1}^{\psi} D .... + s_{(\ell-1)} \sum_{i=1}^{\psi} D = \sum_{j=1}^{\ell-1} s_j \psi D < s_\ell. \tag{10}$$

So, $X_{\ell-1} = X_\ell mod s_\ell = s_1 \sum_{i=1}^{\psi} d_{i1} + \sum_{i=1}^{\psi} d_{i2} + ... + \sum_{i=1}^{\psi} d_{i(\ell-1)}$, and

$$\frac{X_\ell - X_{\ell-1}}{s_\ell} = \frac{s_\ell \sum_{i=1}^{\psi} d_{i\ell}}{s_\ell} = \sum_{i=1}^{\psi} d_{i\ell} = D_\ell. \tag{11}$$

In a similar way, we can show $D_j = \sum_{i=1}^{\psi} d_{ij}$ for $j \in \{1, 2, ..., (\ell - 1)\}$. After the analysis of almost real-time power consumption data $D_1, D_2, ..., D_\ell$ is performed by the CA, it sends a message $m \in G_2$ to HAN users $\{u_i : 1 \leq \psi\}$ in the residential area. It informs them about their power consumption and allows them to reduce their costs. The following are the concrete measures to understand:

1. CA choose $a \in Z_{q'}^*$ randomly and calculates $\overline{M} = (\overline{M_1}, \overline{M_2}, \overline{M_3}, \overline{M_4})$, where

$$\kappa = H(e(P,P)^{\delta a}), \overline{M_1} = E_{\kappa}(m), \overline{M_2} = sP, \overline{M_3} = sy - sR_1, \overline{M_4} = -sR_2. \quad (12)$$

Then CA creates $\varsigma = xh_1(\overline{M}||ID_{ra}||ID_{ca}||T_3)$ as signature and sends $\overline{M}||\varsigma$ to GW.

2. After getting $\overline{M}||\varsigma$ GW verifies $e(P,\varsigma) = e(y, h_1(\overline{M}||ID_{ra}||ID_{ca}||T_3))$. If verification holds then GW broadcasts $\overline{M}$.
3. Upon receiving $\overline{M}$ from GW, every user u_i in HAN uses the key ak_i to get m from $\overline{M}$ as:

$$\begin{aligned}
&\frac{e(\overline{M_2}, \delta P + t_{i1}y)}{e(t_{i1}P, \overline{M_3})e(ti2P, \overline{M_4})e(t_{i1}R_1 + t_{i2}R_2, \overline{M_2})} \\
&= \frac{e(aP, \delta P + t_{i1}y)}{e(t_{i1}P, ay - aR_1)e(t_{i2}P, -aR_2)e(t_{i1}R_1 + t_{i2}R_2, aP)} \\
&= \frac{e(aP, \delta P + t_{i1}y)}{e(t_{i1}P, ay)} e(P,P)^{\delta a},
\end{aligned} \quad (13)$$

$$\kappa = H(e(P,P)^{\delta a}), m = D_{\kappa}(\overline{M_1}). \quad (14)$$

u_i can utilize the recovered information m to calculate when to shift power consumption from peak to non-peak hours in order to maximize efficiency of electricity uses.

5 Security Analysis

The proposed scheme's security characteristics are examined in this section. Our study will focus on how the proposed method can ensure the privacy, data integrity, and source authentication of individual residential users' reports. After this, the CA's respond back in confidential manner by considering the security requirements discussed earlier.

- **Preserving the privacy of each user's report in the suggested model:** In the suggested model, smart meters sense user $U_i's$ data $d_{i1}; d_{i2}; ...; d_{i\ell}$ and form $C_i = g_1^{d_{i1}}.g_1^{d_{i1}}....g_1^{d_{i1}}.r_i^N \; modN^2$, which can be stated implicitly as

$$\begin{aligned}
C_i &= g_1^{d_{i1}}.g_2^{d_{i2}}....g_\ell^{d_{i\ell}}.r_i^N \; modN^2 \\
&= g^{s_1.d_{i1}}.g^{s_2.d_{i2}}....g^{s_\ell.d_{i1}}.r_i^N \; modN^2 \\
&= g^{s_1.d_{i1}+s_2.d_{i2}.......+s_\ell.d_{i\ell}}.r_i^N \; modN^2
\end{aligned} \quad (15)$$

Let w_i be $s_1.d_{i1} + s_2.d_{i2} .... + s_\ell.d_{i\ell}$, and the cipher $C_i = g^{wi}.r_i^N \ mod \ N^2$ which satisfies Paillier Cryptosystem cipher. The semantic security and privacy of the data $d_{i1}, d_{i2} ...., d_{i\ell}$ in w_i are also preserved by the fact that the Paillier Cryptosystem is privacy-preserved and semantically safe against the chosen plaintext attack. As a result, even if the adversary A_d monitors in on C_i, he/she is unable to deduce the contents. The GW will not retrieve each report after collecting all reports $C_1; C_2; ...; C_\psi$ from residential users; instead, it will compute $C = \prod_{i=1}^{\psi} C_i mod N^2$ to conduct report aggregation. As a result, even if adversary A_d gains access to the GW's database, he/she will not be able to obtain the individual reports $d_{i1}; d_{i2}; ...; d_{i\ell}$. Finally, after receiving $C = \prod_{i=1}^{\psi} C_i mod N^2$ from the GW, the CA recovers C as $D_1; D_2; ...; D_\ell$, where a database is used to store each $D_j = \sum_{i=1}^{\psi} d_{ij}$. Even if adversary A_d obtains the data, he won't be able to access the data of every user $U_i, d_{i1}; d_{i2}; ...; d_{i\ell}$ since each $D_j = \sum_{i=1}^{\psi} d_{ij}$ is an aggregated result. As a result of the three factors discussed above, the proposed method protects the privacy of individual users' reports.

- **The proposed approach ensures data integrity and verification for both individual user reports and aggregated reports:** The source authentication and data integrity can be ensured because of the provably secure BLS short signature in the arbitrary oracle model under the CDH issue [19]. As a result, the suggested system can detect adversary A_d's damaging acts in smart grid communications.
- **The advanced scheme also achieves confidentiality of the CA's response:** CA encrypts m as $\overline{M} = (\overline{M_1}, \overline{M_2}, \overline{M_3}, \overline{M_4})$, when CA respond to $U_i's$ in RA. To demonstrate that m's secrecy is guaranteed, we establish that $\overline{M} = (\overline{M_1}, \overline{M_2}, \overline{M_3}, \overline{M_4})$, is semantically safe under DBDH assumption against a chosen-plaintext attack, as mentioned in the following theorem.

Theorem 1 *Under the DBDH assumption, the cipher* $\overline{M} = (\overline{M_1}, \overline{M_2}, \overline{M_3}, \overline{M_4})$, *is semantically safe against chosen-plaintext attack.*

Proof Let $a, b, c \in Z_q$, and $b = \{0, 1\}$ be the variables. Set $W = e(P, P)^{abc}$ if $b = 0$ is true, and set $W \in G_2$ to be a arbitrary element, if $b = 1$ is true. To forecast b given (bP, cP, P, aP), and W is the DBDH problem. Assuming there is a polynomial time adversary A_d with the non-insignificant benefit of breaching semantic security of $\overline{M} = (\overline{M_1}, \overline{M_2}, \overline{M_3}, \overline{M_4})$, in the proposed scheme, To solve the DBDH problem, we might create a new opponent, B_d, who has access to A_d and gains a sizable advantage. Firstly, B_d gets input (aP, bP, cP, P), and W as DBDH instantiation with $W = e(P, P)^{abc}$ when $b = 0$. Then B_d choose $\delta_1, \beta_1, \beta_2 \in Z_q$ adjusts the system parameters at random.

$$Y = xP = aP$$

$$R_1 = aP + \beta_1 P = aP + \beta_1 P, R_2 = \beta_2 P$$

$$e(P, P)^{\delta} = e(aP, bP).e(P, P)^{\delta_1} \quad (16)$$

where $\delta = ab + \delta_1$. The distribution of $(R_1, R_2, Y, e(P, P)^{\delta})$ are unaltered because $\delta_1, \beta_1, \beta_2, aP, bP$ follow randomness. So it is impossible to distinct the real environment from the simulated parameters $(R_1, R_2, Y, e(P, P)^{\delta})$ for A_d.

Now A_d selects m_1 and m_2 from G_2 after getting $(R_1, R_2, Y, e(P, P)^{\delta})$ and send m_1 and m_2 to B_d. Then B_d flips a bit $b^* \in \{0, 1\}$ and produce $\overline{M} = (\overline{M_1}, \overline{M_2}, \overline{M_3}, \overline{M_4})$, where "E" is symmetric encryption (AES/DES).

$$\overline{M_1} = E_{(W.e(cP,\delta^1 P))}(w_{b^*})$$
$$\overline{M_2} = cP$$
$$\overline{M_3} = cY - cR_1 = caP - c(aP + \beta_1 P) = \beta_1 cP$$
$$\overline{M_4} = -cR_2 = -\beta_2 cP. \tag{17}$$

Finally, B_d sends $\overline{M} = (\overline{M_1}, \overline{M_2}, \overline{M_3}, \overline{M_4})$ to A_d. After getting $\overline{M}$, A_d returns b' to B_d. If $b' = b^*$ then B_d guesses $b = 0$. So we have

$$\overline{M_1} = E_{(W.e(P,\delta_1 P))}(w_{b^*})$$
$$= E_{e(P,P)^{abc}.e(cP,\delta_1 P)}(w_{b^*})$$
$$= E_{e(P,P)^{abc+\delta_1 c}}(w_{b^*}) = E_{e(P,P)^{\delta c}}(w_{b^*}) \tag{18}$$

$\overline{M_1}$ becomes a legitimate cipher component after that. In this scenario, A_d will correctly estimate b_* with a chance of $\frac{1}{2} + \epsilon$. As a result, $Pr[success.B_d|b = 0] = \frac{1}{2} + \epsilon$. Due to the unpredictability of W, if $b = 1$, $\overline{M_1} = w_{b^*}.W.e(cP, \delta_1 P)$ is not dependent on b_*. As a result, $Pr[success.B_d|b = 1] = \frac{1}{2}$. In a nutshell, we have the following two scenarios:

$$Pr[success.B_d] = \frac{1}{2}\left(\frac{1}{2} + \epsilon\right) + \frac{1}{2}.\frac{1}{2} = \frac{1}{2} + \frac{\epsilon}{2} \tag{19}$$

The following conclusion contradicts the premise that the DBDH issue is difficult because ϵ is non-negligible. As a consequence, The chosen plaintext attack is semantically safe for the cipher $\overline{M} = (\overline{M_1}, \overline{M_2}, \overline{M_3}, \overline{M_4})$, implying that the CA's answer achieves secrecy in the presented scheme. We can see from the preceding analysis that the provided system is safe and it achieves security goals.

6 Performance Evaluation

This part assesses the proposed scheme's performance in terms of residential user computing complexity, local CG and CA computation complexity, and communication cost of CG-to-CA and user-to-CG communications.

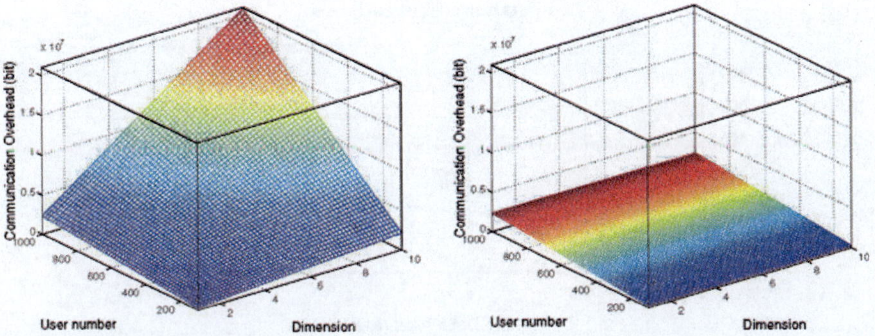

Fig. 2 Illustration of communication cost of traditional design user to central gateway (left) and proposed design user to central gateway (right)

6.1 *Computation Complexity*

If the (residential user) U_i produces an encoded report $C_i||ID_{ra}||U_i||T_1||\varsigma_i$ using the proposed method, it takes $\ell + 1$ times exponential operations in Z_{N^2} to produce C_i, and to generate σ_i, it takes 1 multiplication operation in G_1. The local CG checks the incoming data by executing a batch verification that comprises $\psi + 1$ pairing procedures after obtaining the cipher from ψ users. Furthermore, the CG should combine information from many users so that it can produce a signature for all the combined data. The computational cost of aggregation is minimal since multiplication in Z_{N^2} is regarded insignificant relative to exponentiation and pairing operations, and the multiplication operation in G_1 is required only for creating signature. As for the CA, it uses two pairing operations to verify CG's aggregated data and Paillier decryption to retrieve the data, which involves one exponentiation operation in Z_{N^2}.

The CA then responds to the CG, who then responds to residential users, and CA performs four multiplication operations in G_1 and one exponential operation in G_2 for the production of the response message. The overhead computational costs or expenses for the CG are the two pairing procedures that are used to provide the response m to consumers. All users will incur additional computational expenses as a result of four pairing processes after receiving the response. By e_t, m_t, p_t, and e'_t, respectively, we use to denote the computing expenses of an exponential operation in Z_{N^2}, multiplication operation done in G_1, a pairing operation, and an exponentiation operation in G_2. Then the total computation cost will be $(\ell + 1).e_t + m_t + 4p_t$, $(\psi + 3).p_t + m_t$ and $2.p_t + e_t + 4m_t + e'_t$ for user, CG and CA, respectively. As mentioned in the suggested method,the residential user is able to combine many data sets of multiple dimensions into a single compressed data set. It significantly decreases the time it takes for users to encrypt their data. In this part, we will compare the presented scheme to a conventional technique (described by TRAD), in which each user creates a cipher for one-dimensional data. If we consider ℓ-dimensional data, this option requires the user to produce ℓ ciphers and spends a total of $2.\ell$

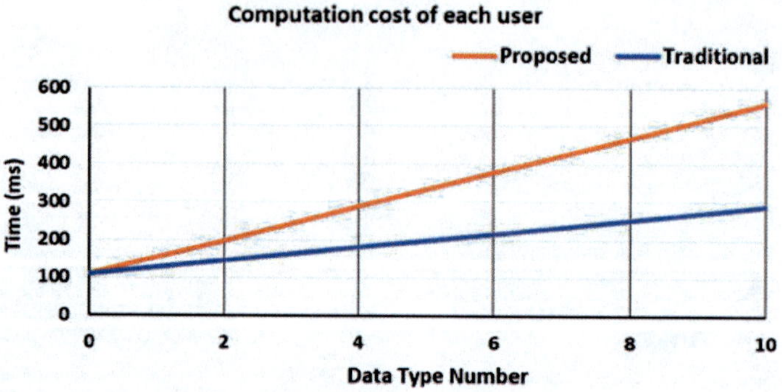

Fig. 3 Illustration of computation cost of each user

exponential operations in Z_{N^2}. The overall computing expenses for a user would be $2\ell.e_t + m_t$, including the one G_2 multiplication operation, used for the signature (see Fig. 3). The CG then sends ℓ ciphers to the CA, each cipher containing the sum of all users' one-dimensional data. The CA verification will need two pairing operations. There are ℓ decryptions carried out by CA, therefore the decryption takes ℓ exponentiation operations in Z_{N^2}. We presume TRAD operates in the same way as the proposed scheme in the response phase. So the computational cost will be $2\ell.e_t + m_t + 4p_t$, $2p_t + \ell.e_t + 4.m_t + e_t'$ and $(\psi + 3)p_t + m_t$ and, for a residential user, the CA and CG, respectively. Moreover, we analyze the operating costs using the PBC [20] and MIRACL [21] functioning on a computer with a 3.0 GHz CPU and 512 MB of RAM are libraries. According to the findings, It takes roughly 12.4 milliseconds to perform one exponentiation in Z_{N^2} ($|N^2| = 2048$), 6.4 ms to execute a single 160-bit G_1 multiplication operation, and 20 ms to execute a single pairing operation.

Table 2 gives an illustration of computation costs of relevant protocols. The symbols of costs are:

- T_e: Exponentiation operation time in Z_{n^2} ,
- T_m: Point multiplication calculation time in ECC,
- T_p: Pairing time,
- T_{et}: exponentiation time in a cyclic group (e.g., Grp_T in [26]),
- T_{log}: Pollard lambda method calculation time,
- $T_{e(ELG)}$: ELGamal encryption time,
- $T_{d(ELG)}$: ELGamal decryption time,

Table 2 Comparision of computational costs

Protocol	User	GW	CS
[26]	$2nT_e + T_m + 4T_p$	$(q+3)T_p + T_m$	$2T_p + nT_e + 4T_m + T_{et}$
[26]	$(n+1)T_e + T_m$	$(q+3)T_p + T_m$	$2T_p + T_e$
[27]	$2T_e + T_m$	$(q+2)T_p + 3T_e + T_m$	$2T_p + T_e$
[28]	$(n+1)T_e + T_p$	$(q+1)T_p$	$T_p + T_m + T_e$
[29]	$T_{eELG} + 2T_e$	$(q+1)T_p + T_e$	$2(q+1)T_p + T_{dELG} + T_{log}$
[30]	$T_e + T_m$	$(q+1)T_e + (q+1)T_m$	$T_e + T_m$
Proposed	$nT_e + T_p$	$(q+1)T_e + T_p$	$T_p + T_e$

6.2 *Communication Cost*

The proposed scheme's communications (see Fig. 2) may be split into two categories: The first category is about user-to-CG communication and the second category is about CG-to-CA communication. We'll start with the first category user-to-CG communication, in which users create data reports and transmit them to the local gateway CG. If 1024-bit N is choosen with 160-bit G_1, the report of the data is produced in the form of $C_i||ID_ra||U_i||T_1||\varsigma_i$ for the user U_i, and it is defined that size should be $Size = 2048 + |ID_{ra}| + |U_i| + 160$. The CG receives reports from a total of ψ users, implying a total communication overhead of $S_{pro} = Size \times \psi$ between users and the CG. Alternatively, each user must produce a cipher of 2048-bit for every dimensional data if the conventional TRAD method is used. In this instance, the user-to-CG communication overhead will grow to $S_{trad} = 2048.\ell + |ID_{ra}| + |U_i| + |T_1|160$. When compared to the TRAD, the suggested method always achieves reduced communication cost. Following that, we look at the proposed scheme and TRAD's CG-to-CA communication. The CG initiates communication in the proposed scheme with the goal of delivering the final collected report to the CA. The report, which is collected, is written in $C||ID_{ra}||ID_{CG}||T_1||\varsigma_g$ format and has a length of $2048 + |ID_{ra}| + |ID_{CG}| + |T_1| + 160$. In TRAD, however, various dimensional data must be aggregated individually, resulting in aggregated reports with a size of $2048 + |ID_{ra}| + |ID_{CG}| + |T_1| + 160$. We also exhibit the communication cost in terms number of users, that is ψ and the data type that is represented by ℓ. It is demonstrated that the proposed technique effectively lowers the CG-to-OA communication cost. According to the results of the foregoing study, the suggested scheme is both computationally and communicationally efficient, making it appropriate for the high-frequency real-time data gathering in smart grid communications.

7 Conclusions

The demonstration of a productive and privacy-preserving gathering technique for a smart grid communications is done in this article. The proposed framework multidimensional data uses super increasing sequence and the homomorphic Paillier cryptosystem. Collection of data is done directly on the cipher at the local gateways without any decryption for transferred data from users to smart grid operating centers, in addition to it, the gathered outcome of the source data may be accessed at the operation centers. Finally, our findings demonstrate that the proposed method is cost-effective in terms of computing and communication, is appropriate for large user groups, and allows for the flexible and quick expansion of residential scales in smart grids.

References

1. Wu D, Zhou C (2011) Fault-tolerant and scalable key management for smart grid. IEEE Trans Smart Grid 2(2):375–381
2. Xia J, Wang Y (2012) Secure key distribution for the smart grid. IEEE Trans Smart Grid 3(3):1437–1443
3. Park JH, Kim M, Kwon D (2013) Security weakness in the smart grid key distribution scheme proposed by Xia and Wang. IEEE Trans Smart Grid 4(3):1613–1614
4. Tsai JL, Lo NW (2015) Secure anonymous key distribution scheme for smart grid. IEEE Trans Smart Grid 7(2):906–914
5. Odelu V, Das AK, Wazid M, Conti M (2016) Provably secure authenticated key agreement scheme for smart grid. IEEE Trans Smart Grid 9(3):1900–1910
6. Kumar P, Gurtov A, Sain M, Martin A, Ha PH (2018) Lightweight authentication and key agreement for smart metering in smart energy networks. IEEE Trans Smart Grid 10(4):4349–4359
7. Garg S, Kaur K, Kaddoum G, Rodrigues JJ, Guizani M (2019) Secure and lightweight authentication scheme for smart metering infrastructure in smart grid. IEEE Trans Ind Inform 16:3548–3557
8. Abbasinezhad-Mood DN (2018) An anonymous ECC-based self-certified key distribution scheme for the smart grid. IEEE Trans Ind Electron 65(10):7996–8004
9. Boneh D, Franklin MK (2001) Identity-based encryption from the Weil Pairing. In: Proceeding of 21st annual international cryptology conference. Advances in cryptology (CRYPTO), pp 213–229
10. Sang Y, Shen H, Tian H (2009) Privacy-preserving tuple matching in distributed databases. IEEE Trans Knowl Data Eng 21(12):1767–1782
11. Zhong S (2007) Privacy-preserving algorithms for distributed mining of frequent itemsets. Inf Sci 177(2):490–503
12. Chen Y, Martínez J-F, Castillejo P, López L (2017) An anonymous authentication and key establish scheme for smart grid: FAuth. Energies 10(9):1354
13. Braeken A, Kumar P, Martin A (2018) Efficient and provably secure key agreement for modern smart metering communications. Energies 11(10):2662
14. Abbasinezhad-Mood D, Nikooghadam M (2018) Design and hardware implementation of a security-enhanced elliptic curve cryptography based lightweight authentication scheme for smart grid communications. Future Gener Comput Syst 84:47–57
15. Palacios-Garcia EJ, Carpent X, Bos JW, Deconinck G (2022) Efficient privacy-preserving aggregation for demand side management of residential loads. Appl Energy 328:120112

16. Li Kunchang, Shi Runhua, Mingxia Wu, Li Yifei, Zhang Xiaoxu (2022) A novel privacy-preserving multi-level aggregate signcryption and query scheme for Smart Grid via mobile fog computing. J Inf Secur Appl 67:103214
17. Zhang Wenzheng, Liu Shiyun, Xia Zhe (2022) A distributed privacy-preserving data aggregation scheme for smart grid with fine-grained access control. J Inf Secur Appl 66:103118
18. Boneh D, Lynn B, Shacham H (2004) Short signatures from the Weil pairing. J Cryptol 17(4):297–319
19. Bellare M, Rogaway P (1993) Random oracles are practical: a paradigm for designing efficient protocols. In: Proceedings of ACM conferences computer and communications security, pp 62–73
20. Lynn B (2012) PBC library. http://crypto.stanford.edu/pbc/
21. Multiprecision Integer and Rational Arithmetic c/c++ Library (2012) http://www.shamus.ie/
22. Zhai Feng, Yang Ting, Zhao Bing, Chen Hao (2022) Privacy-preserving outsourcing algorithms for multidimensional data encryption in smart grids. Sensors (MDPI) 22(12):4365
23. Ming Y, Li Y, Zhao Y, Yang P (2022) Efficient privacy-preserving data aggregation scheme with fault tolerance in smart grid. In: Security and communication networks, Hindawi, vol 2022
24. Singh AK, Kumar J (2023) A secure and privacy-preserving data aggregation and classification model for smart grid. In: Multimedia tools and applications. Springer, pp 1–19
25. Wu Liqiang, Fu Shaojing, Luo Yuchuan, Yan Hongyang, Shi Heyuan, Xu Ming (2023) A robust and lightweight privacy-preserving data aggregation scheme for smart grid. IEEE Trans Dependable Secur Comput
26. Lu R, Liang X, Li X, Lin X, Shen X (2012) EPPA: an efficient and privacy-preserving aggregation scheme for secure smart grid communications. IEEE Trans Parallel Distrib Syst 23(9):1621–1631
27. Shen H, Zhang M, Shen J (2017) Efficient privacy-preserving cube-data aggregation scheme for smart grids. IEEE Trans Inf Forensics Secur 12(6):1369–1381
28. Chen Y, Martínez-Ortega, J-F, Castillejo P, López L (2019) A homomorphic-based multiple data aggregation scheme for smart grid. IEEE Sens J 19(10):3921–3929
29. Zuo X, Li L, Peng H, Luo S, Yang Y (2020) Privacy-preserving multidimensional data aggregation scheme without trusted authority in smart grid. IEEE Syst J 15(1):395–406
30. Mohammadali A, Haghighi MS (2021) A privacy-preserving homomorphic scheme with multiple dimensions and fault tolerance for metering data aggregation in smart grid. IEEE Trans Smart Grid 12(6):5212–5220

A Low-Cost WECS for Remote Area Electrification

S. K. Gupta, R. K. Srivastava, and Meet Kumari

Abstract The Government of India under the Ministry of New and Renewable Energy has taken an initiative to electrify remote areas. Though, due to physical terrain, it is difficult and costly to transmit grid power to all remote areas. A stand-alone generation has ample potential to electrify these locations. In this paper, a low-cost stand-alone wind energy conversion system (WECS) has been designed. The objective of the paper is to make a robust system with improved power quality at a low cost. Appropriate generator topology has been selected with field-weakening technique for voltage regulation at above-rated wind speed. At below-rated wind speed voltage boost has been achieved by interchanging coil connections using smart relays. Internet of Things (IoT) technology can be used to establish communication between a sensor and multiple relays of different generators in the wind farm. The hardware implementation of the technique has been done and results have been presented for further analysis

Keywords Wind energy conversion system · Voltage control · Permanent magnet synchronous generator · Remote area electrification

S. K. Gupta (✉)
Department of Electrical Engineering, Chandigarh University, Mohali, India
e-mail: shailendra.e12031@cumail.in

R. K. Srivastava
Department of Electrical Engineering, IIT BHU, Varanasi, India
e-mail: rksrivastava.eee@iitbhu.ac.in

M. Kumari
Department of Electronics and Communication Engineering, Chandigarh University, Mohali, India
e-mail: meet.e8072@cumail.in

A. Kumar et al. (eds.), *Decarbonisation and Digitization of the Energy System*, Lecture Notes in Electrical Engineering 1099, https://doi.org/10.1007/978-981-99-7630-0_4

1 Introduction

Geography has made some remote locations still un-electrified. In these remote locations, power transfer from the centralized grid is not cost-effective. A stand-alone generation has ample potential to electrify these areas using different forms of renewable energies such as solar, wind, and bio-waste. This paper focuses on tapping wind energy through a wind energy conversion system (WECS) for remote area electrification. A WECS mainly consists of a wind turbine, to convert wind energy to rotational energy; a wind generator, to convert mechanical rotational energy of turbine to electrical energy; and a power electronic converter for power conditioning of the electrical power.

In WECS, the variation in wind speed is high. The high power variation in WECS limits its penetration into the electrical grid [1]. The problem is severe in the stand-alone generation. Therefore, in a stand-alone generation, the excess output power is either stored or regulated [2]. Storing the excess power is a costly affair and therefore should be optimized with suitable power regulation technique. The output power regulation is achieved either from the turbine end or generator end. From the turbine end, the wind intermittence is countered by the blade-pitch control and/or yaw control [3]. From the generator end, generator speed control [4, 5] or/and DC-link voltage/current control [6, 7] technique is used. In the stand-alone generation, the power rating of the system is generally low. In low-power system, the yaw control system and blade-pitch control are considered as complex and costly, and therefore, in this paper, the power regulation is achieved using the generator-end control.

For generator-end control, choosing the right generator topology is essential in optimizing the control system and cost of the plant. In a stand-alone WECS, the permanent-magnet-excited generator is preferred owing to the absence of field winding and the grid [8]. To further reduce the number of active switches in the power conditioner, increase the reliability, and reduce the maintenance cost of the system, a passive rectifier is used at generator end. However, use of passive rectifier looses the freedom of power regulation from generator end.

In this paper, wind intermittence is controlled by producing variable magnetic field by changing the inner geometry of the generator. A dual-stator sandwiched-rotor axial-flux machine is the most suitable topology owing to its geometry [9]. The two stators can be made to be displaced with respect to each other for field weakening of the generator. A detailed description is provided in Sect. 3. The technique of generator voltage control by field weakening frees load from turbine control and generator-end converter to have more robust and reliable solution. Owen et. al have categorized such mechanism under mechanically flux-weakened topology (MFWT) [10]. MFWT is simple in construction, incur low manufacturing cost, and reliable and safe for high-power NdFe magnets in comparison to electrically flux-weakened topology (EFWT) [10].

It has been pointed out that MFWT has only flux-weakening capability, except in topology with air-gap length variation [10], and thus can only buck the voltage at a predefined level for above-rated wind speed. In this paper, to boost the generator voltage, at low wind speed, the winding parallel path is reduced by using high-power smart relays. Internet of Things (IoT) technology can be used to actuate multiple relays associated with different generators in a wind farm at the same instance. IoT establishes communication between the sensor and multiple relays using short-distance communication techniques such as Wi-Fi, ZigBee, Bluetooth, Li-Fi through telecommunication network, or Internet [11].

2 Generator Topology

Choosing a generator topology depends upon the application. As discussed in previous section, a permanent magnet synchronous generator is to be used as generator in the proposed system. Also, it has been highlighted that there should be a provision to shift the inner structure to have magnetic field weakening in the generator. Considering the purpose, a dual-stator/rotor axial-flux permanent magnet synchronous generator (AF PMSG) has been chosen for experimental hardware development owing to its easy construction, ease of mechanical field weakening, higher power density, efficiency, and higher magnet usability factor [9, 12, 13]. However, the chosen generator comes in two variations Kalman type (dual stator and sandwiched rotor) and Torus type (dual rotor and sandwiched stator) as shown in Fig. 1a and b [14].

A Kalman-type dual-stator sandwiched-rotor topology has been chosen in the proposed system owing to its low inertia [15, 16] and ease of changing the inner structure [9]. Owing to the low inertia of the topology, the Kalman-type generator can have higher radius and output power for the same inertia of a Torus-type generator topology. However, the choice of such topology increases the axial length of the generator. Therefore, the magnetic flux path effective length also increases, consequently decreasing magnetic loading and output voltage.

(a) Dual-stator and single-rotor topology of Axial-flux permanent magnet generator.

(b) Single-stator and Dual-rotor topology of Axial-flux permanent magnet generator.

Fig. 1 Topologies of axial-flux permanent magnet generator [14]

3 Generator Design

The typical process followed in designing an axial-flux machine is first to decide upon machine rating such as phase voltage, phase current, output power, magnetic loading, electric loading, and its cooling. Second, standard sizing equation [17, 18] is used to decide upon the dimension of the machine such as outer diameter, inner diameter, axial length, permanent magnet dimension, and air-gap length. Third and final steps include deciding upon winding, teeth, slot, and yoke dimensions.

In this paper, the machine has been designed as per the need of the proposed system, availability of material, and ease of construction. The first step of the design process has been avoided to reduce the cost of machine and due to the fact that fabricated prototype is just to test the operating principle of proposed WECS. In the second step, the machine dimensions have been decided upon by the choice of aspect ratio of AFM. The outer diameter and air gap of the machine are decided upon availability of NdFeB magnet. Before proceeding toward third step a thought process was carried out in deciding stator core material. Pollinder et al. [15] and Capponi et al. [16] have used an air-core Torus-type dual-rotor topology in designing a low-cost and low-weight cogging less generator. On the same line in this paper an air-core topology has been considered for machine fabrication. The construction details have been given in Fig. 2. The specifications of the generator have been provided in Table 1.

In the topology, PMs are pasted on the both sides of the disk-shaped rotor. The rotor is finally put in between the two stators. Figure 3a and b shows the rotor and one of the wounded stators used in the hardware, respectively. Both stators are mounted on the shaft. Pottings, consisting of bearings, are used to support the stators on the shaft. Figure 4a shows the arrangement of mounting stator on the shaft using potting arrangement. Figure 4b shows the rotor mounted on the shaft. To both sides of the rotor, the shaft is stepped. The potting arrangements are placed on the shaft steps which fix the air gap between the rotor and the stators on the both sides.

MFWT in the generator is achieved by having a provision of angular displacement of the rotating stator (RS) with respect to fixed stator (FS). Both stators are equipped with windings that are connected in series. Upon displacement of the RS angularly, a phase displacement is achieved between emf induced in RS winding and in FS

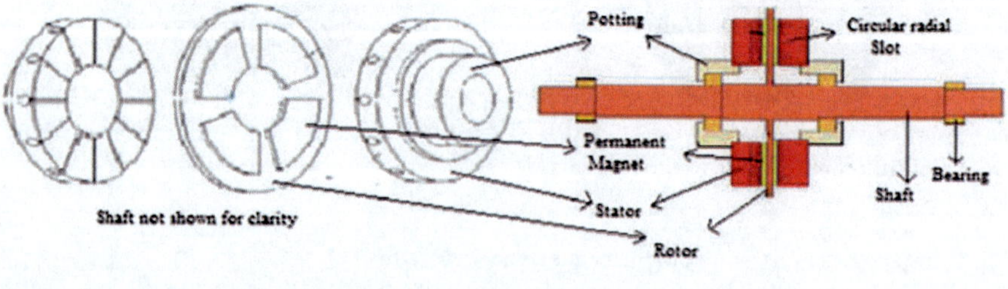

Fig. 2 Construction details of the Kalman-type dual-stator sandwiched-rotor axial-flux permanent magnet synchronous generator

Table 1 Design parameters of the air-core three-phase axial-flux PMSG [19, 20]

Parameter	Value
No. of phases	3
Output phase voltage	15 V per phase per stator
Power	90 W
Rated speed	1000 rpm
No. of poles	4
No. of slots	18
Slot dimension	Rectangular slots 8 mm wide and 10 mm depth
Inner radius of stator	48 mm
Outer radius of stator	80 mm
No. of turns per coil	22
Winding type	Double layer distributed
NdFeB magnet	N35 H
Magnet operating temperature	200 °C
Magnet shape	Arc shape
Magnet dimensions	80 mm outer radius 48 mm inner radius
Magnet arc length	70° mech degrees
Magnet width	3 mm
Residual flux density	1.170 T
Air-gap clearance	1 mm

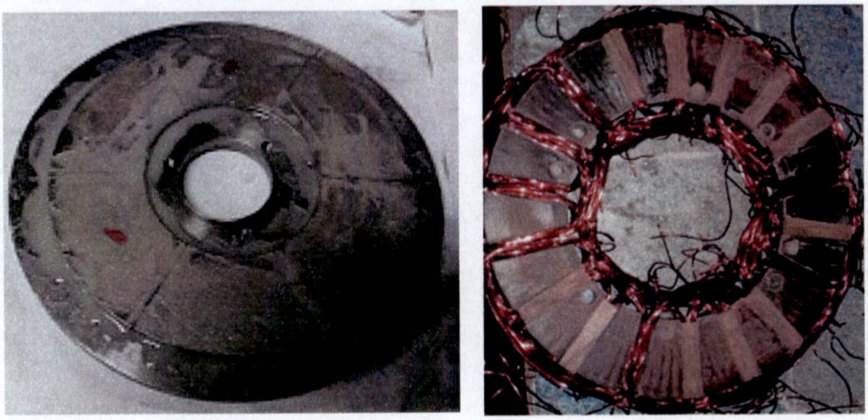

(a) Permanent magnet pasted on the rotor.

(b) Pancake-shaped air-core stator with distributed winding .

Fig. 3 Rotor and stator used in the hardware setup

(a) Arrangement of two stators mounted on the shaft through potting arrangement.

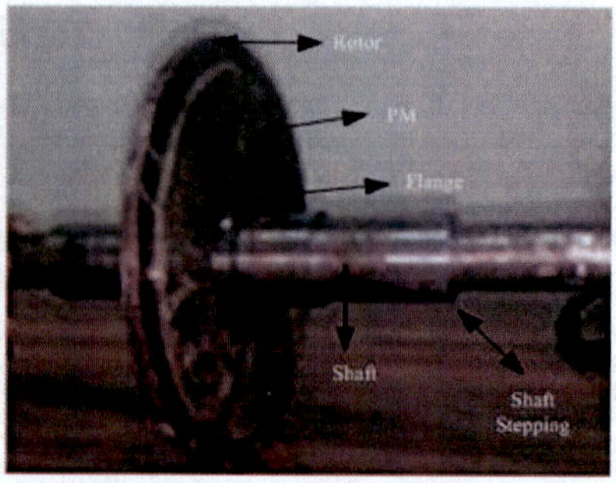

(b) Rotor pinned to the shaft .

Fig. 4 Arrangement of rotor and stators mounting on the shaft

winding. The series output voltage V thus depends upon the angular position of RS with respect to FS. Mathematically, V is given in Eq. 1 and amplitude in Eq. 2. Here, ϕ is the electrical angle by which the RS has been shifted with respect to the FS.

$$V = V + V\angle\phi \tag{1}$$

$$|V| = V cos\left(\frac{\phi}{2}\right). \tag{2}$$

RS is angularly shifted by the use of a worm wheel gear. The worm wheel gear is attached to a stepper motor. Stepper motor receives the command for the rotation of RS to maintain the generator voltage constant with respect to a reference voltage. The setup is shown in Fig. 5. The final topology is capable to shift rotating stator with respect to the fixed stator.

Third and final steps of deciding upon winding and number of stator slots are based upon imparting the generator with voltage boost capability at low wind speed. It has been already stressed upon the fact that flux variations achieved by mechanical means have only weakening capability and, thus, can only be used at above-rated wind speed. In this paper, it is proposed to use a generator with multiple parallel paths. In the generator, the provision would be given to change the winding's number of parallel paths. At low wind speed, number of parallel paths would be decreased to increase the number of coils per phase. This would boost the voltage by the same factor by which the parallel paths have been reduced. Obviously, this would come at a cost of decreased output current rating, and thus load shedding would be a preferred choice in the proposed stand-alone generation system. Load shedding seems more logical if we consider the fact that lower wind speed would generate less power. The proposed system will enable the distribution of power at low wind speed as well.

It is proposed to use a single-phase generator to use all the coils in designing a machine with increased number of parallel paths and number of poles.

Fig. 5 Hardware setup to perform experiments on proposed system

4 Principle of Operation

Operating principle of proposed WECS is illustrated by block diagram in Fig. 6. Block diagram explains control logic applicable to the proposed WECS. The principal control logic is to implement generator output voltage control. The voltage generated is observed with respect to a reference voltage and a voltage error is calculated.

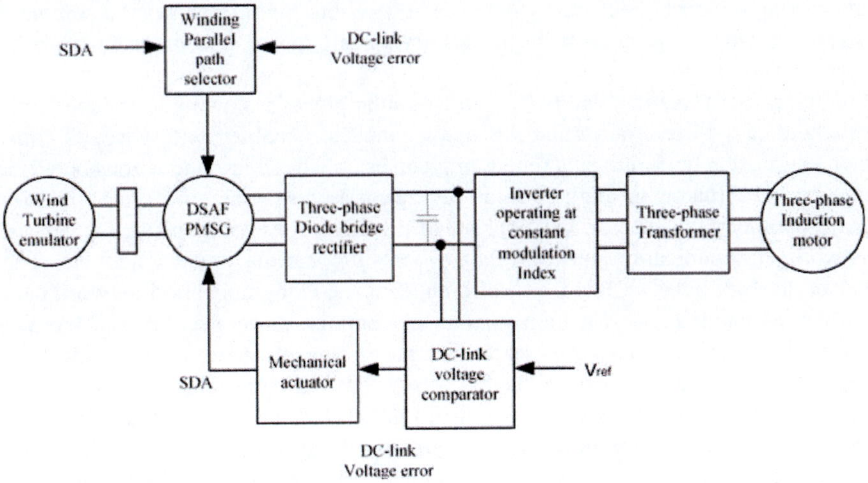

Fig. 6 Block diagram of proposed system

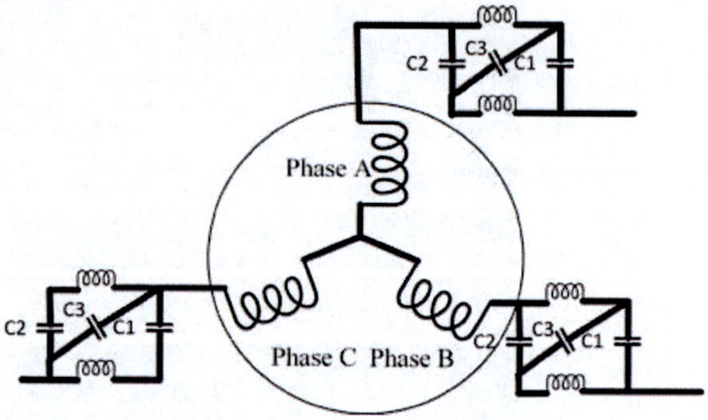

Fig. 7 Circuit showing the details of connecting smart relays in the winding of the wind generator

Based on the voltage error, the stepper motor sends a command to rotate the RS so as to reduce the voltage error. The second control logic is to decide the number of parallel paths of the generator. First the SDA is checked, if SDA is zero and generator voltage is still less than the reference voltage, the winding-parallel-path-selector block generates a command signal to decrease the parallel path of one of the two stators. If then also the generator voltage is less than the reference voltage then the parallel path of the second stator is also reduced to increase the output voltage. Parallel path selector comprises high-power electrical relay connected between coils to switch connection between coils of a single phase. The operation can also be performed using microprocessor-based PLCs [21] or high-power relay-enabled IoT. The circuitry to implement the technique is given by per phase winding diagram in Fig. 7. Figure 7 shows two parallel paths connected by two normally closed contactors, C1 and C2, and one normally open-type contactor, C3. A control circuitry can be designed that closes C1 and C2 and opens C3 for above-rated wind speed and opening C1 and C2 and closing C3 for below-rated wind speed operation. Figure 9 shows the complete hardware setup developed in the laboratory to test the proposed system in both operating conditions.

High-power relay-enabled IoT is more popular nowadays owing to cheaper, reliable, and easy-to-program microprocessor controller, enabled with wireless communication, that performs data processing and establishes communications between sensors and actuators through Internet. Figure 8 presents the architecture of the IoT-based parallel path selector. IoT consists of primarily four components, i.e., sensor network, communication network, data processing, and the actuator network. The sensor network collects data from different sensors. Communication network uses different communication medium such as telecommunication network and Internet to transfer the sensor data to clouds for data processing and control command signal generation. The data processing can also be done by using microprocessor-enabled Arduino board. The processed data/control signal is then communicated to the actuator using different communication mediums as discussed before [11]. The direct advantage of using the IoT-enabled parallel path selector is to synchronize the operation of multiple generators from a wind farm that are interconnected through a single-electric grid. The added advantage of using Internet-connected sensor and

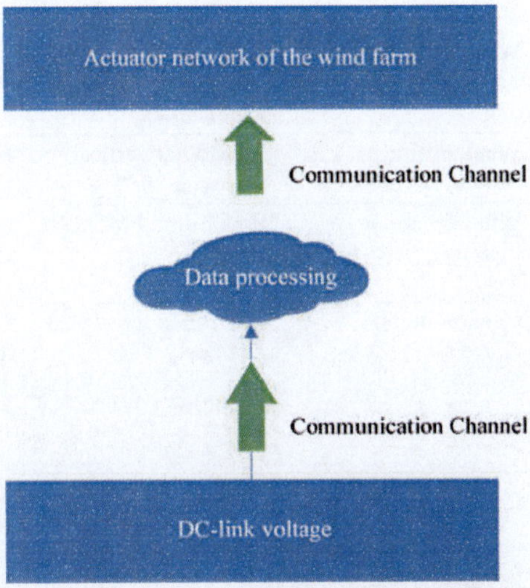

Fig. 8 Architecture of IoT-enabled high-power relay-based winding parallel path selector

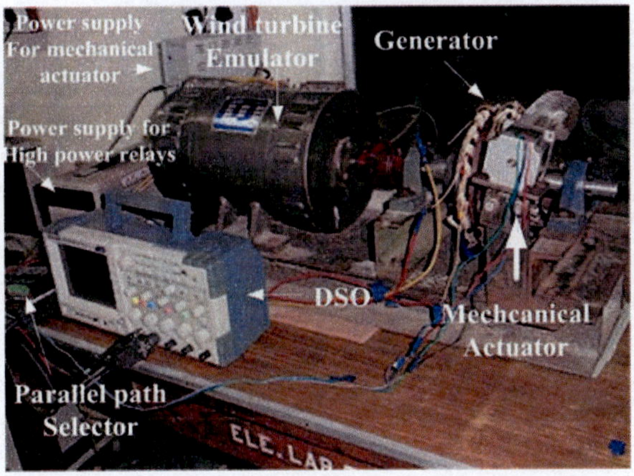

Fig. 9 Complete hardware setup to test the complete system

actuator is the saving of the communication line and, thus, avoiding the interference between the power line with the communication line (Fig. 9).

5 Cost Analysis

Wind turbine, wind generator, generator-end converter, and grid-end converter are the basic components of a WECS. The cost of a WECS depends upon the above-mentioned components. In the proposed system, to reduce the cost of the system, a core-less axial-flux PMSG has been used as wind generator as proposed by Ani et al. [15]. Further, an uncontrolled generator-end converter has been used in the

Table 2 Cost comparison of proposed system with the conventional system

WECS components	Technology used in WECS for remote area applications	Technology used in the proposed system	Cost comparison
Wind turbine	Horizontal/vertical axis wind turbine	Horizontal/vertical axis wind turbine	Similar
Wind generator topology	Radial-flux PMSG	Air-core axial-flux PMSG	Air-core axial-flux PMSG is less costly [15]
Generator-end converter	Controlled three-phase rectifier	Uncontrolled three-phase rectifier	Uncontrolled rectifier is less costly than controlled rectifier. The use of controlled switches and the control circuitry in the controlled rectifier makes it more costly
Grid-end converter	Three-phase inverter	Three-phase inverter	Similar

system that reduces the cost and complexity of the system. The use of generator-end converter results in a variable DC-link voltage. Conventional systems that use an uncontrolled generator-end converter also use a separate DC/DC converter to control DC-link voltage [8]. Instead, a mechanical field-weakening technique has been used in the proposed system for controlling the DC-link voltage at wind speed above-rated value and changing of parallel paths of the generator windings at wind speed below-rated value. Table 2 shows the cost comparison of the proposed system with the conventional system.

6 Results and Discussion

Experimental tests have been conducted based on the principle of operation, as discussed in the previous section. To test the proposed WECS, a wind turbine emulator (WTE) is designed using a DC motor [22]. The WTE emulates the characteristics of a wind turbine and is very useful for testing and validating proof-of-concept machines in the laboratory environment. The digital pulses required for the WTE, parallel path selector, and mechanical actuator have been generated using DSPACE hardware-in-loop system. In present work, DS1104 board has been used. The output signals such as voltage, current, speed, mechanical actuator angle and its processing is done on the real-time interface provided by DSPACE. The real-time interface has been used to collect the data in Excel and is plotted using MATLAB.

To test the proposed proof of concept, three cases have been considered. In the first case, the output voltage regulation capability of the generator, at variable wind speed,

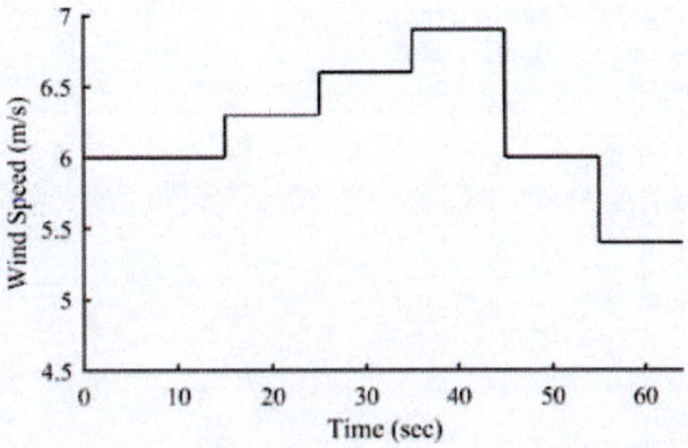

Fig. 10 Wind speed profile for testing the voltage regulation capability of setup in first two cases

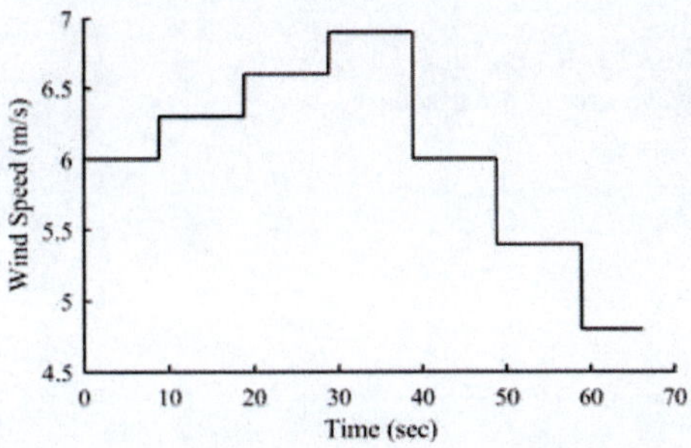

Fig. 11 Wind speed profile for testing the voltage regulation capability of setup in the third case

has been checked for stationary stator having two parallel paths and rotatable stator having one parallel path. In the second case, the stationary stator has one parallel path and the rotatable stator has two parallel paths. In the third case, both the stators have single parallel paths.

Figures 10 and 11 show wind speed waveform for first two cases and the third case, respectively. There is a slight difference between both the waveforms. However, the wind speed and the step size are same for both the waveforms. Figures 12 and 13 show the rectified output voltage and the RS position waveform, respectively, for first case. Figures 14 and 15 show the rectified output voltage and the RS position waveform, respectively, for second case. It is to be noted that the RS position waveform, to maintain the rectified voltage as to be 20 V, in case 2, is little higher than case 1. Therefore, it is inferred that, higher number of parallel paths reduces the capability of RS to regulate the voltage of FS. This is due to higher number of parallel path in case 2. Thus, the per phase voltage in case 2 is half of the per phase voltage in case 1. Therefore, it is inferred that, higher number of parallel paths reduces the capability of RS to regulate the voltage of FS.

Figures 16 and 17 show the rectified voltage and RS position for third case. It is to be noted that the output voltage for RS position to be zero is higher than other two cases. Though the voltage is little higher than expected because of the drive running at higher rotor speed. In this case, drive speed has been found to be 12.91 % higher than other two cases. At wind speed 7.2 m/s, the peak-to-peak voltage of the ripple to controlled DC voltage ratio is found to be 3.2 %. At the same wind speed, in

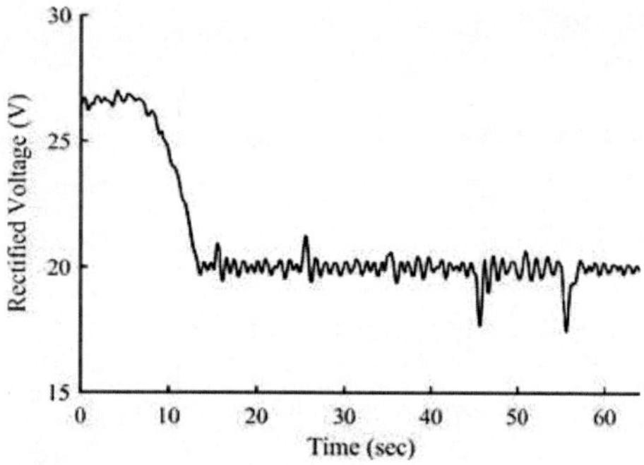

Fig. 12 Rectified output voltage in the first case

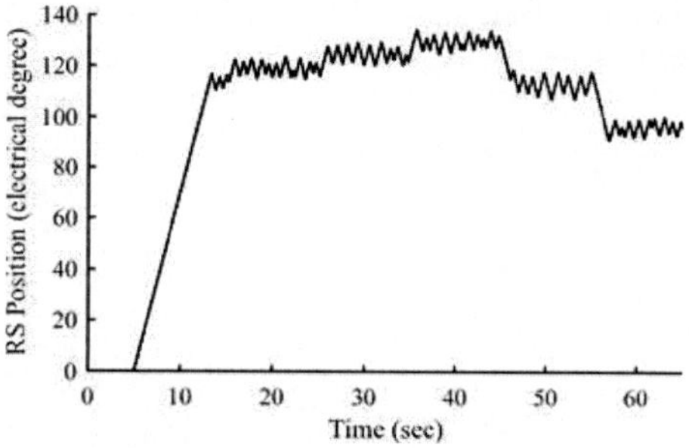

Fig. 13 RS position waveform in the first case

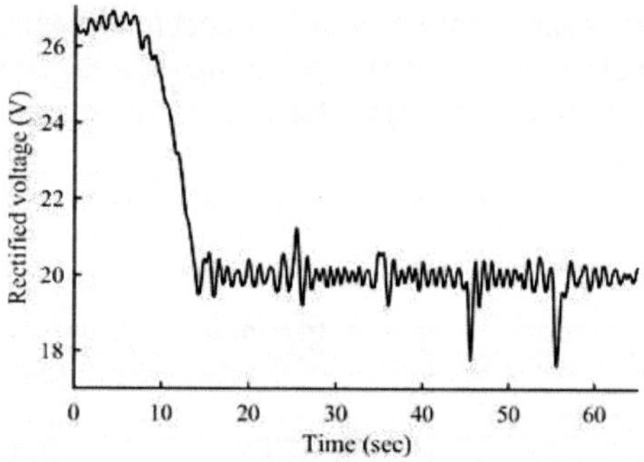

Fig. 14 Rectified output voltage in the second case

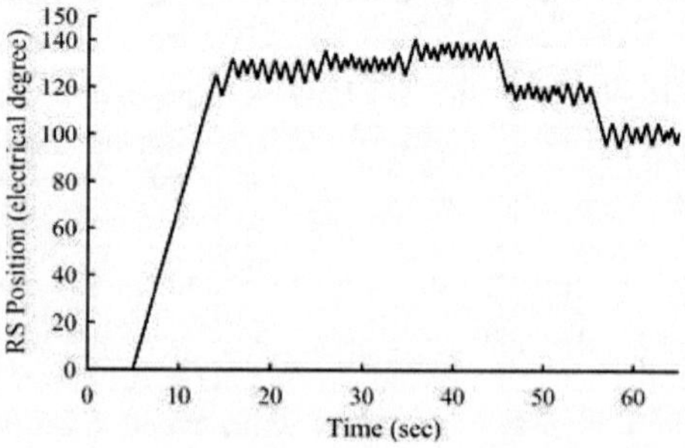

Fig. 15 RS position waveform in the second case

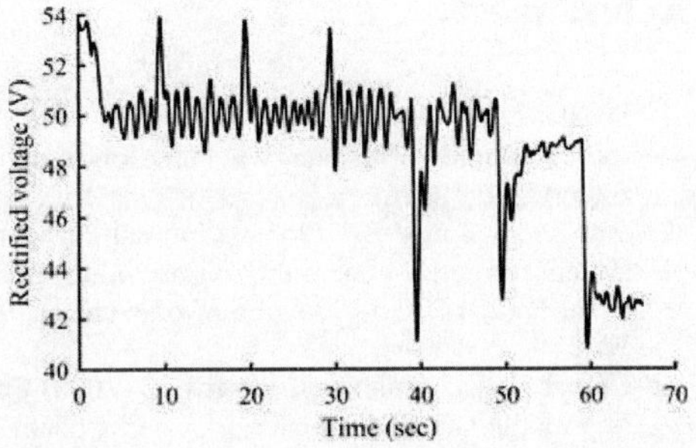

Fig. 16 Rectified output voltage in the third case

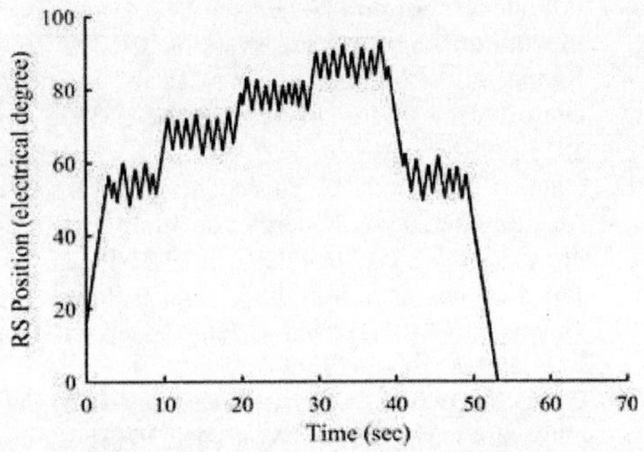

Fig. 17 RS position waveform in the third case

first two cases, the ratio has been found to be around 2.5 %. Higher ripple in case 3 is due to the use of same actuator to angularly displace the RS with respect to FS in all the three cases. Thus, proposed system will have higher amount of ripples in output power when wind speed is high or in other words the boosting capability of the system is not being used. The proposed system will have an added advantage of reducing the output power ripple while boosting the output voltage.

7 Conclusion

In this paper, a low-cost stand-alone wind energy conversion system (WECS) has been designed. The design procedure focuses upon cost, efficiency, and output power quality of the system. As per the requirement of the application, a Kalman-type air-core dual-stator sandwiched-rotor permanent magnet synchronous generator has been designed and proposed to be used as generator. The generator output voltage at above-rated wind speed is regulated by the mechanical field weakening of the generator. At below-rated wind speed, voltage boost has been achieved by decreasing number of parallel paths of the generator by using IoT-enabled high-power relays. Logic block diagram of proposed controller has been presented and explained. Hardware results show output voltage control for above- and below-rated wind speed feeding constant power. Results have shown good agreement with the hypothesis.

References

1. Dalala Z, Al-Omari M, Al-Addous M, Bdour M, Al-Khasawneh Y, Alkasrawi M (2022) Increased renewable energy penetration in national electrical grids constraints and solutions. Energy 246:123361
2. Patel AM, Singal SK (2019) Optimal component selection of integrated renewable energy system for power generation in stand-alone applications. Energy 175:481–504
3. Heier S (2014) Grid integration of wind energy: onshore and offshore conversion systems. Wiley
4. Chinchilla M, Arnaltes S, Burgos JC (2006) Control of permanent-magnet generators applied to variable-speed wind-energy systems connected to the grid. IEEE Trans Energy Convers 21(1):130–135
5. Polinder H, Ferreira JA, Jensen BB, Abrahamsen AB, Atallah K, McMahon RA (2013) Trends in wind turbine generator systems. IEEE J Emerg Sel Top Power Electron 1(3):174–185
6. Ramasamy T, Abdul Basheer A, Tak M-H, Joo Y-H, Lee S-R (2022) An effective dc-link voltage control strategy for grid-connected PMVG-based wind energy conversion system. Energies 15(8):2931
7. Dalala ZM, Zahid ZU, Yu W, Cho Y, Lai J-S (2013) Design and analysis of an MPPT technique for small-scale wind energy conversion systems. IEEE Trans Energy Convers 28(3):756–767. https://doi.org/10.1109/TEC.2013.2259627
8. Jain A, Shankar S, Vanitha V (2017) Power generation using permanent magnet synchronous generator (PMSG) based variable speed wind energy conversion system (WECS): an overview. J Green Eng 7(4):477–504
9. Gupta S, Dwivedi A, Srivastava RK (2015) Fabrication of dual-stator permanent magnet synchronous generator. In: Annual IEEE India conference (INDICON), pp 1–5. https://doi.org/10.1109/INDICON.2015.7443782
10. Owen R, Zhu Z, Wang J, Stone D, Urquhart I (2011) Review of variable-flux permanent magnet machines. In: 2011 International conference on electrical machines and systems, IEEE, pp 1–6
11. Ghasempour A (2019) Internet of things in smart grid: architecture, applications, services, key technologies, and challenges. Inventions 4(1):22
12. Chen Y, Pillay P, Khan A (2005) PM wind generator topologies. IEEE Trans Ind Appl 41(6):1619–1626
13. Sitapati K, Krishnan R (2001) Performance comparisons of radial and axial field, permanent-magnet, brushless machines. IEEE Trans Ind Appl 37(5):1219–1226
14. Parviainen A et al Design of axial-flux permanent-magnet low-speed machines and performance comparison between radial-flux and axial-flux machines
15. Ani SO, Polinder H, Ferreira JA (2012) Low cost axial flux pm generator for small wind turbines. In: IEEE energy conversion congress and exposition (ECCE). IEEE, pp 2350–2357
16. Capponi FG, Terrigi R, Caricchi F, Del Ferraro L (2009) Active output voltage regulation for an ironless axial-flux pm automotive alternator with electromechanical flux weakening. IEEE Trans Ind Appl 45(5):1785–1793
17. Huang S, Luo J, Leonardi F, Lipo TA (1999) A comparison of power density for axial flux machines based on general purpose sizing equations. IEEE Trans Energy Convers 14(2):185–192
18. Spooner E, Chalmers B (1992) 'torus': a slotless, toroidal-stator, permanent-magnet generator. In: IEE proceedings B (Electric Power Applications), vol 139. IET, pp 497–506
19. Gupta SK, Srivastava RK (2020) Performance comparison of AFDS PMSG during voltage regulation by mechanical field-weakening in extended-speed-range 14(4):643–654
20. Gupta SK, Srivastava RK (2017) Comparison of variable-flux PMSG for extended speed-range based on magnet arc-length to pole-pitch ratio. In: 2017 IEEE transportation electrification conference (ITEC-India), 2017, pp 1–5. https://doi.org/10.1109/ITEC-India.2017.8333720
21. Gupta SK, Srivastava R, Mahendra SN (2015) Voltage regulation of dual stator permanent magnet synchronous generator. In: IEEE international transportation electrification conference (ITEC). IEEE, pp 1–6

22. Kumar Gupta S, Kumar Srivastava R (2021) Wind turbine emulator for laboratory environment. In: 2021 IEEE transportation electrification conference (ITEC-India), pp 1–5. https://doi.org/10.1109/ITEC-India53713.2021.9932422

Decentralized Control Strategy for Hybrid Microgrid Based on Coordinate Coefficient

Ashutosh Singh

Abstract This work attempts to establish a control strategy for power management of hybrid microgrid consisting of different voltages of AC and DC subgrids. These subgrids are interfaced through power electronics converter to the common bus which is formed by an energy storage system i.e., battery. The converter interfacing this battery and common bus are bidirectional in nature and controlled by P_{dc}–V_{dc}^2. Bidirectional AC–DC converter (BDAC) for AC subgrid and Bidirectional DC-DC converter (BDDC) for DC subgrid are controlled by droop-controlled techniques. The interaction among these subgrids in a coordinated manner requires communication which leads to the reduced reliability of the system. Hence a decentralized control strategy based on coordinate coefficient is studied in this work to interact among the subgrids in case of change in load. The generating capacity of the DERs connected to the subgrids is considered for the calculation of coordinate coefficient. The system is studied in Simulink for the load changes in AC subgrid with three stages of operations i.e., when the rated load is applied, when the load is increased without increasing power of DERs and when the generating capacity of the DER is increased. A comparative study was also done with coordinate control and without coordinate control.

Keywords Coordinated control · Power management · Multiple subgrid · Distributed control

A. Singh (✉)
IIT (ISM), Dhanbad, India
e-mail: iashutosh09@gmail.com

SOET, IGNOU, New Delhi, India

A. Kumar et al. (eds.), *Decarbonisation and Digitization of the Energy System*, Lecture Notes in Electrical Engineering 1099, https://doi.org/10.1007/978-981-99-7630-0_5

1 Introduction

Power generation is undergoing transformational changes. Depletion of environment, energy insecurity and concerns of location and energy led to increase in research in Distributed energy resources (DERs) [1]. Microgrid can be formed by interconnection of loads, DERs and energy storage system (ESS). Bidirectional converters interface DERs, ESS and loads. Power is managed by controlling these interfacing bidirectional converters. There are various ways to control these bidirectional converters, in this paper decentralized droop control method is used for BDDC and BDAC. A new control technique $P_{dc}-V_{dc}^2$ control technique is used for converter interfacing storage subgrid and common bus. Power management is studied in three stages without and with proposed coordinate coefficient.

1.1 Microgrid Configuration

The system studied consists of three subgrids in which two are DC subgrids namely DC subgrid 1 of 500 V and DC subgrid 2 of 400 V and one AC subgrid of 50 Hz. They are connected to common bus of 700 V formed by energy storage system through BDDC for DC subgrid and BDAC for AC subgrid respectively. These BDAC and BDDC are droop controlled while the interlinking converter is controlled by $P_{dc}-V_{dc}^2$. The common bus provides voltage support to the subgrids. The whole system is standalone and it is not connected to the grid. Loads connected are resistive and Photovoltaic generation is taken as DERs in all subgrids.

1.2 Droop Control of Microgrid

Droop control technique operates without any central controller, it uses local information to determine active and reactive power in microgrid [2]. The inverter can be compared to synchronous generator that operates in large-scale power system. Frequency is reduced when the demand of real power increases [2]. Expression of active power (P_0) and reactive power (Q_0) in case of highly inductive load are as follows:

$$P_0 = \frac{V V_0}{X} \sin\theta \tag{1}$$

$$Q_0 = \frac{V V_0 \cos\theta V^2}{X} \tag{2}$$

where V is the microgrid voltage, V_0 is the inverter's output voltage and X is the output reactance of the inverter, θ is the angle between the inverter's output voltage and the grid's voltage. It can be seen from Eqs. 1 and 2 while the real power depends upon θ, the reactive power depends upon the magnitude of voltage which can be incorporated in the inverter via droop control namely P-f droop and Q–V droop.

$$f_i = f_{\text{rated}} - m(\text{P}_0 - P_{\text{rated}}) \tag{3}$$

$$E_i = E_{\text{rated}} - n(\text{Q}_0 - Q_{\text{rated}}) \tag{4}$$

Equation 3 represents the frequency droop in which f_i represents the output frequency, f_{rated} represents nominal frequency, m represents droop coefficient and P_{rated} is rated real power.

Similarly, Eq. 4 depicts the voltage and reactive power droop as E_i is output voltage, E_{rated} is rated output voltage, Q_{rated} is rated reactive power, while n represents droop coefficient.

2 Literature Review

Ambia et al. [3] described the power management technique based on centralized control and also controls the interfacing converters. It also discusses the control of power electronic converter based on abc/dq0 techniques.

Loh et al. [4] suggest the droop control method for converters combining AC and DC subgrids that depends on the normalized values of voltage and frequency. It also discusses the value of power transfer between subgrids.

Eghtedarpour and Farjah [5] discuss the management of power balancing technique of hybrid microgrid and also gives technique of controlling interlinking converters in the form of ω-Pac and P_{dc}–V_{dc}^2.

Eajal [6] discusses the equal power sharing by using objective function and uses normalized values of voltage and frequency control for interfacing converters.

Wang et al. [7] give the power sharing based on local level within the same grid and also discuss the storage control based on the droop control techniques.

A multistage control of microgrid in standalone mode has been discussed in [8]. It uses Multistage centralized microgrid controller which optimally coordinates selection of DG unit droop characteristics. It aims to increase reliability of microgrid by minimizing load shedding.

To improve the limitations of the conventional droop the proportional integral or derivative coefficient of active power has been studied in [9] and a novel load sharing method has been proposed without using control wire interconnections.

To supply harmonics to the non-linear loads such as fluorescent lamp the virtual impedance technique is given in [10]. It also proposes the current and power droop.

Strategy for managing power for the asymmetrical configuration of AC and DC subgrids in mix type of microgrid is illustrated in [11]. It also analyzes the effect of fault in ILCs during management of power.

3 Structure and Control of Hybrid Microgrid

The DC subgrid 1 of 500 V is connected to DERs (Photovoltaic generation) working in Maximum Power Point Tracking (MPPT) mode which generates 63KW. Connected load to the DC subgrid 1 is 50 kW and allowable variation in voltage is ±25 V. The rated voltage of DC subgrid 2 is 400 V with Photovoltaic generation working in MPPT producing 70 kW and allowable variation in voltage is ±50 V. AC subgrid is 50 ± 1 Hz and the DER (Photovoltaic generation) working in MPPT mode connected to the subgrid is generating power of 60 kW.

3.1 *Control Strategy of AC Subgrid*

Figure 1 shows two loops in AC subgrid namely outer and inner loops. Outer loop calculates the power and generates reference voltage to inner loop. In inner loop three three-phase reference voltage is decomposed into dq component which is subtracted from the converter output voltage and given as input to the Proportional and Integral (PI) controller. PI controller gives the reference current value which is after subtracting from output current of the converter fed to the Proportional (P) controller to generate pulse width modulated (PWM) signal.

$$f_{\text{ref}} = \left(f^* + \Delta f\right) + m\left(P^* - P\right) \tag{5}$$

$$V_{\text{ref}} = V^* + n\left(Q^* - Q\right) \tag{6}$$

where f_{ref} is the reference frequency, f* is rated frequency, Δf is coordinate coefficient and m and n are droop constants. P* and Q* are rated real and reactive power respectively. V_{ref} is reference voltage, V* is rated output voltage, while P and Q are output real and reactive power respectively.

3.2 *Control Strategy of DC Subgrid*

With increase in load, DC reference voltage is decreased.

$$V^* = V_{\text{ref}} - i_0 r \tag{7}$$

where V_{ref} is the reference voltage, i_0 is load current and r is virtual resistance also known as droop constant. Equation (7) shows the value reference DC voltage decrease as we increase the value of load current using virtual resistance known as droop coefficient. Also, droop control is used based on voltage and power sharing given in reference [12].

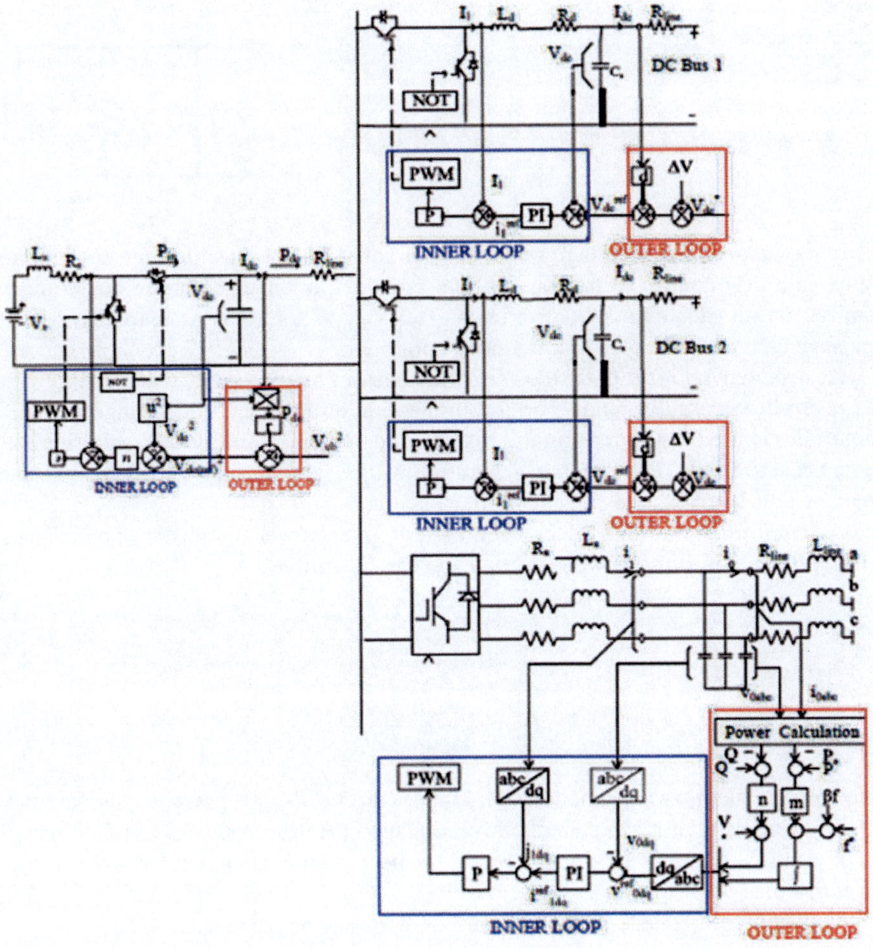

Fig. 1 Control strategy of hybrid microgrid

$$V^{*} = V_{\text{ref}} - m_{p}P \tag{8}$$

where m_p is droop constant and P is output power.

3.3 *Control Strategy of Storage Subgrid*

Energy storage system is interfaced to the subgrid using Bidirectional DC-DC converter (BDDC). This subgrid acts as common bus through which subgrids are linked. BDDC works in boost and buck mode as per requirement of common bus

Fig. 2 Circuit diagram with CPL connected

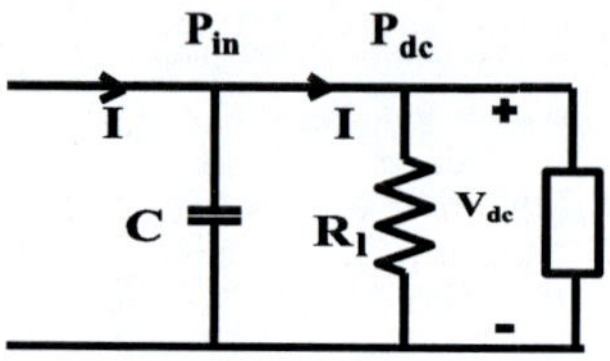

keeping the polarity of voltage same. Conventional i_{dc}–V_{dc} has non-linear relationship, so a new control technique is proposed which involves common bus voltage and dc power at output of converter as P_{dc}–V_{dc}^2. This helps in realizing power management and voltage support among subgrids.

As depicted in Fig. 1, outer loop sets the square of reference DC voltage $(V_{dc}^{ref})^2$. PI controller is used in inner loop for elimination of static error in voltage while P controller is used for increasing the system damping and stability. The subgrids are connected to DERs (Photovoltaic generation) which work in MPPT mode and hence may be considered as the constant voltage sources.

As circuit diagram of Fig. 2, when battery gives power to constant power load (CPL) the relationship between I_{dc} and V_{dc} can be written as.

$$CS\frac{dv_{dc}}{dt} = I_{in} - I_{dc} \tag{9}$$

$$I_{dc} = \frac{V_{dc}}{R_{load}} + \frac{P_c}{V_{dc}} \tag{10}$$

where Cs is the filter capacitance, I_{in} is input current, R_{load} represents passive load and P_c represents CPL's rated real power. Equation (9) shows non-linear relationship of V_{dc} and I_{dc}, hence performance would not be good if conventional V_{dc}–I_{dc} droop is adopted.

Now applying Tellegen's theorem in Fig. 2, it may be written as:

$$\frac{1}{2}C_s\frac{dV_{dc}^2}{dt} = P_{in} - P_{dc} \tag{11}$$

where P_{in} and P_{dc} are input and output power respectively.

$$P_{dc} = \frac{Vdc^2}{R_{load}} + P_{cpl} \tag{12}$$

Equation (12) shows linear relationship between P_{dc} and V_{dc}^2. Based on the equation and analysis of Fig. 2, the control technique may be written as.

$$\left(V_{cb}^{ref}\right)^2 = V_{cb}^2 - r(P_{dc}) \tag{13}$$

where V_{cb}^{ref} denotes reference common bus voltage, V_{cb}^2 is square of common bus voltage, r represents droop coefficient and P_{dc} is the output dc power. This strategy is used for calculations in outer and inner loops and PWM signal is generated through P controller as shown in Fig. 1.

4 Power Management in Hybrid Microgrid

Power management in AC subgrid refers to sharing of active and reactive power between inverters which is connected between DERs and AC grids. In case of DC microgrid, real power is controlled only as no reactive power flows in DC grid. In AC grid, the P-f and V-q droop is used to realize sharing of powers among the parallel connected inverters, while in case of DC grid, V-I droop is used to achieve the same among the parallel connected converters.

Change in frequency in case of AC grid and change in voltage in case of DC grid will represent change in power in respective subgrids while the change in common bus voltage will result in change in power in whole hybrid microgrid [13]. Based on this, the coordinated control strategy power control strategy is formed for regulating frequency in AC subgrids and voltage in DC subgrids.

$$\Delta v = \left(k_{PV} + \frac{k_{iv}}{s}\right)\left(\frac{v_{cb} - v_{cb}^*}{V_{cb\max} - V_{cb\min}} - \beta\frac{v_{dc} - V_{dc}^*}{V_{dc\max} - V_{dc\min}}\right) \tag{14}$$

$$\Delta f = \left(k_{pf} + \frac{k_{if}}{s}\right)\left(\frac{v_{cb} - v_{cb}^*}{V_{cb\max} - V_{cb\min}} - \beta\frac{f - f^*}{f_{\max} - f_{\min}}\right) \tag{15}$$

Above equations depict the control law used for coordinated power management where v_{cb} is common bus voltage, V_{cbmax} and V_{cbmin} are maximum and minimum common bus voltage respectively, Vcb* is the rated common bus voltage, v_{dc} is the dc subgrid voltage, V_{dcmax} and V_{dcmin} are maximum and minimum dc subgrid voltage respectively, V* is rated dc voltage, f is frequency of the subgrid. In Eq. (15) f* is rated frequency of subgrid, f_{max} and f_{min} denote maximum and minimum frequency of the subgrid, k_{pf} is proportional constant, kif is the integral constant, Δv and Δf are the coordinate coefficient that is sent to BDDCs and BDACs, β is called as correction factor whose value depends on the generating capacity of the grids as well as subgrids.

The generating capacity and load connected to subgrids need to be considered while designing the control law this is done by the use of correction factor β given by.

$$\beta = \left(\frac{Pjsum}{PgSUM}\right)^{-1}\left(\frac{PjESS}{Pjsum}\right) \tag{16}$$

where $Pjsum$ is the generating capacity of the DER connected to AC & DC subgrids, PgSUM denotes generating capacity of whole microgrid system and PjESS is the

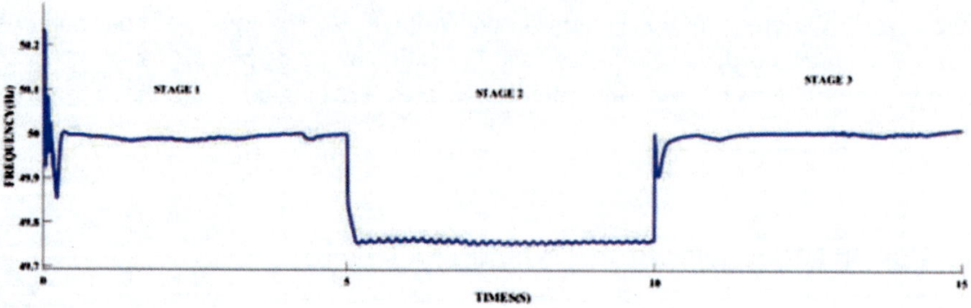

Fig. 3 Change in frequency during three stages of simulation

essential load capacity of the grid. Essential load capacity may be defined for sensitive load for which voltage and frequency should not be changed. If the value of essential loads is higher, then β will be small hence there will be minimum change in voltage and frequency of the subgrid.

5 Simulation Results and Discussions

To draw comparison between the coordinate control and uncoordinated control the load is increased in AC subgrid from 30 to 50 kW and the studied system consists of three stages; Stage 1 when rated load is connected; Stage 2 when the load is increased; Stage 3 when the generation of DER is increased.

5.1 Frequency of AC Subgrid Without Coordinate Coefficient

During Stage1 at rated load there is no change in frequency. During Stage 2 when the load is increased in AC subgrid the frequency decreases due to droop controlled. During Stage 3 when the power of the DER connected to AC subgrid increases, frequency restores to normal value. Frequency change during Stage 2 is abrupt (Fig. 3).

5.2 Frequency of AC Subgrid with Coordinate Coefficient

During Stage 1 in rated load condition the frequency is around 50 Hz and during Stage 2 when the load is increased the frequency decreases. During Stage 3 when the output power of DERs increased with coordinated control the frequency restores to normal value. Abrupt change in frequency is avoided using coordinate coefficient (Fig. 4).

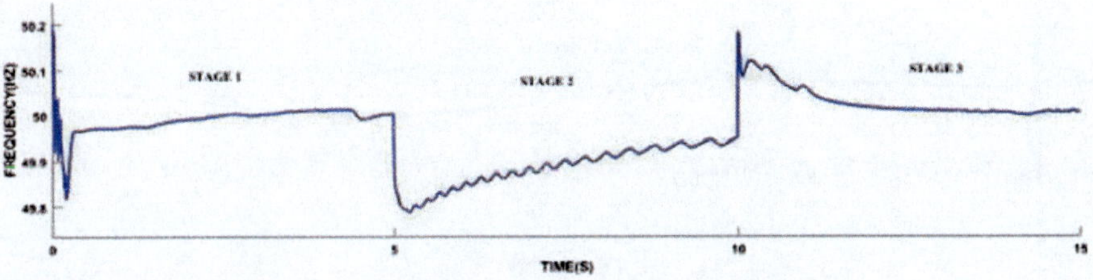

Fig. 4 Frequency of the AC grid during the three stages of operations

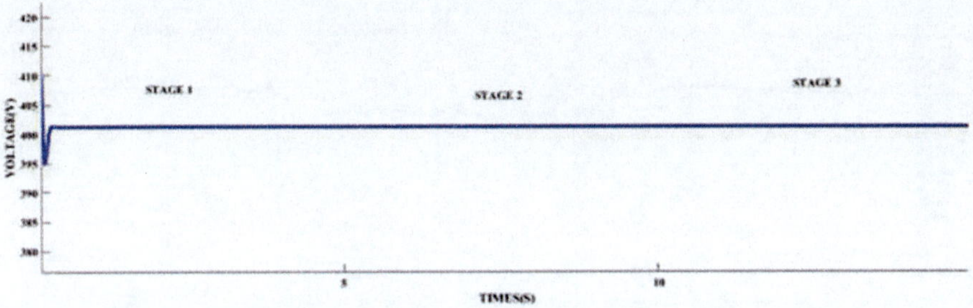

Fig. 5 Output voltage of DC subgrid

5.3 Output of DC Subgrid Without Coordinate Control

Figure 5 shows the change in voltage in DC subgrid 2 when load changes in AC subgrid. The voltage does not change during the three stages of operation. During Stage 2 when load changes in AC subgrid there is no decrease in voltage in DC subgrid 2 showing that there is no coordination between the two subgrids.

5.4 Output of DC Subgrid with Coordinate Control

During Stage 1 the value of voltage is 400 V. During Stage 2 when load demand in AC subgrid is increased the value of voltage decreases representing the support to the AC subgrid due to ΔV in droop control. Hence the two subgrids communicate among themselves during load change and support each other (Fig. 6).

5.5 Output of Common Bus Voltage

Figure 7 shows the common bus voltage when load is changed in AC subgrid. During Stage 2 the common bus voltage also decreases along with frequency of the grid

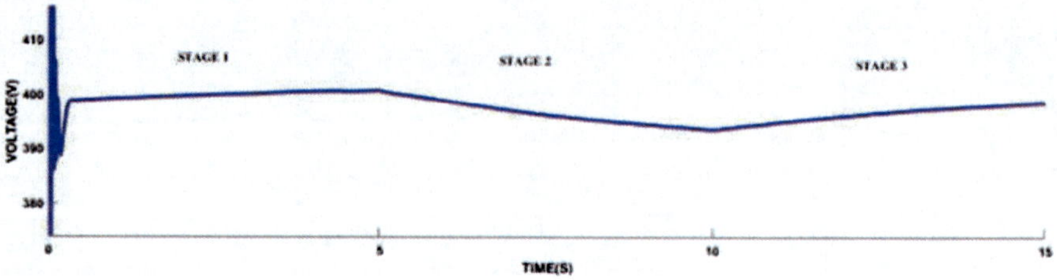

Fig. 6 Output Voltage of DC subgrid with coordinate control

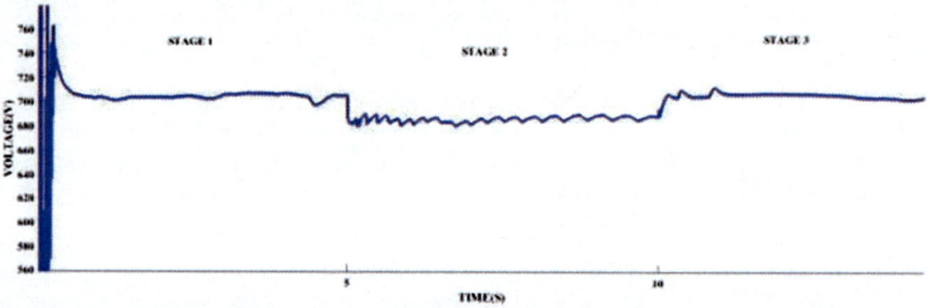

Fig. 7 Common bus voltage

showing that the storage subgrid formed using battery and AC subgrid are working in coordinated manner to share the increased load demand.

6 Conclusion

The power management technique used for hybrid microgrids is studied. The decentralized control technique namely droop control technique for the BDDC and BDAC with coordinate coefficient is analyzed in MATLAB/SIMULINK. The results clearly show that the load change in subgrid makes whole system interact with each other to share the extra power demand.

The interlinking converter between the energy storage and common bus is controlled by P_{dc}–V_{dc}^2 which has a linear relationship. The change in common bus voltage represents the change in overall power demand of the microgrid. This common bus provides voltage support for the subgrid in case of variation in demand. Power management is studied in three stages without and with proposed coordinate coefficient. The power rating of the subgrids is used while calculating the value of coordinate coefficient. This helps in avoiding the drastic changes in voltage and frequency.

References

1. Lasseter RH, Paigi P (2004) Microgrid: a conceptual solution. In: Proceedings of IEEE 35th Annual Power Electronic Specialists Conference. Aachen, Germany, pp 4285–4290
2. Han H, Hou X, Yang J, Wu J, Su M, Gucrrcro JM (2016) 'Review of power sharing control strategies for islanding operation of AC microgrids.' IEEE Trans Smart Grid 7(1):200–215
3. Ambia MN, Al-Durra A, Muyeen SM (2011) Centralized power control strategy for AC–DC hybrid micro-grid system using multi-converter scheme. In: Proceedings of 37th annual conference of the IEEE industrial electronics society (IECON). Melbourne, VIC, Australia, pp 843–848
4. Loh PC, Li D, Chai YK, Blaabjerg F (2013) Autonomous operation of hybrid microgrid with ac and dc subgrids. IEEE Trans Power Electron 28(5):2214–2223
5. Eghtedarpour N, Farjah E (2014) Power control and management in a hybrid ac/dc microgrid. IEEE Trans Smart Grid 5(3):1494–1505
6. Eajal AA, Saadany EFE, Ponnambalam K (2016) Equal power sharing in islanded AC/DC hybrid microgrids. In: Proceedings IEEE electrical power energy conference (EPEC). Ottawa, ON, Canada, pp 1–6
7. Wang P, Jin C, Zhu D, Tang Y, Loh PC, Choo FH (2015) Distributed control for autonomous operation of a three-port ac/dc/DS hybrid microgrid. IEEE Trans Ind Electron 62(2):1279–1290
8. Abdelaziz MMA, Shaaban MF, Farag HE, El-Saadany EF (2014) A multistage centralized control scheme for islanded microgrids with PEVs. IEEE Trans Sustain Energy 5(3):927–937
9. Guerrero JM, de Vicuna LG, Matas J, Castilla M, Miret J (2004) A wireless controller to enhance dynamic performance of parallel inverters in distributed generation systems. IEEE Trans Power Electron 19(5):1205–1213
10. De Brabandere K, Bolsens B, Van den Keybus J, Woyte A, Driesen J, Belmans R (2007) A voltage and frequency droop control method for parallel inverters. IEEE Trans Power Electron 22(4):1107–1115
11. Xia Y, Wei W, Yu M, Wang X, Peng Y (2018) Power management for a hybrid AC/DC microgrid with multiple subgrids. IEEE Trans Power Electron 33(4):3520–3533
12. Dragičević T, Lu X, Vasquez JC, Guerrero JM (2016) DC microgrids—Part I: a review of control strategies and stabilization techniques. IEEE Trans Power Electron 31(7):4876–4891
13. Nejabatkhah F, Li Y (2015) Overview of power management strategies of hybrid AC/DC microgrid. IEEE Trans Power Electron 7072–7708

Optimal Energy Scheduling Using ANT Colony Approach with Consideration of Consumers Preferences in a Residential Smart Home

Degala Isaac and Amit Kumar

Abstract In this paper, an optimization approach is brought forward for optimal energy scheduling in a modern smart home. After the development of smart devices and their integration with smart grid, the end consumers have the chance to schedule their home appliances in order to decrease the overall electricity price and allaying the power Peak-to-Average ratio (PAR). The application pattern of home devices is observed at first for optimal energy scheduling. The home gateway (HG) obtains the demand response (DR) showing the real-time electricity price (RTEP) and later sends it to energy management controller (EMC). With the Demand Response, the EMC results in the essential energy scheduling method that can be sent to each electric appliance by the Home Gateway. The next feature is integrating RTEP with IBR, because if only RTEP is considered the appliances tend to be scheduled at lowest electricity price slot, this may lead to blackout during that time period. By executing this optimization method, our proposed approach could gradually reduce both the electricity price and PAR, thereby, strengthening the stability of the entire Grid system. The optimization technique used to solve this problem was Ant Colony Optimization (ACO) approach as the optimization problem was nonlinear.

Keywords Ant colony optimization · Real time electricity pricing · Energy management system · Inclined block rate · Smart home

1 Introduction

The urge to expand power networks has emerged to implement on-demand electricity production and distribution techniques due to increase in consumer's needs. The impact of smart grids and the beneficial features extracted from it such as dynamic

D. Isaac (✉) · A. Kumar
National Institute of Technology Kurukshetra, Thanesar, India
e-mail: isaac_32114312@nitkkr.ac.in

A. Kumar
e-mail: amitkumar357@nitkkr.ac.in

A. Kumar et al. (eds.), *Decarbonisation and Digitization of the Energy System*, Lecture Notes in Electrical Engineering 1099, https://doi.org/10.1007/978-981-99-7630-0_6

pricing [1] and demand response management have been significant in many aspects. These ideas encourage end consumers to minimize electricity usage at some periods by which they reschedule household appliances to other time period slots. Thus, the distribution of net electricity consumption becomes sustained over time by eradicating the consumption residential peaks over time intervals [2]. Dynamic pricing schemes can be rearranged based on area and consumption time intervals [3]. Many studies showed that dynamic pricing scheme reduces electricity costs by as much as 10% at the level of residential consumers [4]. The main problem encountered here is that the generation and distribution of electricity on demand side and the dynamic pricing of electricity requires real-time monitoring of consumers even under small time intervals. This is resolved by using of smart meters on the demand side and integrating it with Home Gateway (HG) by using IOT devices. Thus, by considering the above parameters the ant colony algorithm is programmed so that the motives of this paper are achieved. The response from these IOT devices ensures that the schedule of household appliances is met. This developed algorithm can send over wireless Trans receivers and data processors can enable the better functioning of the residential optimal scheduling.

2 Application Pattern of Home Devices

Residents generally prefer to operate their shiftable appliances at certain time slots which avoid high pricing. For example, if residents want to have dinner as soon as they arrive home in the night, they must ensure that the electric rice cooker finishes its job before they arrive home. From this aspect, it is necessary for residential consumers to set some parameters for each shiftable appliance which includes the service usage time with their preferable start to end timing. These are treated as customer preferences as shown in Table 1 and considered with utmost care while executing the scheduling algorithm. In Table 1 the customer preferences are taken for an average in New York City with the appliance's application time and service time periods [3].

3 RTEP Combined with IBR

With RTEP, the consumers only pay for the electricity which they use and is worth differently at different times of the day [4] and they can shift their appliances to the off-peak slots which not only decreases the peak-to-average ratio but also their monthly electricity bill. In general, many economists proved that this pricing model will have greater impact in both short term and long-term scheduling. Advancements in research showed that in future with RTEP schemes, the load can be reduced at peak usage times and shifted to off-peak service times with lowest costs resulting a peak in those periods, thereby initiating blackout conditions due to high PAR. The

Table 1 Usage pattern of home appliances

Appliances	Minimum power (kW)	Maximum power (kW)	Application time	Service time
Ventilation	0.0	2.0	00:00–24:00	24
Water hheater	0.0	3.0	00:00–24:00	24
Dish washer	1.40	1.40	08:00–14:00 or 17:30–22:00	2
Washing machine	2	2	08:30–19:00	2 (Wash) 1 (Dry)
Oven	2	2	09:00–14:30 and 17:00–22:30	2 2
Electric kettle	1	1	06:00–08:00	0.5
Vacuum cleaner	0.8	0.8	09:00–13:30 or 16:00–20:30	1

method to distribute the load evenly throughout the day is to set a limit for the service usage at each time slot so that, if the consumption increases by the predefined limit, the consumers should be charged more than the normal tariff scheme at that specified amount of time. This tariff scheme is known as Inclining Block Rates (IBR) as shown in Table 2. Therefore, by combining RTEP with IBR, both the constraints set initially for optimal energy scheduling in a smart residential home is achieved.

Table 2 Hourly market price

Time period	Price ($/MWhr)	Time period	Price ($/MWhr)
1	15.03	13	65.79
2	10.97	14	66.57
3	13.51	15	65.44
4	15.36	16	79.79
5	18.51	17	115.45
6	21.80	18	110.28
7	17.30	19	96.05
8	22.83	20	90.53
9	21.84	21	77.38
10	27.09	22	70.95
11	37.06	23	59.42
12	68.95	24	56.68

4 Problem Formulation

After the nature of appliance is known i.e., whether the appliance is shiftable or non-shiftable the scheduling starts to take place. Only shiftable appliances are scheduled. In this scenario dish washer, washing machine, oven, electric kettle and vacuum cleaner are scheduled. Consider that there are n appliances and $P_1, P_2, P_3 \ldots. P_n$ are the power ratings of the appliances and $P_{r1}, P_{r2}, P_{r3} \ldots P_{r24}$ are the hourly market prices of each time block. j is the time slot from 1 to24 hours where that particular ith appliance operates, it is also termed as application time in this paper.

The cost of ith appliance is given by:

$$C_i = P_i * \left(\sum_{k=1}^{24} P_{rk} \right) \forall k \in j \tag{1}$$

The total cost of all n appliances is given by:

$$C = \left(\sum_{i=1}^{n} C_i \right) \tag{2}$$

The power consumed by ith appliance in each kth time slot is given by:

$$P_{C_{ik}} = P_i k = \{1, 2 \ldots., 24\}, \forall k \in j \tag{3}$$

$$P_{C_{ik}} = P_i k = \{1, 2 \ldots., 24\}, \forall k \notin j \tag{4}$$

The power consumed by all n appliances in each kth time slot is given by:

$$P_{C_k} = \left(\sum_{i=1}^{n} P_{C_{ik}} \right) k = \{1, 2, \ldots, 24\} \tag{5}$$

The peak-to-average ratio (PAR) is given by:

$$PAR = \frac{\max(P_{C_k})}{\frac{(\sum_{k=1}^{24} P_{C_k})}{24}} \tag{6}$$

n = Total number of appliances.
k = {1, 2…., 24}.
j = time period when appliance operates i.e., application time period.
C_i= Cost of ith appliance.
P_i = power rating of ith appliance.
P_{rk}= price at kth hour when the appliance is operated.
C = Total cost of all n appliances.

PAR = Peak-to-Average Ratio.
$P_{C_{ik}}$ = Power consumed by ith appliance in kth time slot.
P_{C_k}= Power consumed by all n appliances in kth time slot.

The scheduling of appliances is done under two conditions namely without considering customer preferences and secondly considering customer preferences. The appliances are scheduled under constraints of decreasing the overall cost and reducing the Peak-to-Average Ratio in order to reduce power outages at over peak periods. The cost formulation is given in Eq. (2) and the PAR formulation is given in Eq. (6). The cost is minimized and the Peak-to-Average Ratio (PAR) made near to unity for Optimal Energy scheduling.

5 Final Aim of Methodology-Optimization Approach

The solution for this problem is found in the following procedure. Initially an Unscheduled algorithm is developed where the appliances are scheduled randomly. In this algorithm, only the customer's demand is considered and no constraints are taken up. This is the generalized algorithm to calculate the electricity bill. Based on customer's demand, the appliances are categorized into 2 categories namely shiftable and non-shiftable appliances. The non-shiftable appliances are not evaluated for scheduling. Secondly, scheduled algorithm with deterministic approach is evaluated with the constraint of minimum cost where the shiftable appliances are shifted to the time slots where the price is minimum value irrespective of any constraints. By this scheduling, the cost is decreased but the Peak-to-Average Ratio soars up [1]. So, an approach which is efficient and gives the best solution in a smaller number of iterations with satisfying both constraints is required. Therefore, an ACO algorithm is adopted to provide better solution.

The considerations made for Ant Colony Optimization algorithm are as follows. The time period of one day or 24 h is divided into 48 time slots i.e., the time resolution is 0.5 h and 1day has 48 slots. The time resolution is set to be 0.5 h as it is feasible to solve the nonlinear problem using ACO and the minimum application time of customer appliances is 0.5 h. The hourly price for each time slot is given in hourly market price table as shown in Table 2. In 48 time slots, the appliances that are active in a particular time slot are indicated as 1 and the others which are inactive are indicated as 0. Thereby, the operating time and the corresponding price for that time period are evaluated more precisely. The power of the appliance can be obtained from the service usage pattern as shown in Table 1. By using price and power of each of the appliances, the cost is calculated [8] as shown in Eq. 2.

In ACO, there are two factors to be considered for evaluation of the probability of choosing a route and the evaporation rate of pheromone. A network routing problem of Peak-to-Average Ratio is evaluated in this problem. The probability reflects the route chosen by the ant for evaluation. The evaporation rate signifies the impact of each route chosen and their impact for other ants to follow up [4]. In this problem consider that there are 'n' number of appliances. Initially the cost and PAR are evaluated using Deterministic approach. Now to decrease the cost as well as PAR, initialize the population (Here population is nothing but price) over the 48 time slots and set limits for the price of each appliance. There are few parameters other than the two to be evaluated through this procedure such as the power consumption at peak periods. An acceleration method is used to set limits for the pheromone concentration such as PAR set near to unity and only the ants with the current best global solutions will be allowed to deposit pheromone. Now the cost is evaluated with random values of time slots within limits and recursion for obtaining better results is performed. Additionally, ranking of the best global fitness can also be done for better optimality [10]. This method provides better solution with a smaller number of iterations.

6 Results and Discussions

The following are the graphs obtained under Unscheduled (Scheduled randomly) approach, Deterministic approach and Ant Colony techniques with consideration of customer approaches.

6.1 Scheduled Randomly with Customer Considerations

In this case the appliances are scheduled with customer preferences as shown in Table 1. Here based on the dynamic pricing model in application time, at different preferred time periods different hourly pricing is valuated. Figure 1 shows the cost distribution at each time period. Figure 2 shows the power distribution at each period with customer preferences. Both the graphs are dynamic in nature.

6.2 Deterministic Approach with Customer Considerations

This is to evaluate the minimum cost among the customer's preferred time periods. The PAR value is also high in this case as all appliances are scheduled in the minimum electric pricing (EP) time periods among customer preferences. Figure 3 shows the daily cost at different time periods and Fig. 4 shows the power consumption at different time periods in the day.

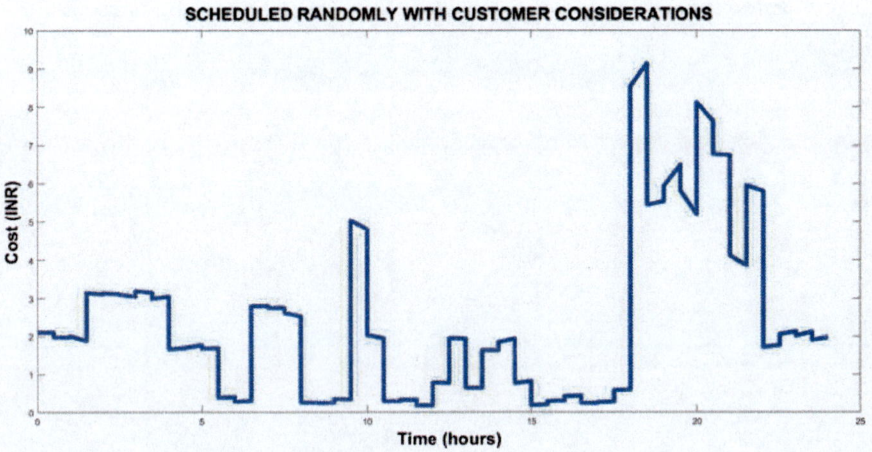

Fig. 1 Graph for unscheduled algorithm (time vs. cost)

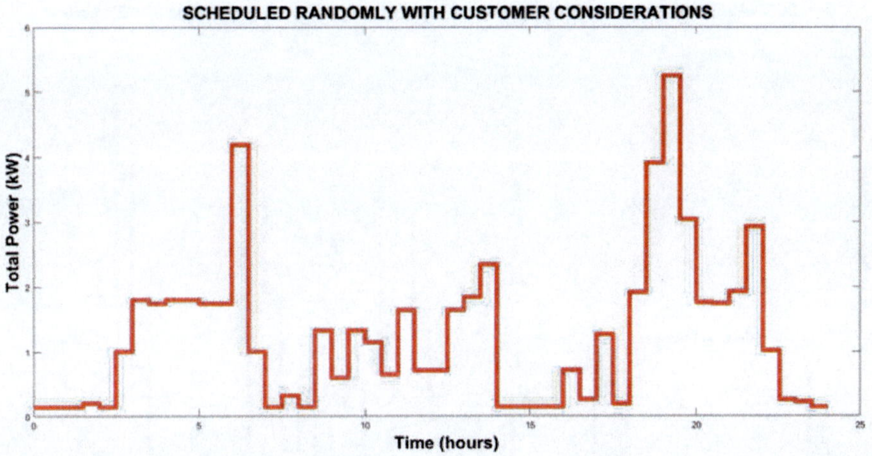

Fig. 2 Graph for unscheduled algorithm (time vs. power)

6.3 Ant Colony Optimization with Customer Considerations

The time periods which are preferred by the customer are considered for evaluation of Ant Colony optimization to provide the optimal PAR. Figure 5 shows the cost variation at different time periods and Fig. 6 shows the power consumption at those time periods.

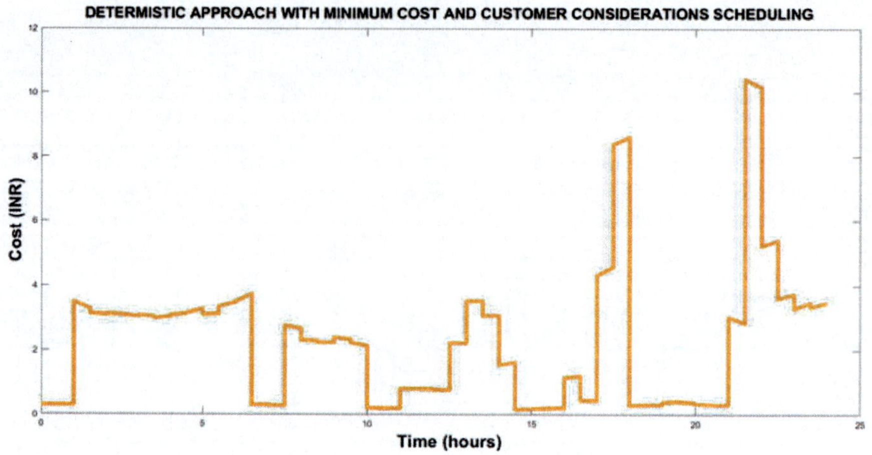

Fig. 3 Graph for deterministic approach with minimum cost (time vs. cost)

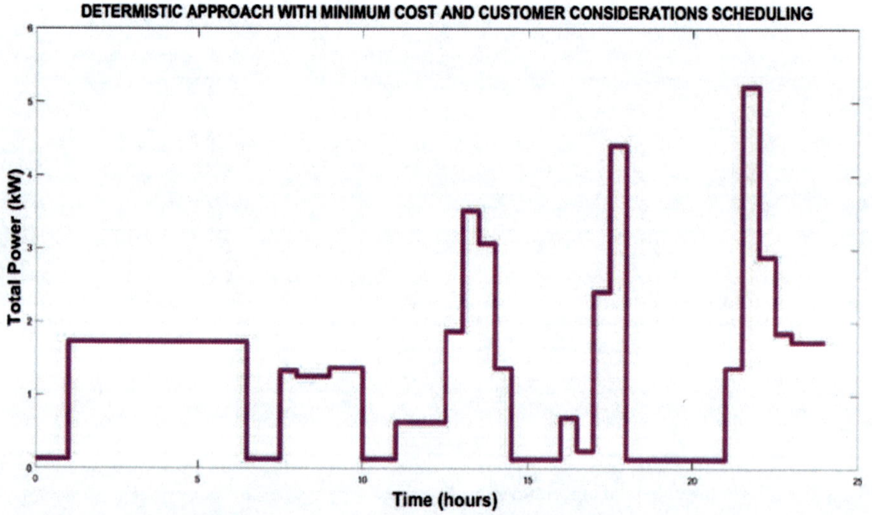

Fig. 4 Graph for deterministic approach with minimum cost (time vs. power)

6.4 Statistical Comparison

The comparison of all 3 approaches is shown in Table 3 with parameters of Cost and Peak-to-Average Ratio (PAR).

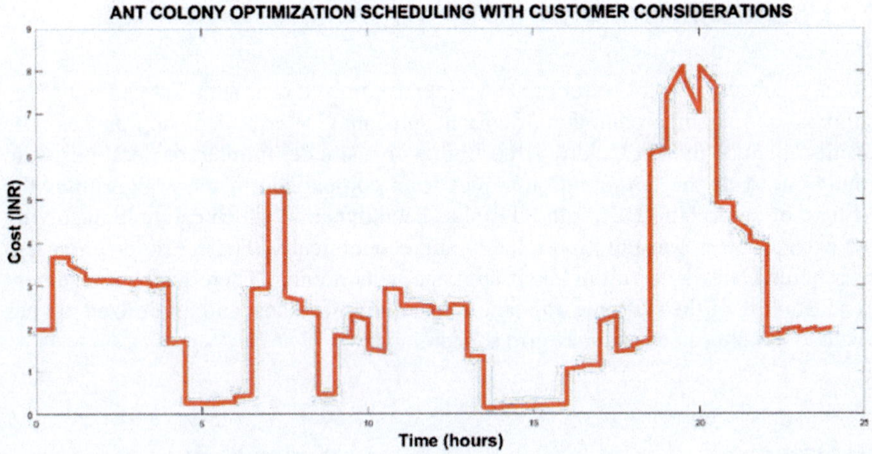

Fig. 5 Graph for scheduled with aco algorithm (time vs. cost)

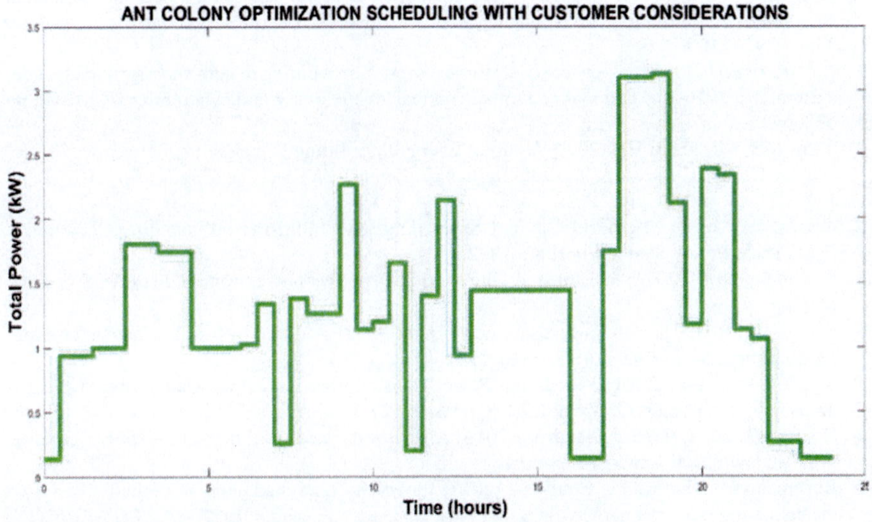

Fig. 6 Graph for scheduled with aco algorithm (time vs. power)

Table 3 Comparison of 3 approaches

Approach	Monthly electricity price ($)	Peak-to-Average Ratio (PAR)
Unscheduled	217.8768*	1.8078*
Deterministic Approach	207.9472	1.9464
ACO Approach	213.0496*	1.5988*

***:** indicates that value is dynamic and the value depends on iterations and number of times the algorithm is run to reach its global best

7 Conclusion

Based on observations a better par and better minimized cost with Ant Colony Algorithm is found compared to that of Unscheduled and Deterministic approaches. The results obtained by Ant Colony Algorithm with a smaller number of iterations were found out to be the best solution. Apart from comparison of these algorithms the method of combining IBR with RTEP has been dynamic which can help in solving the excess power consumption reliably and economically. These results when sent over home Gateway result in better appliance functioning. Thereby, the assignment or scheduling of the shiftable appliances to a minimized cost and minimized par has resulted in betterment of smart grid scheduling.

References

1. Zhao Z, Lee WC, Shin Y, Song KB (2013) An optimal power scheduling method for demand response in home energy management system. In: IEEE Transactions on Smart Grid, vol 4, no 3, pp 1391–1400
2. Lin SK, Chen CR (2016) Optimal energy consumption scheduling in home energy management system. In: 2016 international conference on machine learning and cybernetics (ICMLC), pp 638–643
3. Toosi MB, Mashhadi HR (2020) Optimal Energy Scheduling Considering Consumers Preference in a Smart Home. In: 2020 28th Iranian Conference on Electrical Engineering (ICEE), pp 1–5
4. Khodaei A (2014) Microgrid Optimal Scheduling with Multi-Period Islanding Constraints. IEEE Trans Power Syst 29(3):1383–1392
5. Shahidehpour M (2016) Role of smart microgrid in a perfect power system. IEEE PES General Meeting, pp 1–1
6. Peretto L (2010) The role of measurements in the smart grid era. In IEEE Instrumentation & Measurement Magazine, vol 13, no 3, pp 22–25
7. Kim TT, Poor HV (2011) Scheduling Power Consumption with Price Uncertainty. In: IEEE Transactions on Smart Grid, vol 2, no 3, pp 519–527
8. Haider HT, See OH, Elmenreich W (2016) A review of residential demand response of smart grid. Renew Sustain Energy Rev 59:166–178
9. Begemann M, Thopil GA, Chudy M (2017) Electricity load management potential based on the behaviour of consumers in the South African residential sector. In: 2017 IEEE AFRICON, pp 725–730
10. Mitra J, Vallem MR (2012) Determination of storage required to Meet Reliability Guarantees on Island-Capable Microgrids with Intermittent Sources. IEEE Trans Power Syst 27(4):2360–2367

Optimal Sizing of Grid-Connected Hybrid Renewable Energy System Using the GWO Algorithm and Adapting the Time-of-Use Tariff Rates

Joshi Sukhdev Nirbheram, Aeidapu Mahesh, and Ambati Bhimaraju

Abstract The use of renewable resources like solar and wind has been encouraged by the depletion of conventional fuels and global warming since they are friendly to the environment. Combining these resources with battery storage can produce clean, affordable, and dependable energy. This paper presents the optimization of the solar photovoltaic (PV)-wind turbine (WT)-battery using the gray wolf optimization (GWO) to optimize the levelized cost of energy (LCE). To obtain the operational benefits of the hybrid renewable energy system (HRES), limit the use of gird power, maximize the renewable use of renewable sources, and limit the surplus Energy of the HRES, the restrictions are the power import rate from grid (PIRG) and the Excess energy rate of renewable (EERR). A novel approach to energy management is suggested, offering a variable rate for grid electricity purchases that adapts the Time-of-Use (TOU) price. The variable tariff from the grid enhances the capability and stability of the HRES. The energy management system (EMS) considers the higher cost of the grid when the burden on the grid is more and vice versa. The EMS also balances energy between renewable sources, batteries, and Demand. The proposed study has been investigated in the location of the Kanyakumari district, India.

Keywords Hybrid renewable energy systems · Energy management system · Levelized cost of energy

1 Introduction

In today's world, replacing fossil fuels with renewable sources is critical considering climate change [1]. But, the uncertain and sporadic nature of renewable sources causes an unreliable power supply to the load. The combination of two and more renewable sources is used to improve reliability, like solar PV, WT, bio-gas, etc.

J. S. Nirbheram (✉) · A. Mahesh · A. Bhimaraju
Sardar Vallabhbhai National Institute of Technology Surat, Surat, Gujarat, India
e-mail: sukhdevjoshi58@gmail.com

A. Kumar et al. (eds.), *Decarbonisation and Digitization of the Energy System*, Lecture Notes in Electrical Engineering 1099, https://doi.org/10.1007/978-981-99-7630-0_7

Solar PV gives the maximum power during the day, whereas WT gives the most power at night. So, Due to the complementary nature of solar PV, WT is the best choice for renewable sources [2]. But, a backup storage system is essential whenever the renewables are unavailable to supply the load [3]. The battery storage has a distinct advantage over other storage due to its high specific energy, prolonged storage durations, and quick response. Hence, solar PV, WT, and battery storage are the best configuration for the HRES to fulfill demand [4]. The Ref. [5] presented the optimal sizing of HRES with the solar PV, WT, and battery storage configuration.

However, energy management system is crucial in HRES to minimize the system operating cost and optimal dispatch of renewable sources. The EMS of HRES is a difficult task considering the uncertainty of renewable sources and load power [6]. Ref. [7] EMS is used for the load scheduling to minimize the energy cost and maximize the system's savings. The solar PV-WT-battery hybrid renewable system presents excess power from renewables charging the battery. If the renewables are insufficient to supply the load, grid power supplies the deficit power [8]. In this paper, the EMS offers a variable tariff to increase the stability of the grid and provides economic benefits to the customer. The TOU is used to demonstrate the variable tariff of the gird [9].

As a result, the literature reviewed above concluded that the EMS is crucial to the HRES's ideal size. Therefore, this study suggested a revolutionary energy market system (EMS) that offers a variable pricing for buying and selling energy. Additionally, the Gray Wolf Optimization (GWO)-efficient metaheuristic technique is employed to determine the HRES's ideal size. The paper's key contribution is listed below:

- The optimal sizing of the solar PV, WT, and battery has been performed with the objective function of LCE subjected to constraint PIRG and EERR. The constraint PIRG encourages the system to limit the power from utility. At the same time, the EERR limits the over sizing of renewable sizing, which will help in the overall cost reduction of the system.
- The TOU tariff rates are used to exchange power between the grid and HRES. The load is divided into peak, off-peak, and valley, so different tariffs are entitled to different periods.

2 Modeling of HRES Components

2.1 Modeling of Wind Turbine

Measure the wind speed in the designated region for an hour and compare it to the cut-out speed (u_{cout}), cut-in speed (u_{cin}), wind speed at turbine altitude (u), and rated speed (u_r) to get the average wind power. The following is how the wind turbine's output power is expressed [10]:

$$P_{wt} = \begin{cases} 0, & when, u < u_{cin} \\ P_{wtr} \times (\frac{u^3 - u_{cin}^3}{u_{cr}^3 - u_{cin}^3}), & when, u_{cin} \leq u \leq u_r \\ P_{wtr} & when, u_r < u < u_{cout} \\ 0, & whenu \geq u_{cout} \end{cases} \quad (1)$$

2.2 Modeling of Solar PV

In [11], the solar irradiance, absorption capability, cell temperature, and panel area are all connected to the output power of the PV system $Ppv(t)$.

$$P_{pv}(i) = \frac{I(i)}{1000} \times P_{pvr} \times \eta_{pv} \times f_{dr}[1 - \alpha_T(T_c - T_{c,STC})] \quad (2)$$

where I is Irradaition in (w/m^2), P_{pvr} is rated power in (kW), η_{pv} is efficiency, f_{dr} is a derating factor of solar PV, α_T is temperature co-efficient, T_c is Cell temperature, and $T_{(c,STC)}$ is temperature at standard test conditions.

2.3 Battery Modeling

The battery is in charging or discharging mode, depending on the available renewable energy. When renewable energy is more than the load demand, the surplus power is used to charge the battery. The instantaneous state of charge $SOC(t)$ during the charging process must be determined using the SOC modeling of the battery [12].

$$SOC_{t+1} = SOC_t + \frac{\eta_{charging}(t) \times I_{battery}(t) \times \Delta t}{C(t)} \quad (3)$$

where $I_{battery}$ is the charging current, $\eta_{charging}$ is charging efficiency, and $C(t)$ is the capacity of the battery.

Similar to the charging process, the discharging process of the system is given below:

$$SOC_{t+1} = SOC_t - \frac{\eta_{discharging}(t) \times I_{battery}(t) \times \Delta t}{C(t)} \quad (4)$$

3 Objective Function Formulation for Optimization

3.1 Total Cost of HRES

The overall cost of the system covers a number of charges, including the cost of the initial investment ($Cost_{in}$), the cost of operation and maintenance ($Cost_{om}$), and the cost of replacement ($Cost_{rep}$) [13].

$$Cost_{total} = Cost_{in} + Cost_{om} + Cost_{rep} \tag{5}$$

3.2 Levelized Cost of Energy (LCE)

The LCE is the per unit cost of the HRES and is defined as the cost of the power delivered by the system. The LCE is the ratio of total cost of the HRES to the total energy delivered to load.

$$LCE = \frac{Cost_{total}(\$)}{Energy_{total}(kwh)} \tag{6}$$

where $Cost_{total}$ denotes the total cost of the HRES and $Energy_{total}$ denotes the overall amount of energy supplied by the HRES.

3.3 Objective Function

The main objective function of the HRES is to minimize the LCE considering the constraint and limitations. So, the HRES has the main aim of providing the design of the system with minimal LCE [14].

$$minf = minimize(LCE) \tag{7}$$

subject to constraints of grid-connected system

$$\begin{cases} 0 \leq PIRG \leq PIRG_{max}, & PIRG_{max} = 15\% \\ 0 \leq EERR \leq EERR_{max}, & EERR_{max} = 10\% \end{cases} \tag{8}$$

The ratio of total power acquired from the grid to the total load demand is known as the power import rate from grid.

$$PIRG = \frac{P_{gridpurchased}}{Load_{Total}} \tag{9}$$

The ratio of excess renewable energy generation to total renewable energy generation is known as the excess energy rate from renewables.

$$EERR = \frac{P_{PV} + P_{WT} - Load_{total}}{P_{PV} + P_{WT}} \tag{10}$$

where P_{PV}, P_{WT}, $Load_{total}$, are the generation from solar PV, WT, and total load of the HRES.

4 Energy Management System

The energy management system is an essential part of the HRES in order to balance the energy balance between renewable sources and the load. The novel EMS is offering the variable tariff and reducing the burden from the gird. The grid offering the high cost while the burden on the grid is more and gird offering the less cost while the burden on the gird is less. The EMS is divided into many parts based on the charging and discharging of the battery.

- In the first mode, when the renewable sources are more than the load and battery SOC is more than 0.8, the surplus power from the renewables supply to the grid at a variable tariff.
- In the second mode, when the renewable sources are more than the load and battery SOC is less than 0.8, the surplus power from the renewables supplies the battery to charge the battery.
- In the third mode of operation, when the renewable sources are less than the load and the battery SOC is greater than 0.2, the battery supplies the deficit power of the load.
- In the last mode of operation, when the renewable sources are less than the load and the battery SOC is less than 0.2, the deficit power of the load is supplied by the grid at a variable tariff (Fig. 1).

5 Optimization Techniques

Metaheuristic algorithmic techniques for sizing have been taken into consideration in this research. This study examined scaling methods for the configured system using metaheuristic algorithms. There are many other metaheuristic techniques, but GWO has received the most attention for optimization because of its accuracy in identifying the best optimal solution.

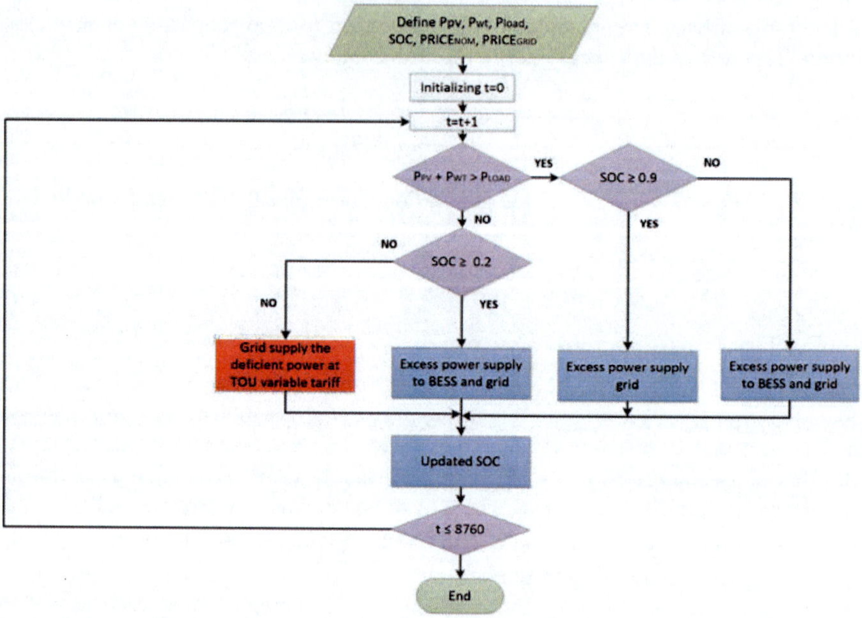

Fig. 1 Energy management system

5.1 *Gray Wolf Optimization*

One of the most current metaheuristics swarms intelligence algorithms is the GWO. Due to its superiority over other swarm intelligence approaches, it has been widely customized for a wide range of optimization issues. This section presents the mathematical modeling of the GWO. The first step of the GWO is to chase and encircle, and the proposed equation is given below [15]:

$$Y(t+1) = Y(t) - B \times S \tag{11}$$

where $Y(t+1)$, $Y(t)$ are the gray wolf's next location and current location, respectively. B and S are the coefficients of the matrix, and the vector depends on the prey location (Y_p) and is calculated by

$$S = |C \times Y_p(t) - Y(t)| \tag{12}$$

where $C = 2 \times r_2$ and r_2 is the randomly generated vector from [0,1].

Using the aforementioned equations, a wolf may migrate to any place in a hypersphere surrounding the prey. However, this is insufficient to approximate gray wolf social intelligence. As previously stated, social hierarchy is important in hunting and group survival. The finest options for simulating social hierarchy are alpha, beta, and delta. Although there may be more than one wolf in each group in nature, in GWO, it is assumed that there is only one solution for each class for the purpose of simplicity. The three best answers thus far are always presumed to be alpha, beta, and delta in GWO. Due to the fact that alpha, beta, and delta are the top solutions in the population, it has been claimed that they have a solid understanding of where the global optimum of optimization problems is located. Other wolves should be required to update their places in the manner described below:

$$Y(t+1) = \frac{Y_1 + Y_2 + Y_3}{3} \tag{13}$$

where Y_1, Y_2, Y_3 is calculated using the below equation:

$$\begin{cases} Y_1 = Y_\alpha(t) - B_1 \times S_\alpha \\ Y_2 = Y_\beta(t) - B_2 \times S_\beta \\ Y_3 = Y_\delta(t) - B_3 \times S_\delta \end{cases} \tag{14}$$

6 Results and Discussion

The optimal sizing of the HRES is performed at the specific site location (https://www.sciencedirect.com/science/article/pii/S2352152X22007873). Figure 2a–c demonstrates the solar radiation, wind speed, and load for 1 year. The PV-wind and battery configuration are considered for optimal sizing using the gray wolf optimization algorithm. The variable grid tariff of this study takes into account the TOU pricing in addition to optimal size. To limit the power taken from the grid and increase the use of the renewable power, import rate from the grid (PIRG) and excess energy rate from renewables (EERR) are utilized. To prevent deterioration, the maximum and minimum states of charge of the battery are regarded as 0.8 and 0.2, respectively. Now, to make the comparison of the results of the optimization of the HRES is performed with TOU tariff rates. In TOU tariff rates, the rates are divided into three parts as per the burden on the grid. When the burden on the grid is more, the tariff is more and period is known as peak period. On the other hand, if the burden on the grid is less, the cost of the grid is less and the period is known as valley. On a similar line if the load is in between the peak and valley the period is known as off-peak. The tariff for selling and purchasing is decided based on load demand and as per the TOU tariff, the load is divided into three parts (1) valley, (2) off-peak, and (3) Peak, and the tariff rates for the time periods are 0.03, 0.07, and 0.17, respectively [9].

For the comparison, the optimal sizing is performed with different algorithms like the sine cosine algorithm (SCA), Seagull optimization algorithm (SOA), and gray wolf optimization (GWO). The maximum number of iterations is 100, and the population size is 50 considered for the algorithm. The results demonstrated that GWO is giving the best optimum result among the given algorithm. Table 1 shows

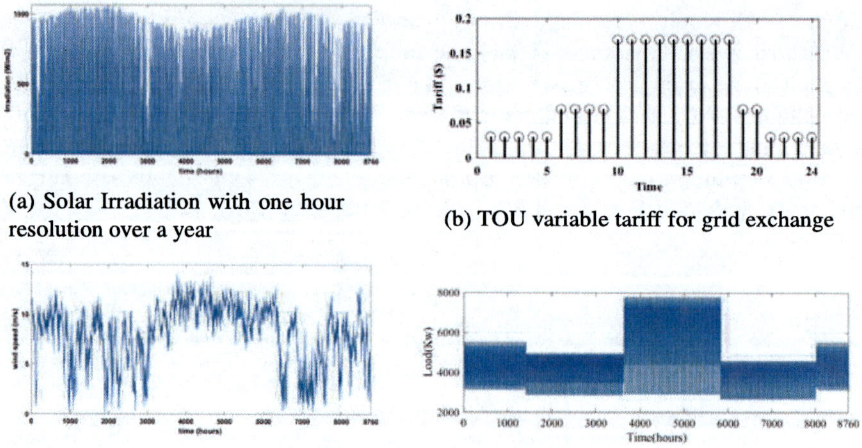

(a) Solar Irradiation with one hour resolution over a year

(b) TOU variable tariff for grid exchange

(c) Wind speed with 1-hour resolution over a year

(d) Seasonal load profile for one year

Fig. 2 Meteorological data

Table 1 Optimization results for the constant grid tariff

Optimization algorithm	N_{PV}	N_{WT}	$N_{battery}$	LCE	$PIRG$	$EERR$
GWO	2976	151	1882	0.1731	0.15	0.1
SCA	2965	157	1855	0.1901	0.1498	0.09
SOA	2970	154	1871	0.1822	0.1499	0.1

Table 2 Optimization results for the variable grid tariff

Optimization algorithm	N_{PV}	N_{WT}	$N_{battery}$	LCE	$PIRG$	$EERR$
GWO	2952	152	1884	0.1584	0.15	0.1
SCA	2960	156	1869	0.1712	0.1499	0.09
SOA	2967	154	1875	0.1752	0.1499	0.1

results at constant tariff with a selling tariff of 0.07 and purchasing tariff of 0.17. The optimal result is $N_{PV} = 2976$, $N_{WT} = 151$, $N_{bat} = 1882$ and $LCE = 0.1713$.

As you can observe from Table 2, the optimal sizing of the HRES considering the variable tariff is demonstrated. The results demonstrate a significant reduction in the LCE. In this case, the GWO optimization gives the best optimum result among the all optimizations given. The optimal results obtained by GWO are $N_{PV} = 2952$, $N_{WT} = 152$, $N_{bat} = 1884$, and $LCE = 0.1584$. Figure 2 shows the power generated by PV and wind, along with that of the SOC of battery, grid power exchange load. The load compensation of the HRES by the renewable sources is demonstrated in Fig. 3.

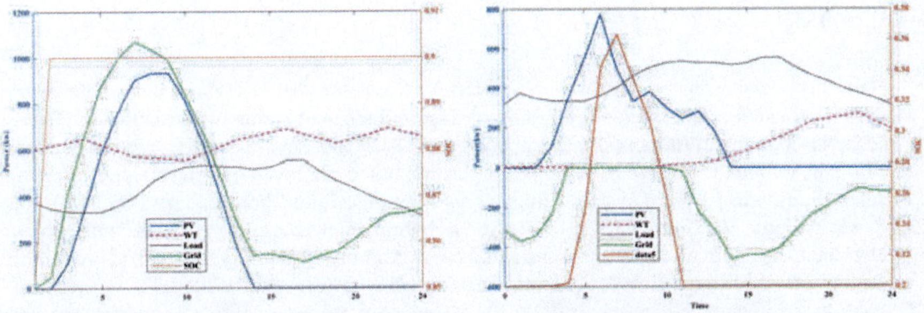

(a) Compensation of load by renewable sources on day-1

(b) Compensation of load by renewable sources on day-6

Fig. 3 Compensation of load by renewable sources

Figure 3a shows that from the hours 0:00 to 1:00, the PV is not sufficiently available. However, excess wind power is available, which supplies the load, and surplus power charges the batteries. From hours 2:00 to 8:00, the excess power is available in both the PV and WT, and the surplus power is supplied to the grid. From hours 14:00 to 24:00, the PV power is unavailable to supply the load. However, the excess WT power is available to supply the load, and the surplus is supplied to the grid. In the same way, Fig. 3b shows the compensation of the load by the various sources. From hours 0:00 to 3:00, neither PV nor WT is available to supply the load. However, the deficit power is supplied by the grid. From hours 4:00 to 6:00, excess PV power is available, which meets the load and charges the battery. From hours 7:00 to 11:00, the PV power is less than the load demand, and the WT power is still zero. However, the battery, along with PV, meets the load.

7 Conclusion

This study demonstrated the techno-economic analysis of the PV, wind, and battery HRES. The location considered for the research has an ample amount of irradiation, and the wind is available. The findings of the comparison demonstrate a considerable decrease in the LCE, and the proposed EMS offers a variable price for the buying and selling of energy. The cost reduction achieved by using the variable tariff is around 7.5% per year. In the optimal sizing of the HRES, the comparison is made between the metaheuristic techniques like GWO, SC, and SOA. It found that GWO is giving the best optimum results compared to other algorithms. The optimal results obtained by GWO are $N_{PV} = 2952$, $N_{WT} = 152$, $N_{bat} = 1884$, and $LCE = 0.1584$. This study is helpful in the design of the distribution system, economic analysis of the system, and various other parts of the power system.

References

1. Bartolucci L, Cordiner S, Mulone V, Rocco V, Rossi JL (2018) Hybrid renewable energy systems for renewable integration in microgrids: in fluence of sizing on performance. Energy 152:744–758. https://doi.org/10.1016/j.energy.2018.03.165
2. Mahesh A, Sandhu KS (2020) A genetic algorithm based improved optimal sizing strategy for solar-wind-battery hybrid system using energy filter algorithm. Front Energy 14(1):139–151
3. Sawle Y, Gupta SC, Bohre AK (2017) Review of hybrid renewable energy systems with comparative analysis of off-grid hybrid system loss of power supply probability loss of load probability. Renew Sustain Energy Rev 1–19. http://dx.doi.org/10.1016/j.rser.2017.06.033
4. Sassi A, Zaidi N, Nasri O, Slama JBH (2017) Energy management of PV/wind/battery hybrid energy system based on batteries utilization optimization. In: International conference on green energy and conversion systems, GECS 2017
5. Ma T, Javed MS (2019) Integrated sizing of hybrid PV-wind-battery system for remote island considering the saturation of each renewable energy resource. Energy Convers Manag 182:178–190. https://doi.org/10.1016/j.enconman.2018.12.059
6. Goud BS, Rekha R, Jyostna MR, Sarala S, Rao BL, Reddy CR (2020) Energy management and power quality improvement in hres grid-connected system. In: Proceeding - 1st FORTEI-international conference on electrical engineering, FORTEI-ICEE 2020, pp 174–178
7. Radhakrishnan A, Selvan MP (2015) Load scheduling for smart energy management in residential buildings with renewable sources. In: 18th national power systems conference. NPSC 2014
8. Torreglosa JP, Ferna LM, Garcia P, Jurado F (2013) Optimal energy management system for stand-alone wind turbine/photovoltaic/hydrogen/battery hybrid system with supervisory control based on fuzzy logic, vol 8
9. Chaudhari K, Ukil A (2016) TOU pricing based energy management of public EV charging stations using energy storage system. In: Proceedings of the IEEE international conference on industrial technology, pp 460–465
10. Amrollahi MH, Bathaee SMT (2017) Techno-economic optimization of hybrid photovoltaic/wind generation together with energy storage system in a stand-alone micro-grid subjected to demand response. Appl Energy 202:66–77. http://dx.doi.org/10.1016/j.apenergy.2017.05.116
11. Diemuodeke EO, Addo A, Oko CO, Mulugetta Y, Ojapah MM (2019) Optimal mapping of hybrid renewable energy systems for locations using multi-criteria decision-making algorithm. Renew Energy 134:461–477. https://doi.org/10.1016/j.renene.2018.11.055
12. Escobar LA, Meeker WQ (2006) A review of accelerated test models. Stat Sci 21(4):552–577
13. Mohammed AQ, Al-Anbarri KA, Hannun RM (2020) Optimal combination and sizing of a stand-alone hybrid energy system using a nomadic people optimizer. IEEE Access 8:200 518–200 540
14. Hamanah WM, Abido MA, Alhems LM (2020) Optimum sizing of hybrid PV, wind, battery and diesel system using lightning search algorithm. Arab J Sci Eng 45(3):1871–1883. https://doi.org/10.1007/s13369-019-04292-w
15. Mirjalili S, Mirjalili SM, Lewis A (2014) Grey wolf optimizer. Adv Eng Softw 69:46–61. http://dx.doi.org/10.1016/j.advengsoft.2013.12.007

Optimal Dispatch of Renewable Sources Under Virtual Power Plant

Pranjali Kumari, Gautam Kumar, and Sanjay Kumar

Abstract The main grid integrated into renewable energy source can be challenging due to their intermittent and unpredictable nature. However, incorporating battery energy storage systems (BESS) into renewable energy systems can provide several benefits. To fully exploit the advantages of BESS, an optimal energy management strategy is required. The Concept of a Virtual Power Plant (VPP) can combine the strength of renewable energy sources and simplify management for a decentralized energy market. The proposal introduces a new structure for optimal energy management of a Virtual Power Plant. The framework considers important aspects, such as managing the risk associated with renewable energy sources, reducing operating costs, and managing voltage levels within the system. Additionally a stochastic simulation with two stages is created to address the issue of unpredictable Power bills and RES generation. The simulation results indicate that the proposed framework can enhance the system by ensuring energy balance accuracy, reducing energy and maintaining system security. By incorporating the VPP approach and optimal energy management strategies, the integration of renewable energy sources and BESS can be streamlined and made more effective.

Keywords Distributed generations · Battery energy storage system · Optimal energy management · Virtual power plant · Renewable energy system · Stochastic method · Deterministic method

P. Kumari (✉) · G. Kumar · S. Kumar
Department of Electrical Engineering, National Institute of Technology, Jamshedpur, Jharkhand, India
e-mail: pranjali2408@gmail.com

S. Kumar
e-mail: skumar.ee@nitjsr.ac.in

A. Kumar et al. (eds.), *Decarbonisation and Digitization of the Energy System*, Lecture Notes in Electrical Engineering 1099, https://doi.org/10.1007/978-981-99-7630-0_8

1 Introduction

Environmental concerns have led to a rise in the use of renewable energy sources (RESs), such as solar, wind power, bio-gas, etc., along with rising demand for energy and a decreasing distribution of fossil fuels with high carbon content. Many Countries have created laws regarding RES as well as standards to encourage the use of RES [1], and as a result, there is a large growth in other sectors of power systems, including renewable Distributed Generators [2]. However, because RES is unpredictable and variable, managing and controlling it in a typical electricity system is challenging [3]. It has been noticed that integrating RESs and various other traditional distributed generators and BESS like thermal plants, micro-CHP, etc. can significantly address the balancing of power and regulating voltage fluctuations [4].

(BESS) have a significant role in modern power systems, as they can directly address the renewable energy intermittency, provide technical support for power systems, and develop smart grid technology [5]. BESS have been researched in a wide range of renewable energy systems to improve the integration of renewable energy, from small-scale systems like distributed renewable systems and micro-grids to large-scale freestanding hybrid renewable energy systems and renewable energy power plants [4, 6]. The reduction of transmission network congestion, assistance with voltage and frequency management, and the postponement of transmission network renovations and expansions are other significant applications of battery storage in power systems that require attention [7].

As mentioned in, VPP has two variants, referred to as commercial Virtual Power Plant (cVPP) and technical Virtual Power Plant (tVPP) [8]. A tVPP is made up of a few renewable DGs in the same geographic area and considers the actual impact of the local network on the whole portfolio. A commercial Virtual Power Plant works as a market operator and maximizes its profit without taking the conditions of the network into consideration [9]. VPP energy management differs from that of traditional power systems, which depends on the type, the level of penetration of RES, and the strategy for market participation. Many things have been done to solve the problem of how to manage VPP energy [10]. RES and BESS are two parameters taken into account by the electrical energy management employed for Virtual Power Plants. They proposed a theoretical energy management methodology for Virtual Power Plants with storage systems [11] and thermal and electrical generators, and introduced a regulation strategy for micro-CHP and wind that decreases operating price [12, 13]. But none of the above works have taken wind power and ESSs into account in a precise manner for tVPP energy management [14].

This paper presents an optimal Virtual Power Plant energy management issue model with wind energy systems, theoretical energy storage systems, and DGs inside the coming day and balancing business. The aim of the prototype is to maintain the energy balance of the Virtual Power Plant and system security while minimizing operational costs for the VPP, including those associated with operation, maintenance, and power market charges. To acknowledge the uncertainty in wind power gains and electricity pricing, a two-stage theoretical technique is presented.

The following are the paper's main contributions:

- A cutting-edge VPP energy management model that takes into account a number of crucial elements, such as the operation of dispersed wind and traditional generators, the higher perception of wind energy, and voltage regulation assurances of power quality.
- A double-mode stochastic optimization model to take into account different VPP uncertainty.
- Depending on earlier records of data sets proper simulations are presented, and the outcome demonstrates that the speculative strategy performs significantly more in lowering operating costs while preserving system dependability.

2 System Modelling

2.1 *Virtual Power Plant Model*

The suggested model concludes that the Virtual Power Plant is made up of Energy Storage Systems units, wind energy systems (Wind Turbines), micro-CHP units, and stable loads. Wind turbines reduce carbon emissions while the entire efficiency is produced by micro-CHP units [15]. Wind power outcome uncertainty is decreased with the coordination of battery storage and micro Combined Heat and Power(micro-CHP) [16]. When the actual generation of wind differs from that of the predicted total, batteries help to compensate for the imbalance by charging or discharging [17]. The Virtual Power Plant is assumed to be run by a central system, from which the holder receives predicted details and has direct authority over each subsystem. The system operating cost can be made low with the help of the controller. A distribution line connects the VPP units that are kept together which are linked to the major grid. The operational time is denoted by t divided by T time slots [18].

2.2 *Model for Energy/Electricity Market*

The Virtual Power Plant configuration is suggested to compete in the energy market in a manner same to the Nordic Energy Market [19], which is an aggressive double-settlement energy market comprised of an actual balancing market. Based on the operational restriction and evaluation of the entire profile, the top authority put their proposal for consumption or generation to the future market [20]. The actual energy/power transfer may differ from the power schedule due to uncertainty. To maintain power balance, the central controller must regulate the absolute-time power/energy transfer. Furthermore, if an unbalancing problem arises, an ascending or descending rule penance will be submitted in the balancing market [21].

3 Problem Interpretation

The issue is expressed as a stochastic prototype that has two stages. Energy Proposal and Unit commitment status of DG units in the future market are among the first-step opinion variables. The second stage consists of DG unit active and reactive power outcome, battery charge cycle, flexible load management, and absolute-time energy transfer in the electricity market.

3.1 Governing Function

$$min\sum_i\sum_t(\alpha_i p_{i,t} + \beta_i) \tag{1}$$

$$\sum_t[p_t^{DA}\lambda_t^{DA} + (p_t^{BL} - p_t^{DA})\lambda_t^{BL} - \phi_t|p_t^{BL} - p_t^{DA}|] \tag{2}$$

$$\sum_k\sum_t C_k\left(\frac{e_{k,t}^-}{\eta_k^-} + e_{k,t}^+\eta_k^+\right) \tag{3}$$

Equation (1) represents the operational cost of a virtual power plant, where $\alpha_i p_{i,t}$ denote the micro-CHP unit generation cost i for t time and the unit β_i is cost for no-load. When unit i is turned on, the zero-load cost is incurred.

Equation (2) mentions the VPP's market cost. In a two-settlement market, the hourly bid and real-time power exchange are denoted by P_t^{DA} and P_t^{BL}, respectively. When P_t^{DA} is negative or positive, the Virtual Power Plant has purchasing/selling energy in the future/coming market at a cost of λ_t^{BL}. The Virtual Power Plant is purchasing/selling energy/power on cost λ_t^{BL} in a balancing market when $P_t^{BL} - P_t^{DA}$) is negative/positive. The higher or lower regulation is given by $\phi_t|P_t^{BL} - P_t^{DA}|$, whenever the real-time power exchange and day-ahead power schedule are not equal then regulation ϕ_t is paid.

Equation (3) shows the price of battery storage k, where $e_{k,t}^+$ is rate of charging having efficiency η_k^+ and $e_{k,t}^-$ is the rate of discharging having efficiency η_k^-.

3.2 Constraints

(1) *Equations of Power Flow*: swing bus is present at Bus 1 and linked to the major grid. The reactive power injection and active power injection are given in Eqs. (4) and (5):

$$P_{j,t} = \sum_i P_{i,t} - p_t^{BL} - p_{j,t}^{LD} + \sum_n w_{n,t} + \sum_k(e_{k,t}^- - e_{k,t}^+), \forall t \tag{4}$$

$$Q_{j,t} = \sum\nolimits_i q_{i,t} - q_t^{BL} - q_{j,t}^{LD}, \forall t \tag{5}$$

The difference between load consumption and the active power generation is known as net active power injection $P_{j,t}$. $P_{j,t}^{LD}$ denotes the consumption of the load active power on bus j, and $W_{n,t}$ represents the output of wind power for unit n. Similarly, the difference between consumption of load reactive power and generation of load reactive power is known as net reactive power injection $Q_{j,t}$. The equations of linear power flow are given by:

$$P_{j,t} = (2V_{j,t} - 1)G_{j,j} + \sum\nolimits_{o(o \neq j)} G_{j,o}(V_{j,t} + V_{o,t} - 1) + B_{j,o}(\theta_{j,t} - \theta_{o,t}), \forall t \tag{6}$$

$$Q_{j,t} = -(2V_{j,t} - 1)B_{j,j} + \sum\nolimits_{o(o \neq j)} - B_{j,o}(V_{j,t} + V_{o,t} - 1) + G_{j,o}(\theta_{j,t} - \theta_{o,t}), \forall t \tag{7}$$

$$P_{j,o}^t = G_{j,o}(V_{j,t} - V_{o,t}) + B_{j,o}(\theta_{j,t} - \theta_{o,t}), \forall t \tag{8}$$

$$Q_{j,o}^t = B_{j,o}(V_{o,t} - V_{j,t}) + G_{j,o}(\theta_{j,t} - \theta_{o,t}), \forall t \tag{9}$$

where $P_{j,o}^t$ is the active power flow over branch (j,o) and $Q_{j,o}^t$ is the reactive power flow over branch (j,o). $V_{j,t}$ denotes the magnitude of voltage and $\theta_{j,t}$ denotes the phase angle at bus j. $G_{j,o}$ is real parts of the Virtual Power Plant admittance matrix at branch (j,o) and $B_{j,o}$ is the imaginary parts of the Virtual Power Plant admittance matrix at branch (j,o).

(2) *Constraints in Network.* To secure the system, the network congestion as. well as phase angle and magnitude of voltages at each bus are needed to be in a secure operational period. Therefore, Eq. (10) defines an upper and lower limit for voltage whereas Eq. (11) represents the power flow over branches.

$$V_j^- \leq V_{j,t} \leq V_j^+, \theta_j^- \leq \theta_{j,t} \leq \theta_j^+, \forall j, \forall t \tag{10}$$

$$P_j^- \leq P_{j,o}^t \leq P_j^+ Q_j^- \leq Q_{j,o}^t \leq Q_j^+, \forall j, \forall t \tag{11}$$

(3) *Constraints in Battery Storage.* Since battery power lies in a specified range, a higher and lower limit for batteries' size $E_{k,t}$ is governed by following two equations respectively to avoid the rate of over-charging or over-discharging.

$$E_k^- \leq E_{k,t} \leq E_k^+, \forall k, \forall t \tag{12}$$

$$E_{k,t+1} = E_{k,t} + \left(\frac{e_{k,t}^-}{\eta_k^-} + e_{k,t}^+ \eta_k^+ \right), \forall k, \forall t \tag{13}$$

4 Test Analysis

4.1 Data

The proposed Virtual Power Plant model (Fig. 1) is based on a modified IEEE 13-bus distribution test feeder [11]. Buses 6, 7, and 11 have three micro-CHP units, On bus 9, there is one wind turbine, and on buses 2 and 10, respectively, there are two battery storage systems. In the actual test feeder, the transformer and switch are replaced by distribution lines [7]. The base values voltage magnitude is to be set at 4.16 kV and apparent power base values are to be set at 5000 kVA. The VPP central control is assumed to be able to receive the information to be forecasted and to have straight control over entire unit. The micro-CHP units parameters are displayed in Table 1, while Table 2 displays data from the battery storage system. Historical details are collected to address the power demand, market prices and the wind power, and to meet the required install capacity demand, these data are scaled down.

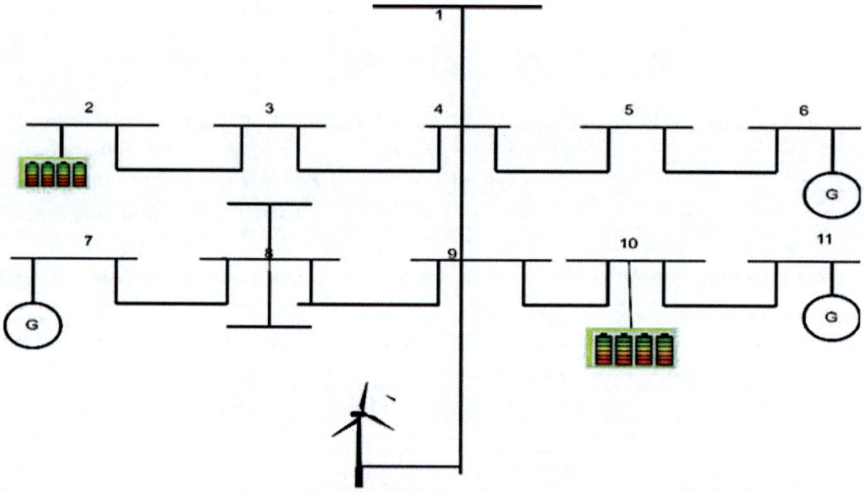

Fig. 1 IEEE 13—bus distribution test feeder

Table 1 Parameters for micro-CHP units

Unit	α_i($/kWh)	β_i($)	P_{cap}(kW)
1	0.15	30	1500
2	0.30	50	1000
3	0.50	70	1500

Table 2 Parameters for storage

E_k^-(kWh)	E_k^+(kWh)
50	Max = 100
$e_{k,t}^-(kW)$	$e_{k,t}^+(kW)$
100	100

4.2 Result

(1) *Deterministic vs. Stochastic*:- The suggested deterministic strategy and a stochastic strategy to determine the Virtual Power Plant power management problem are firstly compared. Simulation result of proposed model can observed in Fig. 2. The figure shows the comparison between the stochastic and deterministic approaches for the 25 iterations. It is noted that growing wind penetration level benefits both approaches by lowering system operating costs. This happens as wind turbines are expected to have lower generation costs. Furthermore, the results show that the stochastic strategy performs better than the deterministic strategy, implying that the stochastic operation is more capable of dealing with RES uncertainties.

(2) *Voltage Control*:- To measure the power quality, voltage regulation one of component and in the use of RES, power quality is one of the concern. The higher and lower limits of phase angle are set to be 0.03 rad and the higher and lower limits for magnitude of voltage are set to be 0.5 p.u., in this case, and penetration of wind is set to be 20%. Both magnitude of voltage and phase angle are in the safe operating range as shown in Figs. 3 and 4 of the simulation results.

(3) *Associated with storage:*- Effectiveness of ESS in relieving ambiguity is assessed here. The use of battery storage systems and those without storages is

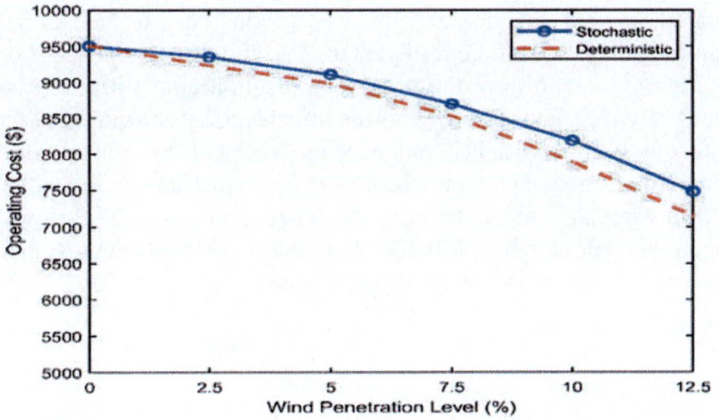

Fig. 2 Stochastic versus deterministic approach operating cost having different wind penetration

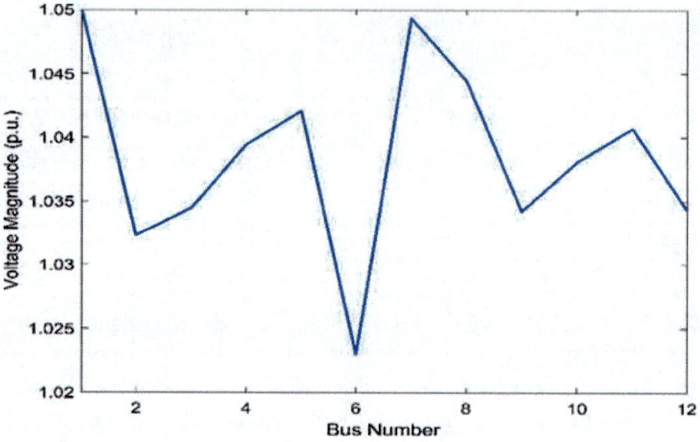

Fig. 3 Each bus voltage magnitude

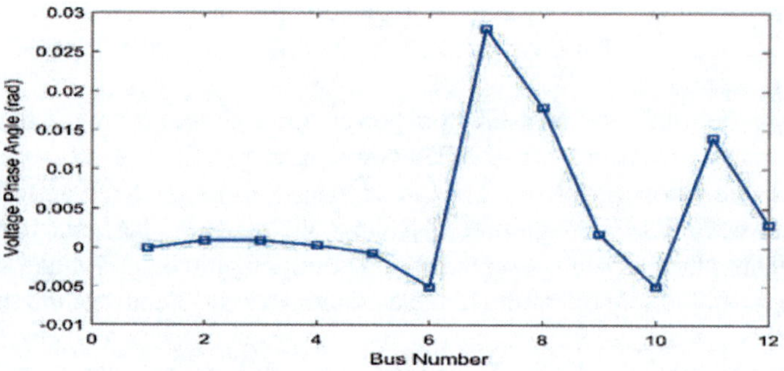

Fig. 4 Each bus voltage Phase

compared, and per day storage capital cost is included to the storage system. The storage of energy efficiency is set at 75%, and the wind penetration is set at 20%. Figures 5 and 6 show the simulation result. Figure 5 illustrates the effectiveness of storages in reducing power imbalance by showing that the system running cost is lower with ESSs added to the VPP than it would be without them. Additionally, Fig. 6 shows that the storage system's most efficient capacity is 200 kWh when the per capital cost of storage is included. Therefore, building storage capacity exceeding 200 kilo-Watt hour is not financially viable because the cost of the storage will exceed the revenue.

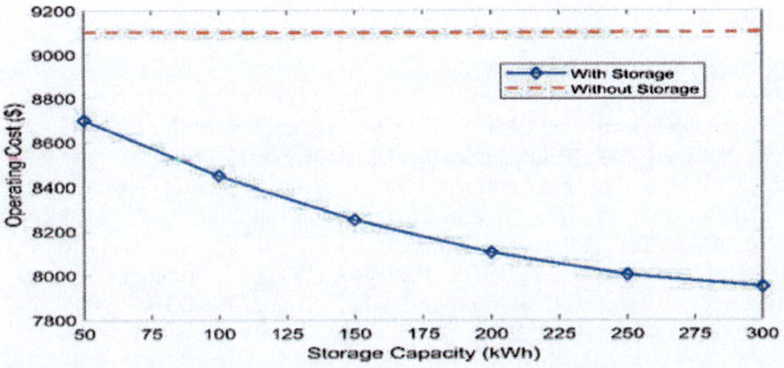

Fig. 5 Cost versus storage

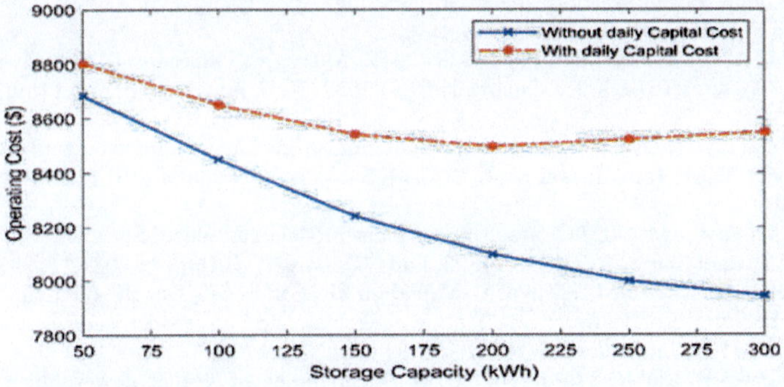

Fig. 6 Cost versus storage

5 Conclusions

This research provides an innovative energy management paradigm for VPP that includes energy storage and renewable power generation. The issue with A two-stage stochastic optimization problem is used to describe the task of minimizing the operational costs of the VPP in electricity markets. Numerous simulations have been run, and the results demonstrate how the framework can save operating costs for the VPP while managing power quality and security of system. Further scope of this work includes taking into account bidding technique in stochastic optimization and also expanding the quantity of input cases to evaluate results' stability and tackle many subtasks concurrently.

References

1. Chen Z, Liu Q, Xiao X, Liu N, Yan X (2013) Integrated mode and key issues of renewable energy sources and electric vehicles charging and discharging facilities in microgrid. IET Conf. Publ., vol. 2013, no. 623 CP. https://doi.org/10.1049/cp.2013.1759
2. Wang Z (2020) Modelling and Analysis of Grid Integrated TSCAOI Generators for Renewable Distributed Generation Systems. CPSS Trans. Power Electron. Appl. 5(4):372–387. https://doi.org/10.24295/CPSSTPEA.2020.00031
3. Lee JT, Henriquez-Auba R, Poolla BK, Callaway DS (2022) Pricing and Energy Trading in Peer-to-Peer Zero Marginal-Cost Microgrids. IEEE Trans. Smart Grid 13(1):702–714. https://doi.org/10.1109/TSG.2021.3122879
4. Shaqsi AZ, Sopian K, Al-Hinai A (2020) Review of energy storage services, applications, limitations, and benefits. Energy Reports, vol 6. Elsevier Ltd, pp. 288–306. https://doi.org/10.1016/j.egyr.2020.07.028.
5. Mbungu NT, Bansal RC, Naidoo RM, Bettayeb M, Siti MW, Bipath M (2020) A dynamic energy management system using smart metering. Appl. Energy, vol 280, p 115990. https://doi.org/10.1016/j.apenergy.2020.115990.
6. Ross M, Abbey C, Bouffard F, Joos G (2018) Microgrid Economic Dispatch with Energy Storage Systems. IEEE Trans. Smart Grid 9(4):3039–3047. https://doi.org/10.1109/TSG.2016.2624756
7. Li F, Bo R (2007) DCOPF-based LMP simulation: Algorithm, comparison with ACOPF, and sensitivity. IEEE Trans Power Syst 22(4):1475–1485. https://doi.org/10.1109/TPWRS.2007.907924
8. Naval N, Yusta JM (2021) Virtual power plant models and electricity markets - A review. Renew Sustain Energy Rev 149:111393. https://doi.org/10.1016/j.rser.2021.111393
9. Bhuiyan EA, Hossain MZ, Muyeen SM, Fahim SR, Sarker SK, Das SK (Oct.2021) Towards next generation virtual power plant: Technology review and frameworks. Renew Sustain Energy Rev 150:111358. https://doi.org/10.1016/J.RSER.2021.111358
10. Yu S, Fang F, Liu Y, Liu J (Apr.2019) Uncertainties of virtual power plant: Problems and countermeasures. Appl Energy 239:454–470. https://doi.org/10.1016/J.APENERGY.2019.01.224
11. Yang Q, Wang H, Wang T, Zhang S, Wu X, Wang H (Jul.2021) Blockchain-based decentralized energy management platform for residential distributed energy resources in a virtual power plant. Appl Energy 294:117026. https://doi.org/10.1016/J.APENERGY.2021.117026
12. Kumar A, Alaraj M, Rizwan M, Nangia U (2021) Novel AI Based Energy Management System for Smart Grid with RES Integration. IEEE Access 9:162530–162542. https://doi.org/10.1109/ACCESS.2021.3131502
13. Guo W, Liu P, Shu X (Jan.2021) Optimal dispatching of electric-thermal interconnected virtual power plant considering market trading mechanism. J Clean Prod 279:123446. https://doi.org/10.1016/J.JCLEPRO.2020.123446
14. Rahimi M, Ardakani FJ, Ardakani AJ (May2021) Optimal stochastic scheduling of electrical and thermal renewable and non-renewable resources in virtual power plant. Int J Electr Power Energy Syst 127:106658. https://doi.org/10.1016/J.IJEPES.2020.106658
15. Yavuz L, Önen A, Muyeen SM, Kamwa I (2019) Transformation of microgrid to virtual power plant - A comprehensive review. IET Gener Transm Distrib 13(11):2077–2087. https://doi.org/10.1049/iet-gtd.2018.5649
16. Ghasemi Olanlari F, Amraee T, M. Moradi-Sepahvand, Ahmadian A (2022) Coordinated multi-objective scheduling of a multi-energy virtual power plant considering storages and demand response," IET Gener. Transm. Distrib., vol 16, no. 17, pp 3539–3562. https://doi.org/10.1049/gtd2.12543
17. van Summeren LFM, Wieczorek AJ, Bombaerts GJT, Verbong GPJ (May2020) Community energy meets smart grids: Reviewing goals, structure, and roles in Virtual Power Plants in Ireland, Belgium and the Netherlands. Energy Res Soc Sci 63:101415. https://doi.org/10.1016/J.ERSS.2019.101415

18. A smart-grid energy management system using a multi objective control of residential appliances Abstrac
19. Khodadadi A, Herre L, Shinde P, Eriksson R, Söder L, Amelin M (2020) Nordic Balancing Markets: Overview of Market Rules. Int. Conf. Eur. Energy Mark. EEM, vol. 2020
20. Capper T et al (Jul.2022) Peer-to-peer, community self-consumption, and transactive energy: A systematic literature review of local energy market models. Renew Sustain Energy Rev 162:112403. https://doi.org/10.1016/J.RSER.2022.112403
21. Li J, Fu Y, Xing Z, Zhang X, Zhang Z, Fan X (2019) Coordination Scheduling Model of Multi-Type Flexible Load for Increasing Wind Power Utilization. IEEE Access 7:105840–105850. https://doi.org/10.1109/ACCESS.2019.2932141

Variation of Cost in Scheduling of Residential and Commercial Appliances with the Inclusion of Energy Price Tag of EV and Household Battery

Ajay Kumar Prajapat and Sandeep Kakran

Abstract Due to the rising use of the electric vehicle (EVs) in today's transportation systems, the analysis of integration of EVs with the electricity grid has become a necessity of current time. This paper considers a residential and commercial area with some constant and variable load. The main purpose of this article is energy management of sources, i.e., EVs, photovoltaic cells (PVs), household batteries, and grid, along with scheduling of the variable appliances to reduce the peak demand and cost of the electricity bill. In this cost of the energy storage devices, i.e., energy price tag (EPT) is considered and variation in total cost is observed with and without EPT. This paper presents an algorithm to evaluate the variation in the cost of energy consumption with and without considering the EPT. For scheduling of the appliances, two cases are considered, i.e., appliances are scheduled randomly and user preferred duration. For the optimization, mixed-integer linear programming is used, and in that DICOPT solver is used to solve the equations in GAMS software.

Keywords Electric vehicle · PV · Energy price tag · Energy scheduling · GAMS · DICOPT

1 Introduction

Recently with the increment of pollution, use of electric vehicles (EVs) and renewable energy sources (RES) has increased exponentially. Renewable sources include solar, wind, geothermal, and many other which reduces pollution, and these sources are not going to be diminished after a certain interval of time and EVs are a convenient source of energy; therefore recent research has been focused on how to use RES and EVs for fulfilling demand of different sectors and reduce pollution. In [1], EVs and emerging technologies for future application have been elaborated. In this, methods to improve efficiency, reliability, and safety have been explained. In [2], EVs, battery

A. K. Prajapat · S. Kakran (✉)
NIT Kurukshetra, Haryana 136119, India
e-mail: skakran@gmail.com

A. Kumar et al. (eds.), *Decarbonisation and Digitization of the Energy System*, Lecture Notes in Electrical Engineering 1099, https://doi.org/10.1007/978-981-99-7630-0_9

technology trends, and charging methods have been reviewed. In this standard that is available for the charging process, power control, battery technologies, and battery energy management have been mentioned. In [3], a review of electric vehicles in developing countries has been mentioned. In [4], a survey of the impact of fast-charging stations on the existing utility grid has been done. In [5], for the modeling of a charger at a particular location deep reinforcement learning (DRL) based on charging and discharging algorithm has been used. The effectiveness of DRL has been analyzed in two ways, i.e., charging cost and the effect of load shifting. From the results, it has been seen that the charging cost has been reduced and the load has been flattened. In [6], electric vehicle charging station has been implemented which operates in both directions with the combination of photovoltaic and battery which has been interconnected with commercial sector using the intelligent optimization technique which operates in four steps, and from the result, it has been seen that customer satisfaction cost has been reduced, and higher tolerability has been provided toward the uncertainties. In [7], a model of single residence has been proposed using mixed-integer linear programming and residential appliances have been optimally operated, and energy received from distributed energy resources is managed. In this day, ahead time of use price has been used, and energy management and load scheduling system have been embraced. In [8], research during the last five years on cooling techniques with phase change materials (PCM) was reviewed, and it has been observed that 20% in PV efficiency has been achieved. Heat regulation has been improved as well as temperature uniformity has been achieved with two PCM. In [9], the technology of solar PV has been reviewed and various incentive schemes have been introduced by the government to attract consumers from different sectors. The government of India gives a 30% subsidy to general category states and a 70% subsidy in Uttarakhand, Sikkim, Himachal Pradesh, Jammu Kashmir, and Lakshadweep [10]. In [11], optimization methods that address the power scheduling problem in the smart home have been reviewed. In this classification has been made, i.e., exact and metaheuristic algorithms. In [12], for the realization of the uncertainty of renewable energy, electrical load tracking and thermal load tracking have been used. In [13], to do the energy scheduling, the Cuckoo search algorithm has been used and it has been compared with other evolutionary algorithms, and the result shows that the technique used in the paper gives the best result. In [14], the energy price tag (EPT) of all storage devices connected with smart homes has been introduced. A smart home contains a plug-in electric vehicle, household battery, and photovoltaic, which fulfills the demand of a smart home. To set the priority between PEV, household battery, and utility based on EPT, rule-based priority has been used.

Section 2 contains problem formulation. Section 3 contains results, and data related to the case study are shown. Section 4 presents the conclusion of the paper.

2 Problem Formulation

2.1 Storage Device Real Time Price

In this paper, different appliances of residential and commercial sectors are considered and their scheduling is done with the consideration of storage energy costs, i.e., EPT of EV and battery. Sources considered in this are EV, PV, battery, and utility grid. Sources that take and give away energy are EVs and batteries. Equations (1) and (2) denote the charging and discharging power of EVs.

$$Po^{elv}(h)Po(h)\ gte(h) + Po^{bte}(h) + Po^{pvte}(h)\ (charging) \tag{1}$$

$$Po^{elv}(h) = Po^{etl}(h) + Po^{etb}(h)\ (discharging) \tag{2}$$

$$P^{gte}(h), Po^{bte}(h), Po^{pvte}(h), Po^{etl}(h), Po^{etb}(h) \geq 0$$

To charge the EV, power given by grid and battery are $Po^{gte}(h)$ and $Po^{bte}(h)$ respectively, and after meeting the load demand, the charging power provided by PV is $Po^{pvte}(h)$. $Po^{etl}(h)$ and $Po^{etb}(h)$ denote the power provided by the EV to the load and for the charging of the battery respectively. Equations (3) and (4) denote the charging and discharging power of battery.

$$Po^{b}(h) = Po^{gtb}(h) + Po^{etb}(h) + Po^{pvtb}(h)(charging) \tag{3}$$

$$Po^{b}(h)Po^{btl}(h) + Po^{bte}(h)(discharging) \tag{4}$$

$$Po^{gtb}(h), Po^{bte}(h), Po^{pvtb}(h), Po^{btl}(h), Po^{bte}(h) \geq 0$$

To charge the battery, power given by grid and EV are $Po^{gtb}(h)$ and $Po^{etb}(h)$ respectively and after meeting the load demand the charging power provided by PV is $Po^{pvtb}(h)$. $Po^{btl}(h)$ and $Po^{bte}(h)$ denote the power given by the battery to the load and for charging the EV respectively. In this, EPT is used by which exchange of energy takes place among all the storage devices and the price of storage devices is in (ct/kWh). Equations (5) to (9) denote the stored energy of EV.

$$SOC_e(h) = SOC_e(h-1) + Ef_{ec} \times CC_e \times R_{ec} - \frac{1}{Ef_{ed}} \times DD_e \times R_{ed} \tag{5}$$

$$P^{ec} = Ef_{ec} \times R_{ec} \tag{6}$$

$$P^{ed} = Ef_{ed} \times R_{ed} \tag{7}$$

$$SOC_e^{mi}(h) \leq SOC_e(h) \leq SOC_e^{ma}(h) \tag{8}$$

$$CC_e + DD_e \leq 1 \tag{9}$$

Charging and discharging of EV is denoted by (6) and (7) [19]. $SOC_e(h)$ and $SOC_e(h-1)$ are the EV energy at present and before the present time. R_{ec} and R_{ed} are the rates of charging and discharging. Equation (8) shows the maximum and minimum energy limits of EV. Charging and discharging binary variables are denoted by CC_e and DD_e. Ef_{ec} and Ef_{ed} are the charging and discharging efficiencies, whereas P^{ed} and P^{ec} are the charging and discharging power of the EV. Equations (10) and (11) show the EPT of EV.

During charging

$$EPT^E(h) = \frac{EPT^E(h-1) \times energy^E(h-1) + Pg(h) \times cg(h) + P^{bd} \times EPT^E(h-1)}{energy^E(h)} \tag{10}$$

During discharging

$$EPT^E(h) = \frac{EPT^E(h-1) \times energy^E(h-1) + P^{ed} \times EPT^E(h-1)}{energy^E(h)} \tag{11}$$

$$SOC_b(h) = SOC_b(h-1) + Ef_{bc} \times CCC_b \times R_{bc} - \frac{1}{Ef_{bd}} \times DDD_b \times R_{bd} \tag{12}$$

$$P^{bc} = Ef_{bc} \times R_{bc} \tag{13}$$

$$P^{bd} = Ef_{bd} \times R_{bd} \tag{14}$$

$$SOC_b^{mi}(h) \leq SOC_b(h) \leq SOC_b^{ma}(h) \tag{15}$$

$$CCC_b + DDD_b \leq 1 \tag{16}$$

Battery energy at present and before the present time are denoted by $SOC_b(h)$ and $SOC_b(h-1)$. The charging and discharging rates of the battery are R_{bc} and R_{bd} respectively. Charging and discharging binary variables are denoted by CCC_b and DDD_b, whereas the charging and discharging efficiency of the battery are denoted by Ef_{bc} and Ef_{bd}. P^{bd} and P^{bc} are the discharging and charging power of the battery. Equations (17) and (18) show the EPT of the battery.

During charging

$$EPT^E(h) = \frac{EPT^E(h-1) \times energy^E(h-1) + Pg(h) \times cg(h) + P^{bd} \times EPT^E(h-1)}{energy^E(h)} \tag{17}$$

During discharging

$$EPT^B(h) = \frac{EPT^B(h-1) \times energy^B(h-1) + P^{bd} \times EPT^B(h-1)}{energy^B(h)} \tag{18}$$

Equation (19) shows the evaluation of PV power with the help of PV radiation [19].

$$P_{pv}(h) = P_{rr} \times P_d\left(\frac{R_d(h)}{R_{re}}\right) \tag{19}$$

where

P_{rr}: PV panel rated power.

P_d: Derating factor.

Rd: Solar radiation.

R_{re}: Reference solar radiation.

2.2 *Objective I*

In this work, four different appliances of residential and commercial areas are considered. The main objective of this research is to implement the scheduling of loads for the minimization of consumer cost with minimal impact on consumer preference. Equation (20) shows the objective function.

$$Cos_{min} = \sum_{h=1}^{24} \sum_{apll=1}^{4} A_{apll}^{h} \times Pri_g \tag{20}$$

$$A_{apll} = \left[A_1^1, A_2^2, \ldots A_4^4\right] \tag{21}$$

$$\sum_{h=1}^{24} A_{apll}^{h} = A_{apll}^{total} \tag{22}$$

Cos_{min} : Optimized cost

A_{apll}^{h} : Energy consumption of each appliance during the interval.

$$A^{h}_{apll} = f_{h,apll} \times W^{mx}_{apll} + \left(1 - f_{h,apl}\right) \times W^{mn}_{apll} \tag{23}$$

$$\sum_{h=1}^{24} f_{h,apll} \times g_{apll,h} = u_{apll} \tag{24}$$

$f_{apll,h}$: Binary variable which indicates the ON or OFF moment of appliance

$g_{apll,h}$: Binary variable which indicates the duration for which appliance is ON or OFF

u_{apll} : Overall slots when the appliance remains ON

$$\sum_{n \in M} x_{n,apll} = 1 \tag{25}$$

x is a binary variable, and 'M' is the total number of schedules of any appliances, and Total number of schedules = Total number of consumer preferred duration—appliance ON duration + 2.

$$A^{h}_{apll} = \sum_{n \in A} x_{n,apll} \times A_{n,h} \tag{26}$$

$A_{n,h}$: Energy consumption during the hth hour of nth schedule.

$$W^{mi}_{apll} \leq A^{h}_{apll} \times g_{apll,h} \leq W^{ma}_{apll} \tag{27}$$

$$\sum_{a=1}^{4} A^{h}_{apll} \leq H^{mx} \tag{28}$$

H^{mx}: Energy consumption limit during each hour.

2.3 Objective II

The objective is to reduce the total cost of buying electricity from the grid while lowering the EPT of EV and domestic batteries at the same time. The total cost is given in (29).

$$cost_{min} = \sum_{h=1}^{24} Cg(h)Pg(h) + EPT^{EV}(h) * energy^{EV}(h) + EPT^{B}(h) * energy^{A}(h) \tag{29}$$

The first term of the objective denotes the total cost of purchasing energy from the utility, whereas the other terms indicate the charging and discharging cost of EV as well as battery.

3 Result

In this, to see the variation in cost with EPT different cases are considered. Firstly, scheduling of appliances is done randomly without EPT. Secondly, scheduling of appliances is done considering the user preferred duration without EPT. Thirdly, the random scheduling of appliances is done with EPT, and, lastly, the appliances are scheduled considering the consumers preferred duration with EPT. The assumption considered in this is that EV stays at house for complete duration. Figure 1 shows the static load of residential and commercial sectors during each hour, and Table 1 shows the different variable appliances of residential and commercial sectors. Figure 2 and Fig. 3 show the solar irradiation and utility real time tariff respectively [17, 18]. Different parameters used by storage devices are shown in Table 2.

Figures 4 and 5 show the scheduling of different appliances randomly and considering consumer preferred duration without EPT. These figures show the power used by different sources to meet the fixed demand and operation of variable appliances for the minimization of cost. From the figure, it is observed that when the appliance is scheduled randomly charging and discharging of EV and battery are done randomly whereas discharging of EV and battery is done during the hours when power demand is at peak and charged during the hours when power demand is normal in the case of appliances scheduling considering the consumer duration.

Figures 6 and 7 show the scheduling of different appliances and consumption of different sources randomly and considering consumer preferred duration with EPT. These figures show the power used by different sources to meet the fixed demand and operation of variable appliances for the minimization of cost. From the figure it is observed that when the appliance is scheduled randomly charging and discharging of EV and battery are done randomly whereas discharging of EV and battery is

Fig. 1 Static load of the residential and commercial sector

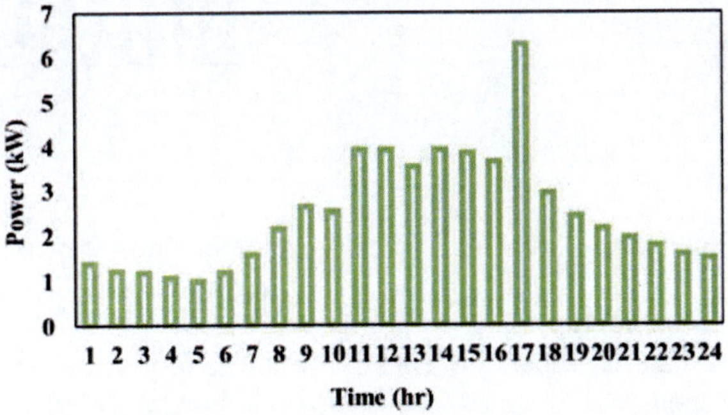

Table 1 Shiftable residential and commercial appliances consumer preferred duration

Appliance	Energy consumed (kWh)	Consumers preferred duration
Shiftable residential appliances		
Geyser	1	7–11
Water Pump	0.75	13–15
Toaster	1.5	6–8
Washing machine	1.3	5–10
Shiftable commercial appliances		
Ice crusher	0.4	7–9
Grinder	1.5	7–9
Deep Fryer	2	9–13
Dishwasher	1.2	12–14

Fig. 2 Irradiation of solar

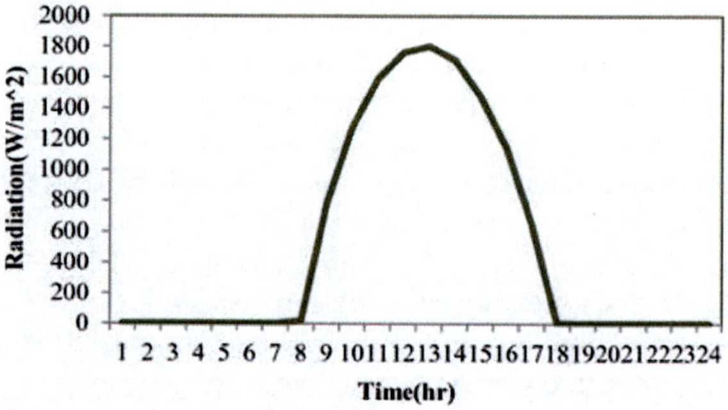

Fig. 3 Grid tariff

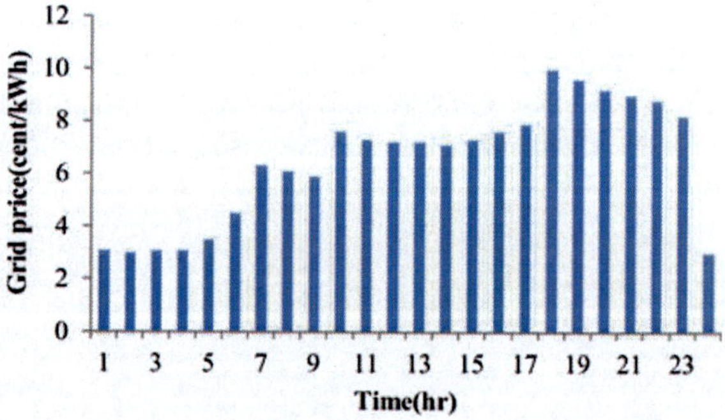

done during the hours when power demand is at peak and charged during the hours when power demand is normal in the case of appliances scheduling considering the consumer duration.

The discrepancy of EPT of EV and battery with random and user preferred duration scheduling of appliances is shown in Figs. 8 and 9.

Table 2 Storage device parameters

Parameter	Value	Parameter	Value
EVs initial EPT	3.5(ct/ kWh)	Battery's initial EPT	2(ct/ kWh)
Storage of EV	20 kWh	Storage of battery	10kWh
EVs preliminary SOC	40%	Battery's preliminary SOC	35%
EVs minimum SOC	35%	Battery's minimum SOC	30%
EVs maximum SOC	80%	Battery's maximum SOC	85%

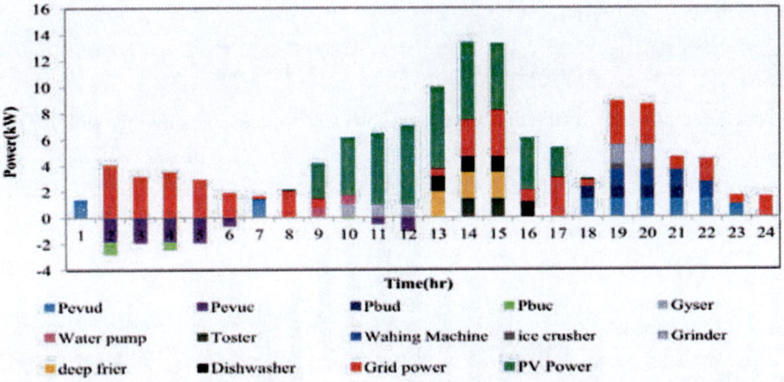

Fig. 4 Scheduled power of different sources and shiftable appliances randomly without EPT

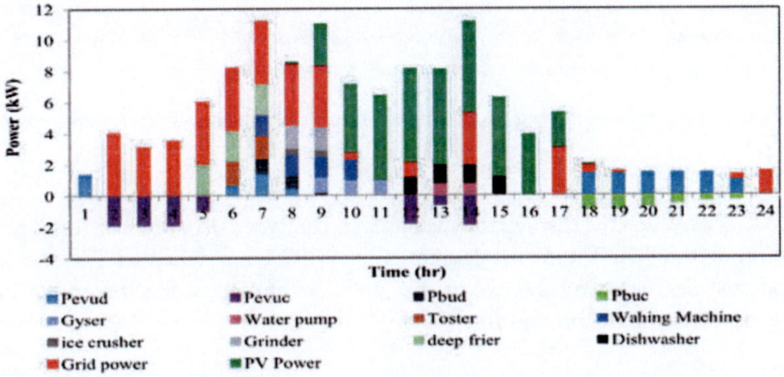

Fig. 5 Scheduled power of different sources and shiftable appliances in user preferred duration without EPT

The discrepancy in cost for different cases is shown in Table 3. From the table, it can be seen that when the appliances are scheduled randomly then the discrepancy in cost with the use of EPT is 157.22% and when the appliances are scheduled considering consumer's preference then with the use of EPT, discrepancy comes out

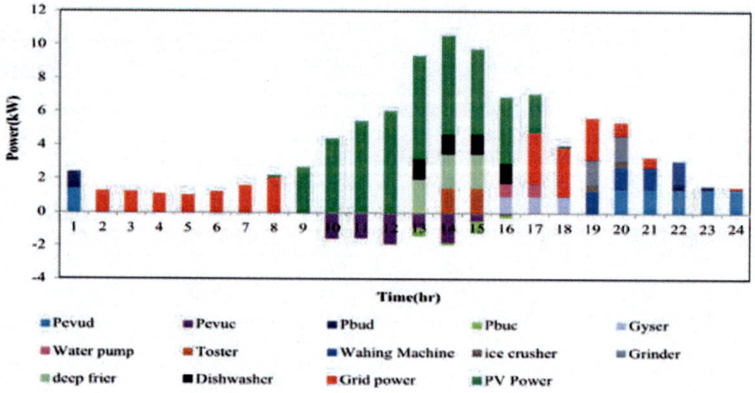

Fig. 6 Scheduled power of different sources and shiftable appliances randomly with EPT

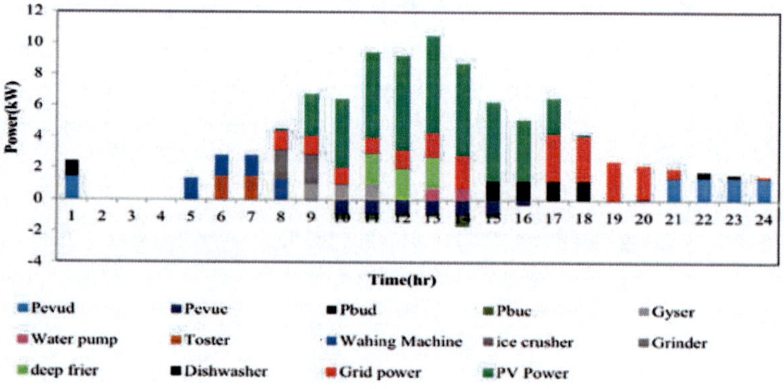

Fig. 7 Scheduled power of different sources and shiftable appliances in user preferred duration with EPT

to be 207.27%. When the discrepancy is observed between user preferred and random scheduling without EPT, the cost decreases by 17.99% and when EPT is considered then the cost decreases by 2.03%. In the table, '+' indicates the increment in cost and '–' indicates the decrement in cost.

4 Conclusion

In this paper EV, PV, battery, and grid have been used to meet the demand of consumers of residential and commercial areas. In addition to scheduling of appliances, charging and discharging costs of EV and battery have been evaluated using EPT. To see the effect of EPT on energy scheduling cost, different cases have been

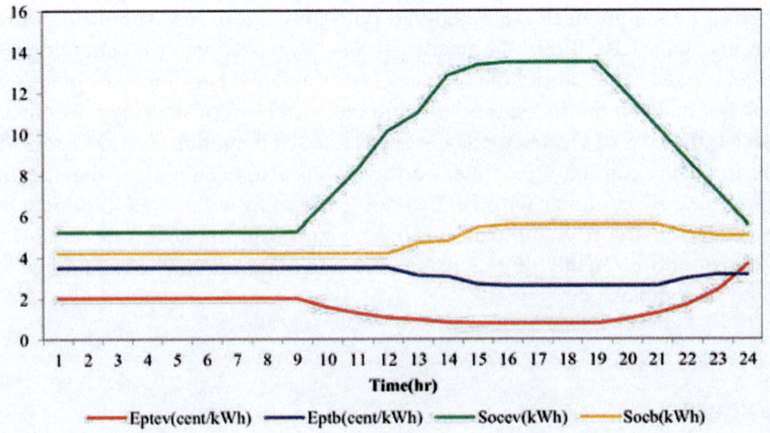

Fig. 8 Variation of EPT and SOC of EV and battery in case when the appliance is scheduled randomly

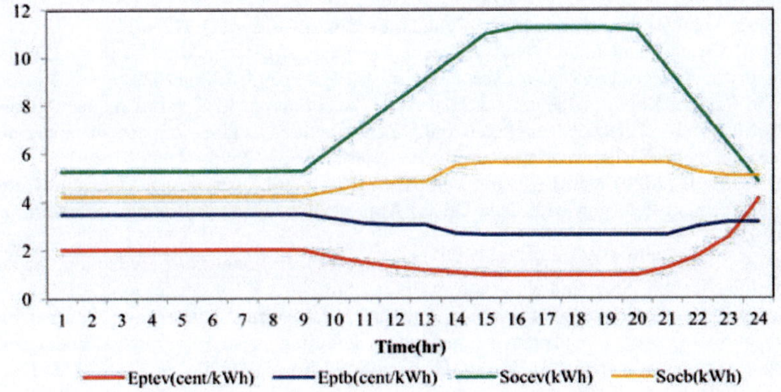

Fig. 9 Variation of EPT and SOC of EV and battery in case when the appliances are scheduled considering the user preferred duration

Table 3 Discrepancy in cost for different cases

	Random scheduled	Consumer preferred duration schedule	Discrepancy (%)
Without EPT	468.429	384.142	–17.99
With EPT	1204.919	1180.362	–2.03
Discrepancy (%)	+ 157.22	+ 207.27	

considered, i.e., appliances are scheduled randomly and in user preferred duration without and with EPT. From the results, it has been observed that charging of EV and battery takes place during the off-peak hours and discharging takes place during the peak hours. With the scheduling of appliances and by considering the consumers preference, the cost of electricity has been reduced and appliances have been scheduled during the minimum price time. Further, it has also been noticed that electricity consumption cost is higher with EPT in comparison to without EPT, which shows that the cost of using EVs and batteries has a significant impact on the cost of electricity consumption. In the future, cost of EVs and batteries can be considered while doing the overall cost analysis.

References

1. Sun X, Li Z, Wang X, Li C (2019) Technology Development of Electric Vehicles: A Review. Energies 13(1):90
2. Sanguesa JA, Torres-Sanz V, Garrido P, Martinez FJ, Marquez-Barja JM (2021) A Review on Electric Vehicles: Technologies and Challenges. Smart Cities 4(1):372–404
3. Rajper, Sarmad Zaman, and Johan Albrecht. Prospects of Electric Vehicles in the Developing Countries: A Literature Review. Sustainability 12. MDPI AG: 1906 (2020)
4. Mohd Rizwan Khalid, Mohammad Saad Alam, Adil Sarwar, M.S. Jamil Asghar,A Comprehensive review on electric vehicles charging infrastructures and their impacts on power-quality of the utility grid, eTransportation, Volume 1 ,100006,ISSN 2590–1168 (2019)
5. Lee J, Lee E (2020) Kim J Electric Vehicle Charging and Discharging Algorithm Based on Reinforcement Learning with Data-Driven Approach in Dynamic Pricing Scheme. Energies 13:1950
6. Yan, Q., Zhang, B., & Kezunovic, M. Optimized Operational Cost Reduction for an EV Charging Station Integrated with Battery Energy Storage and PV generation (2018)
7. Paramvir Singh, Sandeep Dhundhara, Yajvender Pal Verma, Nisha Tayal,Optimal battery utilization for energy management and load scheduling in smart residence under demand response scheme,Sustainable Energy, Grids and Networks,Volume 26 ,100432,ISSN 2352–4677 (2021)
8. Hafiz Muhammad Ali, Recent advancements in PV cooling and efficiency enhancement integrating phase change materials based systems – A comprehensive review, Solar Energy,Volume 197 ,Pages 163–198,ISSN 0038–092X (2020)
9. Dambhare, Mugdha&Butey, Bhavana &Moharil, Sanjiv. Solar photovoltaic technology: A review of different types of solar cells and its future trends. Journal of Physics: Conference Series. 1913. 012053 (2021)
10. https://mnre.gov.in/solar/schemes/
11. Sharif Naser Makhadmeh, Ahamad Tajudin Khader, Mohammed Azmi Al-Betar, Syibrah Naim, Ammar Kamal Abasi, Zaid Abdi Alkareem Alyasseri, Optimization methods for power scheduling problems in smart home: Survey, Renewable and Sustainable Energy Reviews, Volume 115 ,109362,ISSN 1364–0321 (2019).
12. Tang, Yanmei et al. "Robust optimization for energy scheduling of CCHP systems under electrical load tracking and thermal load tracking." *2019 IEEE Sustainable Power and Energy Conference (iSPEC)*: 536–540 (2019).
13. Antonyraj, S & Samuel, G.Giftson. Optimal Energy Scheduling of Renewable Energy Sources in Smart Grid using Cuckoo Optimization Algorithm with Enhanced Local Search. IOP Conference Series: Earth and Environmental Science. 312. 012014 (2019).

14. Aznavi, Sima& Fajri, Poria& Asrari, Arash& Harirchi, Farshad. Realistic and Intelligent Management of Connected Storage Devices in Future Smart Homes Considering Energy Price Tag. IEEE Transactions on Industry Applications. PP. 1–1. https://doi.org/10.1109/TIA.2019.2956718 (2019).
15. National Renewable Energy Laboratory (USA), www.nrel.gov/homer [last accessed January 2022].
16. Real time electricity prices. https://hourlypricing.comed.com/live-prices/ [last accessed January, 2022]
17. Kakran S, Chanana S (2019) Operation management of a renewable microgrid supplying to a residential community under the effect of incentive-based demand response program. Int J Energy Environ Eng 10:121–135
18. Lal Karn A.K., Kakran S. Operation Management of Microgrid Supplying to the Residential, Industrial and Commercial Community using Different Demand Response Techniques. 4th International Conference on Energy, Power and Environment (ICEPE), 2022, pp. 1–6, (2022)

Comparative Analysis of Load Forecasting by Using ANN, FUZZY Logic and ANFIS

Jaya Shukla and Rajnish Bhasker

Abstract With the paradigm shift of power transmission and distribution system towards decentralized control, there has been a great upsurge in forecasting of load which would be effective for economic utilization of resources. In this paper, there is a comparative study about load forecasting using various Artificial Intelligence techniques which includes Neural Network, Fuzzy logic as well as adaptive neuro-fuzzy system to indicate an effective solution for optimum utilization of renewable sources which is incorporated in the existing grid conditions. The AI techniques have helped to utilize information like the data of historical load, weather conditions and temperature difference such that a decision can be taken when to integrate how much renewable energy source in the present system such that optimum results are obtained. The comparison reveals that the ANFIS system is more effective to pass decision on when to integrate renewable energy sources into the grid. Load Forecasting at various places is critical for efficiently utilizing energy potential. With this in mind, intelligent models based on fuzzy logic, ANN, and ANFIS are created and presented for Load prediction. The suggested Review can readily integrate climate variables' uncertainty and nonlinearity. A comparison of the aforementioned models has also been performed. It is explored if the findings produced by the ANFIS model for load prediction are better and more accurate. As a result, the ANFIS model may be useful in anticipating load and optimizing resource consumption.

Keywords Power demand · Load forecast · AI techniques

1 Introduction

For the design of an effective energy management system load forecasting is an effective tool for utilities and power system operators such the demand can be met effectively. It requires various decisions during the operation of the system which includes economic distribution of generation, scrutiny of the security system as well

J. Shukla (✉) · R. Bhasker
Department of Electrical Engineering, VBSPU, Jaunpur, India
e-mail: jayashuklaee@gmail.com

A. Kumar et al. (eds.), *Decarbonisation and Digitization of the Energy System*, Lecture Notes in Electrical Engineering 1099, https://doi.org/10.1007/978-981-99-7630-0_10

as planning of the maintenance which hugely depends on load forecast [1]. The forecasting of load is stereotypically programmed beginning for a limited small time (in minutes) to an hour forward, otherwise to a larger value like for 20 years ahead. The forecasting of the electrical load is classified into four types: long term, medium term, short term and very short term [2]. Energy is the most often used resource in today's globe. Throughout our daily lives, we use energy in a variety of ways, including electricity, refined oils, LPG, solar energy, wind energy, chemical energy in the form of batteries, and many more. We can be flamboyant at times and cautious at others. Yet, the goal is to provide customers with an uninterrupted supply of electricity, and in order to achieve this goal, current and future power demand must be properly evaluated. That is why we want the Load Forecasting approach to inform us about customer demand and the exact capability to create power. Amongst these, the most significant process of load forecasting required for supervisory control of the production of power is short-term load forecasting because it is needed for the determination of the power plants' work plan in addition to the decision of taking out the best production group which may either exist from current grid connection or by integration of new renewable energy source which includes generation from solar or PV integrated into the grid [3]. Energy establishment units face many economic and technical problems such as operating, planning and controlling of electricity [4]. Short-term load forecasting aids the service providers to formulate decisions about energy production as well as its purchasing, switching to renewable sources as well as switching of the load. Subsequently, it becomes extremely significant to effectively forecast the load with accuracy aimed at both service providers as well as prosumers in the upcoming scenario which will be a more viable market [5].

2 Background of AI in Load Forecasting

During the previous few years, there has remained a lot of development in the techniques of forecasting load which ranges from manual operating approaches to controlling automated models which include artificial intelligence-based techniques [6]. The conventional technique of load forecasting hugely depends on statistical models. These models had data of an hour or a week and information about specific events which was performed offline by using time functions [7]. Nevertheless, subsequently, the forecasting of load is non-linear, this conventional arithmetical approach delivers imprecise results as compared to AI-based techniques [8]. In general, there are conventional methods with a consequential outcome from the Box Jenkins thereafter it was advanced by means of Multiple Linear Regression (MLR), adaptive models besides general exponential smoothing (GES) procedures [9]. Doubts in addition to the upsurge in the density of present loads, the forecast errors through these methods have also augmented. Worthwhile Solutions can be thus frequently consequential from Artificial Intelligence practices including Evolutionary Algorithms, Fuzzy Logic, artificial neural networks (ANNs) etc. as they contain promising features [10]. By means of ANN models, there is no requirement of embracing the functional relationships amongst the input parameters as well as the outputs and can be performed both online as well as offline by using various topologies [11]. Individually online as well as offline ANN predicting models have efficaciously provided improved forecast results, which is at an advanced level of accuracy, as compared to the conventional methods. This is accomplished over inclusion of entirely essential inputs besides self-learning algorithms [12]. In [13], the authors offered an expert system through the concept of fuzzy set theory for load forecasting. The expert structure has its aim at the Updating function. Short-term forecasting is to be achieved in addition to the estimation that was done on the power system of the country Taiwan. Later, [14] expressed a fuzzy linear programming model which was intended for scheduling the electric generation, instead of uncertainties in the forecast in addition to the input data by means of fuzzy set notation. With the evolution of smart grids, specific and correct load forecasting has now become ever more important since it might aid the utilities for improvement in scheduling of the lord besides decreasing disproportionate electricity production. Authors in [15] used numerous linear as well as non-linear machine learning algorithms and then gave preference to the best as reference, indicating the best features by means of packaging and embedded feature assortment approaches then lastly with the help of genetic algorithm (GA) to determine optimal time lags in addition with the quantity of layers for LSTM model projecting the optimization of its performance.

Authors in [16] stated that electrical load as well as its charge are the utmost significant aspects within this sector. For refining reliability, regulation along with management of operations, a precise estimate of the day (in advance) load is a significant prerequisite.

3 Forecasting Using ANN Model

Here, we have constructed a typical feed forward back propagation (BP) ANN which was used for producing the forecast representations, by means of a fully connected structure consisting of a sole hidden layer [17]. The load data obtained is from UPSLDC (Uttar Pradesh State Load Despatch Centre) for the time span of 2017 and 2018. These data are divided into two parts one which is used as a training set data and the other used for verifying the forecasting process. Both the input as well as output data were normalized between −1 to +1, and it was also done before training so as to scale the input within the given range. It follows the load graph as represented in Fig. 1.

We have considered L_s as the scaled data element and L as the main data element. For calculations, we have obtained the maximum and minimum quantity (L_{max} and L_{min}). The scaled data will be given by the formula:

$$\mathrm{Ls} = \frac{(Ls\mathrm{max} - Ls\mathrm{min})(L - Lmin)}{Lmax - Lin} + \mathrm{Lsmin} \quad (1)$$

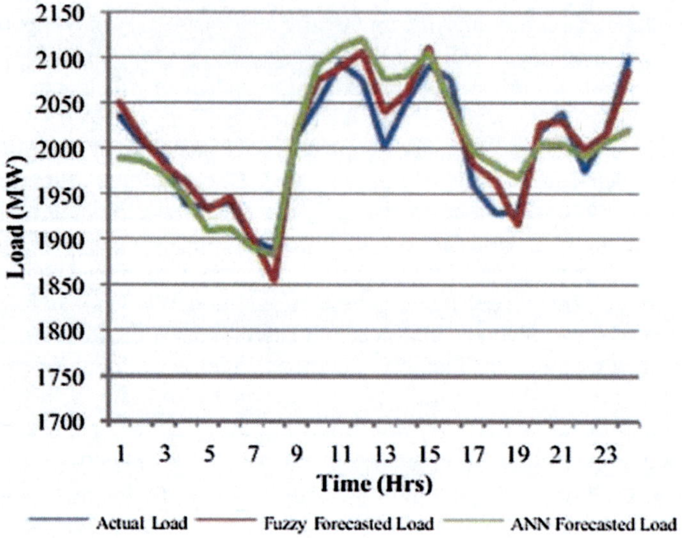

Fig. 1 Demand met graph for 3 days consecutively [18]

The ANN model uses seven inputs including load at days 1st 2nd, 5th 6th, 9th, 10th, 14th and 15th as well as considering only one output node as a representation for the day ahead load forecast for the training data set. For analysis of the model, we have considered the next fifteen days of the month such that error can be obtained.

The absolute mean error was then calculated to obtain the error using formula:

$$\text{AME} = \frac{1}{n}\sum_{t=0}^{t=n}\frac{Fa - Ff}{Fa} \tag{2}$$

Here, F_a is the value of the known load and F_f is the value of forecasted load to obtain the error which is forecasted by the artificial neural network model. The structure of ANN model is shown in Fig. 2.

4 Forecasting Using FUZZY Model

The process of a load forecasting system is done using fuzzy logic method that is verified using statistics from the Uttar Pradesh State Load Despatch Centre for various day's categories which is used for training a load forecasting of load. Here we have used real time data of hourly load demand of a week and weather data in terms of temperature and humidity. For the fuzzy membership of these input values, we have used triangular membership function [19]. The load which was forecasted is then equated with the known and correct load to obtain an average percentage error. Table 1

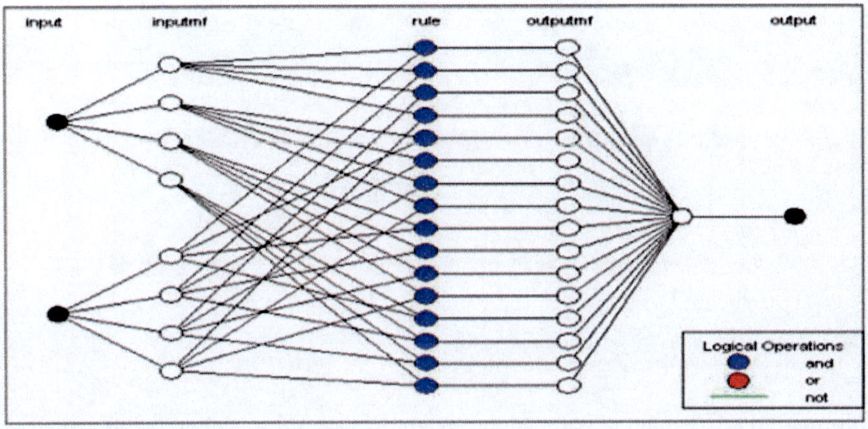

Fig. 2 Structure of ANN model

Table 1 Relation of humid condition as well as temperature with respect to load

Temperature humidity	Extremely hot	Very hot	Hot	Slightly hot	Pleasant
Very humid	Very high	Very high	High	Medium	Low
Low humid	Very high	High	Medium	Medium	Low
Least humid	High	High	Low	Low	Very low

shows the development of the rule base constructed on the general linguistic terms, in addition to the main source of the prediction model. The prototypical yields merely rely on this, besides subsequent initial analysis of the data set, the consequential production rules is formed, nevertheless within equivalent sources they might be diverse for alternative set of data. The rule base which is applied for this forecast is.

For forecasting to be done, the use of a rule base is crucial which is developed according to the available knowledge base from the data obtained from the state load despatch centre of Uttar Pradesh. The resultant membership function for the forecasted load is shown in Fig. 3.

5 Forecasting Using ANFIS Model

The Neuro-Fuzzy Inference System includes the amalgamation of fuzzy logic as well as ANN in addition to the generation of planning structure amongst input parameters as well as output consequences [20]. This structure is proficient with definite information as per the expert system. At that time, the rules are designed along with the equivalent outputs. As equated with a neural network, in this case, the training method remains furthermore difficult. In supervised learning ANN structure, the

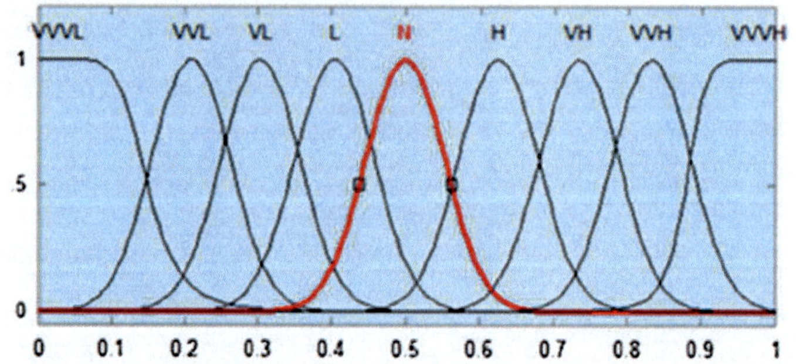

Fig. 3 Membership function for forecasted load

information sets remain articulated through numerous subsets of entry and exit. The transfer function along with the term called bias is an essential requirement in the development of this prototype. Nevertheless, within these fuzzy interference systems, there exists a corresponding requirement to explain the procedures of fuzzification and defuzzification [21].

In this paper, we have used ANFIS aimed at the forecasting of the electrical load. The working of the fuzzy logic archetypal is meanwhile equated to that of the ANFIS model with the help of the existing set of data. The historic load, difference existing in the temperature along with the cyclical data of various seasons form the inputs of ANFIS, whereas the output is the hourly forecasted load. The data for the year 2018–2019 are designated for training; meanwhile, the data for the year 2019 is utilized in testing [9].

The membership function aimed at historical data is represented by Gaussian membership function and trapezoidal membership function. It uses seven membership functions with linguistic terms as very very Low, Very low, Low, None, High, Very high and Very very High and is depicted as shown in Fig. 4.

The Membership function for temperature difference is as shown in Fig. 5 which indicates the usage of Gaussian and trapezoidal membership functions. There are

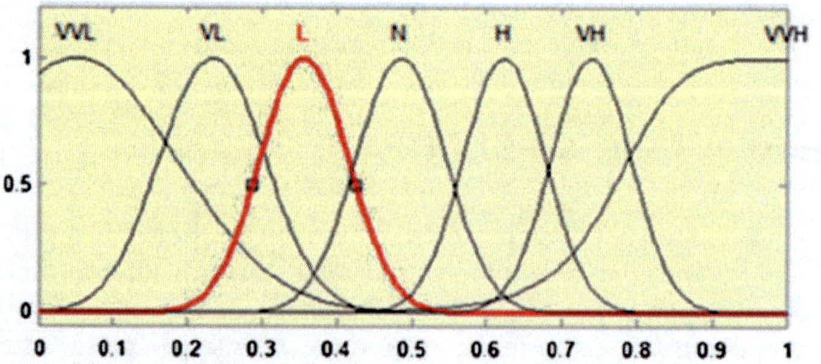

Fig. 4 Membership function for historical data

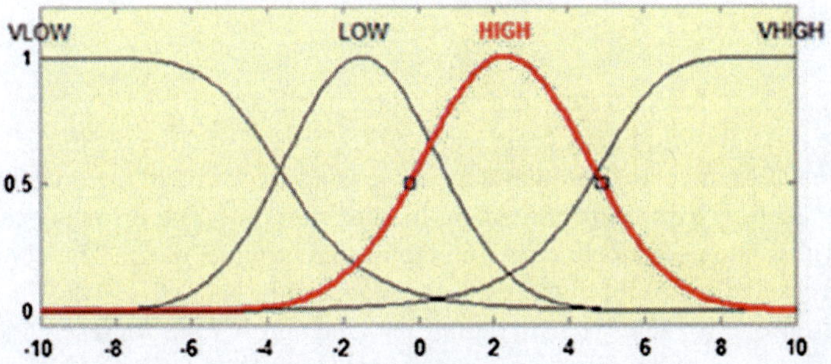

Fig. 5 Membership function for temperature difference

four membership functions which include the linguistic standards as very low, low, high and very high. The range of the membership function varies from the change in temperature within the span of ±10 °C.

The Membership function for indication of various seasons is as shown in Fig. 6, where we have used the Gaussian membership function for the representation of all four seasons that are summer, winter, spring as well as autumn, and thus the range of time span covered is all the twelve months. The range of membership functions lies between 1 and 12 [22–24].

The structure of model consists of five layers, where the primary layer is the input layer for this system which contains its input value given to the system that includes historical data of load, temperature difference of the system as well as the weather conditions [25]. Layer two contains the input membership function and the third layer consists of the membership function of the output. The next layer consists of the layer of weighted sum as well as the last layer which is the resultant of load forecast.

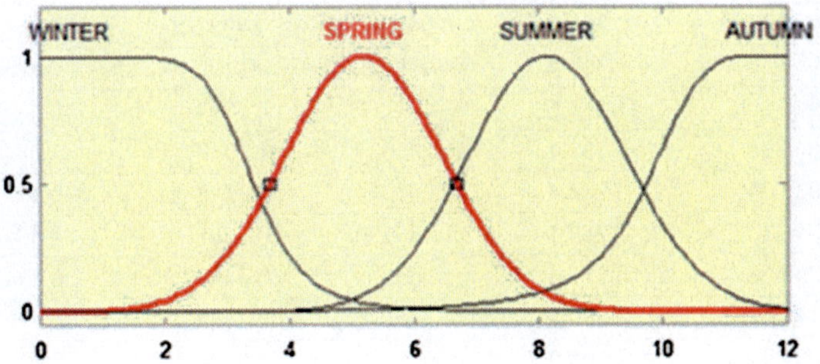

Fig. 6 Membership function for season

6 Results

This study of load forecast has compared three methods of load forecasting which include Neural network, Fuzzy logic and ANFIS system, and the results of load forecasted by each of them are compared by comparing the value of error in forecasting of load by each of them with the actual load of that day. The error is calculate by taking the absolute percentage error and then calculating its mean. The comparisons of load forecasted by all three methods are specified in Table 2.

Furthermore, error rates remain quite truncated and ANFIS representations yield improved results in comparison with fuzzy logic results as well as ANN outcomes. Table II shows the comparison of errors in forecasting load. The accomplishment of ANFIS is owed for an improved exhibition of the previous years' loads. The reason for this improvement in the result is that ANFIS entails fuzzy logic along with artificial neural network approach that includes a capability to learn. Numerous alterations can be made on the way to decrease the forecast error. The preceding and subsequent days with respect to holidays consume diverse load trends. Such types of days might be obtained by using an alternative daytime category. Furthermore, day types including the diurnal in addition to the time of year (e.g., summer-Friday category) can be used. The rate of error can be curtailed with the help of accomplishment of the changes, but then again the intricacy of the classification grows and the arrangement of the archetypal also becomes more difficult.

Table 2 Comparison of error in forecasting of load

S. no.	Forecasted load in (M.W.)	Actual load	Error by ANN	Error by fuzzy	Error by ANFIS
1	2051.11	322.6	−0.8278	−0.6668	0.4731
2	2013.74	380.1	−0.0916	0.1803	0.1431
3	1981.40	410.3	0.4423	0.1122	0.1088
4	1953.11	386.9	0.0430	0.2921	0.3131
5	1942.60	366.4	−0.4953	−0.2510	−0.1856
6	1926.50	353.1	−0.6319	0.5794	0.4646
7	1845.38	338.4	1.2420	1.1932	1.0841
8	2028.24	283.4	2.3630	2.3142	0.3333
9	2065.22	288.4	1.8001	0.8661	0.8042
10	2048.68	311.4	1.6642	0.4492	0.4342
11	2146.62	300.2	−1.9244	1.2632	0.4724
12	2039.94	305.9	−0.5564	−0.0812	0.0265
13	2051.00	332.2	−0.1252	−0.9005	1.0012
14	2171.56	292.2	0.4542	−0.3265	−0.2008
15	2033.73	270.4	0.9272	0.2946	0.0272

7 Conclusion

Short-term load forecasting by means of NN, Fuzzy modelling and Fuzzy-Neural network has been employed in this work. The outcomes attained for the test data set were then compared with those obtained from all the three models. The outcomes attained with the Fuzzy-Neural network remained far better as compared to that of ANN-only and fuzzy-only models. The consequences illustrate that the ANN is appropriate to interpolate amid the training data set to deliver forthcoming load patterns. The cohesive Fuzzy-Neural network supplementary improves the output in addition to the minimization of the error. Meanwhile, the neural network merely interpolates between the training data, and thus it will provide us with more error for the test data which is not nearby to the training data set. ANFIS as well as fuzzy logic approaches give quite close results to the actual information and then yield better forecasting consequences. Thus, we conclude that for further improved performance in the ANFIS system, the training data might be well-spaced in upcoming representations in demand to deliver more better and precise outcomes.

References

1. Chartoniuk W, Chen M-S (2000) Very short term load forecasting using ANN. IEEE Trans Power Syst 15:263–268
2. Hasan Hu¨seyin C¸ evik. Mehmet C¸ Unkas (2014) Short-term load forecasting using fuzzy logic and ANFIS. Nat Comput Appl Forum
3. Gohil P, Gupta M (2014) Short term load forecasting using fuzzy logic. Proc Int J Eng Dev Res 127–130
4. Kalaitzakis K, Stavrakakis GS, Anagnostaki EM (2002) Short-term load forecasting based on artificial neural networks parallel implementation. Electr Power Syst Res 63:196–281
5. Soliman SA, Al-Kandari AM (2010) Electrical load forecasting modeling and model construction. Elsevier
6. Chen JF, Wang WM, Haung CM (1995) Analysis of an adaptive time-series autoregressive moving average (ARMA) model for short term load forecasting, vol 34, pp 187–196
7. Borges CE, Penya YK, Fernandez I (2011) Optimal combined short term building load forecasting. CE Borges, Spain
8. Al-Hamadi HM, Soliman SA (2005) Long-term/mid-term electric load forecasting based on short-term correlation and annual growth. Electr Power Syst Res 74(3):353–361
9. Abedinia O, Amjady N, Zareipour H (2017) A new feature selection technique for load and price forecast of electrical power systems. IEEE Trans Power Syst 32(1)
10. Jang J-SR, Sun C-T, Mizutani E (1997) Neuro-fuzzy & soft computing—a computational approach to learning and machine intelligence. ISBN 0-13-261066-3. Prentice Hall Inc
11. Ghasemi A, Shayeghi H, Moradzadeh M, Nooshyar M (2016) A novel hybrid algorithm for electricity price and load forecasting in smart grids with demand-side management. Appl Energy 177:40–59
12. Bouktif S, Fiaz A, Ouni A, Serhani MA (2018) Optimal deep learning LSTM model for electric load forecasting using feature selection and genetic algorithm: comparison with machine learning approaches. Energies 11(7):1–20
13. Taifour Ali A, Tayeb EBM, Shamseldin ZM. Short term electrical load forecasting using fuzzy logic. Int J Adv

14. Engineering Technology, Management and Applied Science (IJAETMAS)" Volume 03 - Issue 11, 2016.
15. Kaur J, Brar YS (2014) Short term load forecasting using fuzzy logic of 220kV transmission line. Int J Eng Res Technol (IJERT) 3(9)
16. Banda E (2006) Short term load forecasting using artificial intelligence techniques. Department of Electrical Engineering, University of Cape Town, Student Thesis
17. Data Source: Uttar Pradesh state Load dispatch centre
18. Struzikiewicz G, Sioma A (2020) Evaluation of surface roughness and defect formation after the machining of sintered aluminum alloy. AlSi10Mg Mater 13:46–62
19. Konar A (2005) Computational intelligence: principles, techniques and applications. ISBN 3-540-20898-4. Springer, Berlin Heidelberg New York
20. Alfares HK, Nazeeruddin M (2002) Electric load forecasting: literature survey and classification of methods. Int J Syst Sci 33(1):23–34
21. Jose D, Mathew M, Krishnan A (2016) Weather dependency of electricity demand: a case study in warm humid tropical climate. In: 3rd international conference on electrical energy systems 2016
22. Songmene V, Khettabi R, Zaghbani I, Kouam J, Djebara A (2011) Machining and machinability of aluminum alloys. In: Kvackaj T (ed) Aluminium alloys, theory and applications
23. Benardos PG, Vosniakos GC (2003) Predicting surface roughness in machining: a review. Int J Mach Tool Manuf 43:833–844
24. Koenigsburger F (1964) Design principals of metal cutting machine tools. Peragamon, Oxford
25. Melkote S, Grzesik W, Outeiro J, Rech J, Schulze V (2017) Advances in material and friction data for modelling of metal machining. In: CIRP annals—manufacturing technology, vol 66, pp 731–754. Elsevier

A Practical Approach to Volt Var Optimization in an Unbalanced Radial Distribution System

G. Yesuratnam, A. Sriker, G. Jahnavi, and Srikanth Gollapudi

Abstract This paper explains and analyzes a new practical rule-based approach for Voltage VAR Optimization (VVO) which is one of the most popular techniques to improve voltage profiles and minimize feeder losses. VVO uses power system control equipment like capacitor banks and Transformer Taps to calculate new optimal operational state for these devices. The proposed algorithm first optimizes Feeder Capacitor Banks and then optimizes On Load Tap Changing (OLTC) Transformers. Typical VVO objective functions are reactive power compensation, minimization of feeder power losses, Minimum Number of Controls and Minimum Shift or combinations of these. Typical constraints include voltage at each bus within its specified limit and current flows on each feeder section within the specified limit. In this paper, the objective function used is the minimization of feeder power loss with voltage at each bus as a constraint. Feeder capacitor banks and OLTC's are considered as control devices. The proposed algorithm is tested on 55 bus unbalanced Radial network and the results show that proper adjustments of OLTC transformers and Capacitor banks significantly reduce power losses while satisfying all the operational constraints. The results offered by the proposed algorithm are compared with those offered by the conventional Interior point method, and the superiority of the proposed algorithm is demonstrated.

Keywords Volt var optimization · On load tap changer · Unbalanced radial distribution system · Capacitor banks

G. Yesuratnam · A. Sriker (✉) · G. Jahnavi
Osmania University, Hyderabad, India
e-mail: sannangi@yahoo.com

S. Gollapudi
NIT, Jamshedpur, India

A. Kumar et al. (eds.), *Decarbonisation and Digitization of the Energy System*, Lecture Notes in Electrical Engineering 1099, https://doi.org/10.1007/978-981-99-7630-0_11

1 Introduction

The purpose of the Volt Var Optimization (VVO) application is to provide an optimized set of switching recommendations for the operation of distribution capacitor banks and voltage regulators. The optimal solution is achieved by taking into account device states, operational constraints, and expected load conditions. VVO application depends on Power Flows to provide both current load states and load forecasts.

The VVO application is implemented in the form of a generalized optimization application, which tries to minimize a defined performance criterion while respecting various operating limits in the form of constraints. In general, the distribution system Voltage and VAR Optimization application tries to achieve the best system for the current operating condition by judiciously adjusting available controls, subject to various operating constraints.

2 VVO Methodology

Before attempting to solve the VVO problem, the following terms should be understood in detail.

1. Operation Modes
2. Controls
3. Constraints
4. Objective Function
5. Violation Penalty
6. Performance Index

2.1 Operation Modes

There are 2 Operation Modes where an operator can select while solving the Volt Var Optimization problem.

- Shunt operation mode
- Transformer Tap operation mode

Shunt operation Mode: When selecting this mode only Capacitor banks on the Feeder and Substation are used by the VVO program. This mode is selected by the operator to minimize the losses on the feeder. Switching On and Off Feeder capacitors will help minimize the losses. However, Switching On and Off Substation capacitors will help improve the power factor at the substation.

Transformer Tap operation mode: When selecting this mode, the Transformer Taps at the substation transformer are operated by the VVO program. Typically, this mode is selected by the Control Centre operator to minimize the voltage violations after the VVO program completes solving SHUNT operation mode for minimizing the losses.

2.2 Controls

Controls are devices used by the VVO program to meet the Constraints and help in Minimizing or Maximizing the Objective function.

Controls considered for this application include:

Capacitor (Shunt) banks: These are considered whenever the SHUNT Operation Mode is selected by the Control Centre Operator. This mode is selected whenever the losses are to be minimized.

Transformer Taps: These are considered whenever TRANSFORMER TAP Operation mode is selected by the Control Centre Operator. Normally this mode is selected by the operator whenever the Voltage violations still exist after SHUNT Operation Mode is solved.

2.3 Constraints

Typical constraints for distribution network include:

Voltage at each Node (Vimin ≤ Vi ≤ Vimax): Vimin is the Minimum Voltage Limit at the Bus (i), Vi is the current value of the voltage at Bus (i) computed by Power Flow Program and Vimax is the Maximum Voltage Limit at the Bus (i).

2.4 Objective Function

Objective Function used in this work is Minimizing the MW Loss of the distribution system. Mathematically this is represented as

$$Minimize\ Ploss = \sum_{i=1}^{i=N} \left(Iij^2 * R\right) \tag{1}$$

where I_{ij} is the Branch current flowing between Bus i and Bus j. R is the resistance of the Feeder section. The initial losses in the system are computed by the Power Flow program and these losses will be minimized by the VVO program using the controls explained in Sect. 2.2 and meeting the constraints explained in Sect. 2.3.

2.5 Violation Penalty

Violation is the term we use when the voltage at a bus deviates from its limit. In practice, while the Voltage limits should be respected as much as possible, none of them are really hard constraints. One way of handling soft constraints is to incorporate them into the objective function as penalties. This penalty measures the severity of system violations. Mathematically Violation Penalty is represented as shown below

$$\text{Violation Penalty} = (\text{Volt Violation Penalty})$$

Mathematically, this is represented as per Eqs. (2) and 3 below

$$\text{Volt Violation Penalty} = \text{Wv} * \sum_{i=1}^{i=N} (Vi - Vilimit) \tag{2}$$

$$\text{Volt Violation Penalty} = \text{Wv} * \sum_{i=1}^{i=N} (Vi - Vilimit) \tag{3}$$

where Wv is the Voltage Penalty Weights for the Voltage Violation. A value of 100 is considered in this paper while computing the Violation Penalty. When there is no violation on a limit then the corresponding term is made as zero.

By changing the values of penalty factors, it is possible to change the priority of how the constraints are satisfied. A higher penalty factor gives priority to that constraint and minimizes its violation.

2.6 Performance Index

Performance index is a unique index to measure the system's performance. It is a combination of Objective Function and Violation Penalty. By measuring this performance index, the control effectiveness with respect to the objective function and enforced constraints can be evaluated efficiently.

$$\text{Performance Index} = \text{Objective Function} + \text{Violation Penalty}$$

Control effectiveness is its contribution to the improvement of the defined system performance index. This allows unambiguous comparison of different control alternatives and facilitates judicious selection of the best control scheme.

Mathematically, this Performance Index is represented as per Eq. (4) below where loss minimization is the objective function considered, and the constraint is the voltage deviation at each bus. Wv is the weighting factor, and we have considered 100 as the value in this thesis. In case the actual voltage computed is within the high and low voltage limits for a particular node/bus then this term is forced to zero during the performance Index calculations.

$$Performance\ Index = \sum_{i=1}^{i=N}(Iij^2 * R) + \mathrm{Wv} * \sum_{i=1}^{i=N}(Vi - Vilimit) \quad (4)$$

3 Capacitor and Transformer Modeling

3.1 Capacitor Modeling

Capacitors will be modeled as constant Impedance devices while executing the Power Flows and Volt Var Optimization. The Per Unit (PU) Impedance of the capacitor is calculated as explained below.

Steps to calculate the Per Unit (PU) Impedance of the capacitor

1. From the given kvar and voltage rating of the capacitor calculate the reactance of the capacitor
2. From the given Base MVA and Base kV calculate the System Impedance
3. Divide the reactance obtained from Step 1 with System Impedance calculated in Step 2 to calculate the Per Unit (PU) value of the capacitor. This PU value of the capacitor is kept constant throughout the Power Flow program

3.2 Tap Changing Transformer Modeling

Tap Changing Transformer is modeled by changing the input voltage at the feeder head. Normally, when the Power Flows are executed, the input voltage is assumed to be 1 PU at the feeder head. The assumption here is that the Tap Changing Transformer is at its nominal position. However, during optimization, we need to change the tap positions from its nominal value. A simple example below will help understand the modeling of Tap Changing Transformer.

4 Volt Var Optimization Algorithm

The Volt Var Optimization algorithm works by optimizing the state of control devices viz. Capacitor banks on the Feeder and Substation and Tap Changing Transformers inside a Substation. The Feeder where Volt Var Optimization has to be executed is chosen by the user.

The analysis optimizes capacitors and regulators separately as they act very differently on the system. The capacitors are optimized first in order to reduce the losses. Then regulators are optimized to set the voltages to the best level. Applying capacitors first allows more flexibility for the regulators.

4.1 Main Algorithm

Steps

1. Select a Substation Transformer and select all the Feeders under this Transformer
2. Get the list of control devices (capacitor banks) on each Feeder
3. Run the Initial power flow solution using the proposed MDA method
4. Calculate the Performance Index explained in Sect. 2.6. Store this value in a variable PrevPerfIndex
5. Optimize Capacitors (see Sect. 5.2)
6. Run power flow solution using the proposed MDA method
7. Optimize Transformer Taps (see Sect. 5.3)
8. Run the final power flow solution using the proposed MDA method
9. Report results:
   - Control device changes
   - Loss reduction was seen between Steps 3 and 8
   - Voltage profile/violations for Steps 3 and 8
   - Performance Index
   - Branch currents on each feeder section
   - Input Complex Power (Fig. 1).

4.2 Capacitor Optimization

Steps

1. Select a Feeder connected to the Distribution Transformer.
2. Select the Capacitor bank at the downstream of the selected Distribution Feeder. If there are multiple capacitor banks at the same distance on the downstream of the feeder, then select the capacitor bank which has highest rating.
3. Make a note of number of individual capacitors in that bank.
4. Take one capacitor at a time in that bank and change the status of the capacitor i.e., if its initial state is closed open it. If it is open close it.
5. Run the Power Flow using the proposed MDA method with this capacitor status change.
6. Check for Power Flow convergence.
7. Compute the performance index explained in Sect. 2.6 and store this value.
8. Restore the capacitor back to its initial state.

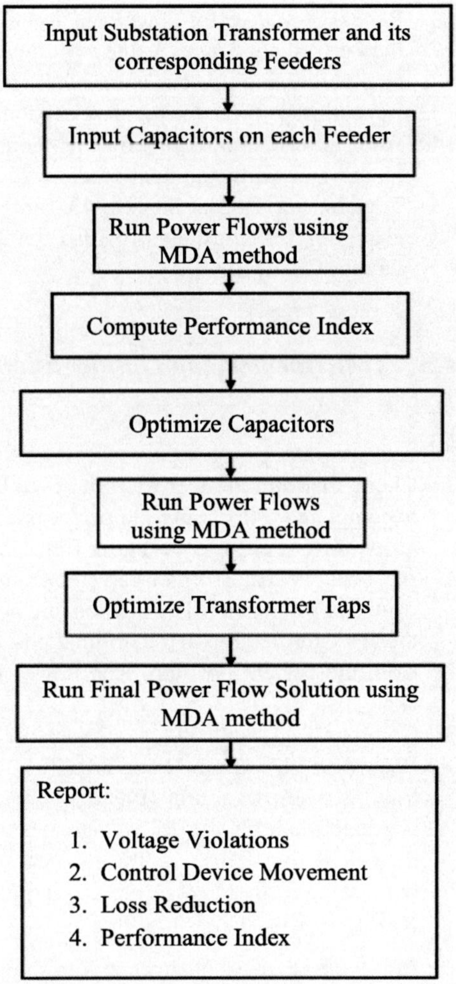

Fig. 1 Main optimization algorithm

9. Repeat Steps 3–7 for all the available capacitors in the same bank.
10. Pick the Capacitor which has the least Performance index value and compare this Performance index value with the base performance index PrevPerfIndex (Step 4, Sect. 5.1). Check whether all the capacitors in the bank are considered. If yes, please proceed to Step 13. Else check whether the Performance index value is less than PrevPerfIndex. If yes proceed to Steps 11 and 12. Else proceed to Step 13.
11. Implement the Capacitor which has made a substantial difference to the performance index.

12. Repeat Steps 4–9 for other capacitors in the same bank. If there is no change in the Performance index value as compared to its previous implementation, then stop processing this capacitor bank and go to the next bank.
13. Repeat Steps 3–11 for the next capacitor bank
14. After all capacitor banks for this feeder are traversed store the results for this feeder and go to the next feeder which is connected to the same Substation Transformer. Repeat Steps 1–13. Store the results. Start implementing the Tap changing Transformers provided the TRANSFORMER TAP operation mode flag is "On" (Fig. 2).

4.3 Transformer Tap Optimization

Steps

1. Check the total number of voltage violations after the Capacitor Optimization is completed. Also, note the performance index (PI_{cap}) and ($PI_{FeederNumber}$ is an array of PI of each feeder) after Capacitor Optimization is completed.
2. Increase one Transformer Tap step from the current position. Run Power Flows using the proposed MDA method and note the Performance Index (PI_{incr}). Bring the Transformer Tap to the original position. Decrease one Transformer Tap step from the current position. Run Power Flows and note the Performance Index (PI_{decr}). Bring the Transformer Tap to the original position.
3. Compare PI_{incr} with PI_{decr}.. Choose the Minimum of the two (PI_{min}). Compare PI_{min} with PI_{cap} (Step 1) and min($PI_{FeederNumber}$). If PI_{min} is lesser than PI_{cap} and PI_{min} is lesser than min ($PI_{FeederNumber}$) proceed to Step 4 and Step 5. If PI_{min} is greater than PI_{cap} or PI_{min} is greater than min ($PI_{FeederNumber}$) stop performing the calculation. This condition indicates that Transformer Tap does not help in optimizing further.
4. If ($PI_{incr} < PI_{decr}$).

   (a) Increase one Transformer Tap step from the current position
   (b) Run Power Flows using the proposed MDA method and note the performance index of each feeder ($PI_{currFeederNumber}$) and PI_{curr} (which is the minimum of ($PI_{currFeederNumber}$).
   (c) Compare this performance index (PI_{curr}) with the previous value (PI_{cap}) and min ($PI_{FeederNumber}$)
   (d) If the performance index is reduced, then copy PI_{curr} to PI_{cap} and $PI_{currFeederNumber}$ to $PI_{FeederNumber}$ and repeat Steps a to c.
   (e) Continue till the performance index does not change between two successive tap change positions.

5. If ($PI_{decr} < PI_{incr}$)

   (a) Decrease one Transformer Tap step from the current position

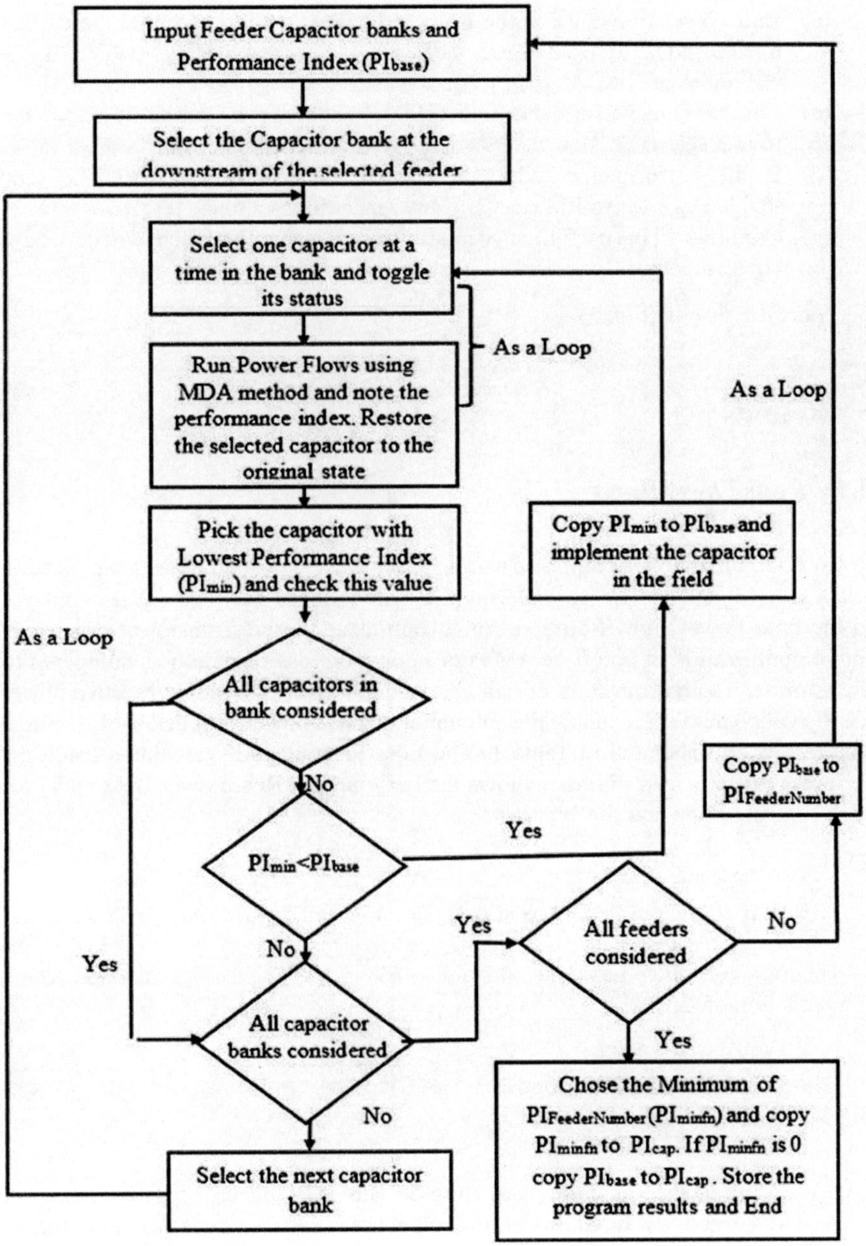

Fig. 2 Shunt optimization algorithm

(b) Run Power Flows using the proposed MDA method and note the performance index of each feeder ($PI_{currFeederNumber}$) and PI_{curr} (which is the minimum of ($PI_{currFeederNumber}$)).
(c) Compare this performance index (PI_{curr}) with the previous value (PI_{cap}) and min($PI_{FeederNumber}$)
(d) If the performance index is reduced, then copy PI_{curr} to PI_{cap} and $PI_{currFeederNumber}$ to $PI_{FeederNumber}$ and repeat Steps a to c.
(e) Continue till the performance index does not change between two successive tap change positions.

6. Store the results (Fig. 3).

5 Results

5.1 *Real Power Loss*

Volt Var Optimization is computed using 55 Bus Data [1] with the steps explained in Sects. 4.1, 4.2 and 4.3. Losses on each phase are reduced due to VVO as compared to the base Power Flow Solution. Shunt optimization and Transformer Tap with Shunt optimization option have a greater impact on loss reduction as compared to Transformer Tap optimization option since Shunts typically supply reactive power locally which helps in reducing the amount of reactive power supplied by the source. The results are tabulated in Table 1. The Loss Percentage is calculated using the formulae given below. Figure 4 shows the bar graph of Real Power Loss (kW) for each control option and for 3 phases

$$\text{Loss Percentage} = \\ (\text{Real part of Complex Power Loss})/(\text{Real Part of Input Complex Power})$$

The loss percentage has reduced from 7.46% to 5.99% post optimization which is a reduction of 2.61%.

5.2 *Input Complex Power*

Volt Var Optimization is computed using 55 Bus [1] with the steps explained in Sects. 4.1, 4.2 and 4.3. Input Power on each phase is reduced by VVO as compared to the base Power Flow Solution. Shunt optimization and Transformer Tap with Shunt optimization option have greater impact on reducing the Input Power as compared to Transformer Tap optimization option since Shunts typically supply reactive power locally which helps in reducing the amount of reactive power supplied by the source. The results are tabulated in Table 2. Figure 5 shows the bar graph of kVA Feeder

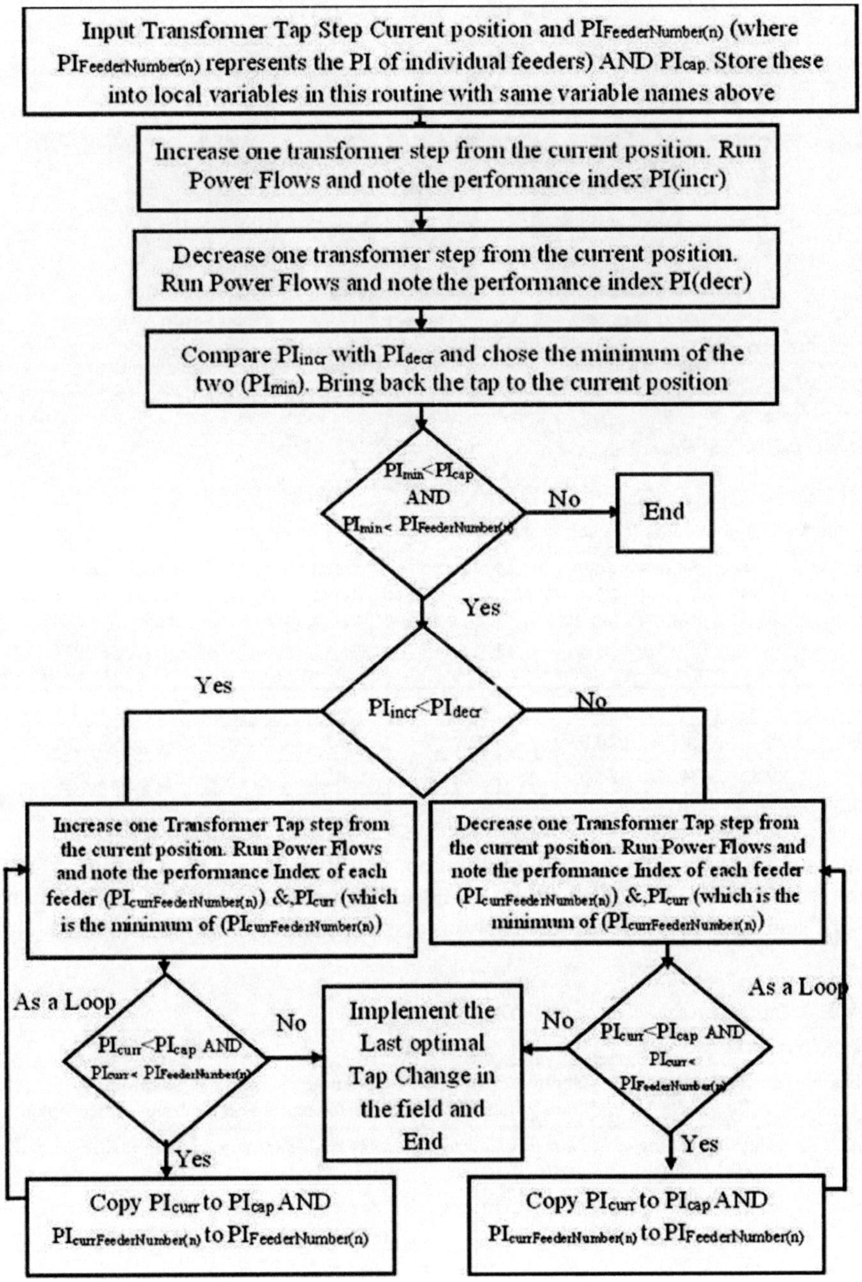

Fig. 3 Transformer tap optimization algorithm

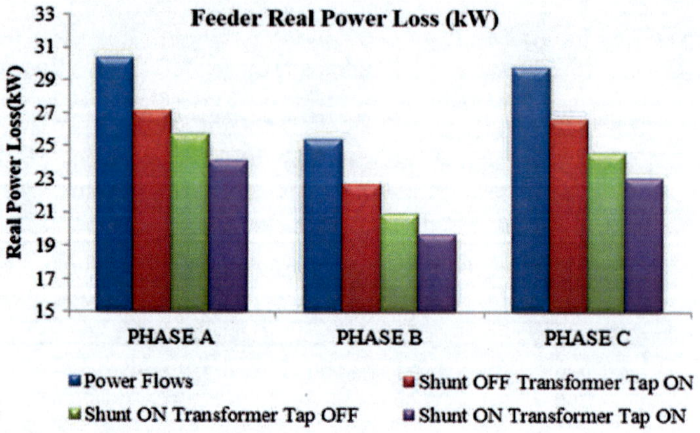

Fig. 4 Feeder real power loss (kW)

Table 1 Real power loss for each phase (kW)

Phase	Power flows (X)	% Loss in power flows	Shunt on transformer tap off (Y)	Shunt off transformer tap on (Z)	Shunt on transformer tap on (M)	% Loss with Shunt on transformer tap on	% Reduction in real power loss IM-XVM
A	30.4	7.46%	25.7	27.1	24.1	5.99%	2.61%
B	25.4	6.88%	20.9	22.7	19.7	5.41%	2.89%
C	29.8	7.49%	24.6	26.6	23.1	5.9%	2.9%

Input of each phase for each control option, viz. Base Power Flows, Shunt "ON" and Transformer Tap "OFF," Shunt "OFF" and Transformer Tap "ON" AND Shunt "ON" and Transformer Tap "ON" options respectively.

Table 2 Input power for each phase (kVA)

Input power (kVA)				
Phase	Power flows	Shunt on transformer tap off	Shunt off transformer tap on	Shunt on transformer tap on
A	Mag = 452 Ang = 25.75	Mag = 411 Ang = 10.76	Mag = 448.69Ang = 25.75	Mag = 408 Ang = 9.96
B	Mag = 411 Ang = 26	Mag = 370 Ang = 9.12	Mag = 407.82Ang = 26	Mag = 369 Ang = 8.22
C	Mag = 440 Ang = 25.54	Mag = 399 Ang = 10.052	Mag = 436.93 Ang = 25.57	Mag = 397 Ang = 9.22

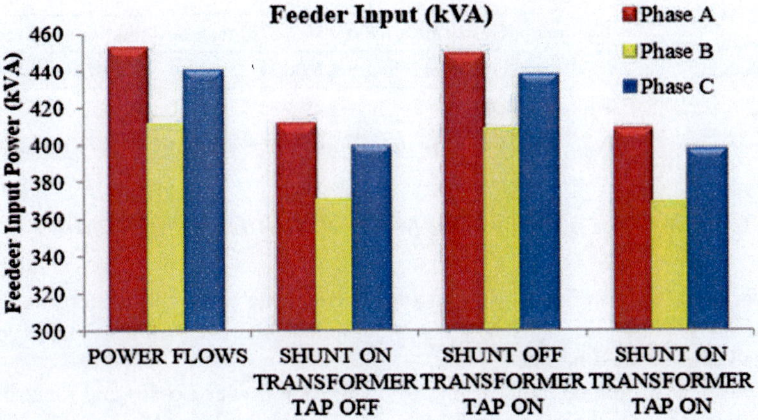

Fig. 5 Feeder input of 3 phases

5.3 *Bus Voltage Violations*

Volt Var Optimization is computed using 55 Bus Data [1] with the steps explained in Sects. 4.1, 4.2 and 4.3. After VVO is computed the Bus Voltage Violations are obtained as explained below.

In this paper, we have considered ±6% as the range with respect to base voltage as voltage deviation i.e., for an 11-kV base line to line voltage the low voltage limit is 10.34 kV and high voltage limit is 11.66kV. The voltage across each bus obtained after successful execution of VVO is then compared with this limit. The Buses with Voltage values outside this range are treated as violated buses. The total number of buses with Low Voltage Violations across each phase is tabulated in Table 3.

Number of Voltage Violations on each phase is reduced to 0 as compared to a huge number of violations in base Power Flow Solution. Transformer Tap optimization and Transformer Tap with Shunt optimization option have a greater impact on reducing the voltage violations as compared to Shunt Optimization option.

Table 3 Buses with voltage violations on each phase

Bus voltage violations				
Phase	Power flows	Shunt on transformer tap off	Shunt off transformer tap on	Shunt on transformer tap on
A	46 buses	39 buses	0 buses	0 buses
B	39 buses	32 buses	0 buses	0 buses
C	46 buses	39 buses	0 buses	1 buses

Table 4 Performance comparison with interior point method

Measurement index	Proposed method	Interior point method
Performance index	0.0666	0.0667
Iterations	20	28
Time	0.3657 s	8 s

5.4 Comparison of Proposed Method with Existing Method

Volt Var Optimization is computed using 55 Bus Data [1] with the steps explained in Sects. 4.1, 4.2 and 4.3. After VVO is computed, the Bus Voltage Violations are obtained as explained below.

The key performance indices, Input Power, Real Power Losses and Controls are compared with the interior point method [36]. The Interior Point Method Algorithm is simulated with MATLAB toolbox. The performance index indicates that the proposed method is more efficient than the interior point method. In terms of accuracy, both the methods have produced the same output. The results are tabulated in Table 4.

5.5 Controls Summary

Volt Var Optimization is computed using 55 Bus Data ([1]) with the steps explained in Sects. 4.1, 4.2 and 4.3. The total number of controls is obtained once the VVO problem is solved as explained in Step 9 of Sect. 4.1. The controls for various configurations and on each phase are tabulated in Table 5. The number of capacitors moved is 4 (2 on each capacitor bank) and 1 Transformer Tap (1.025 PU) from its nominal value (1.0 PU) for Transformer Tap with Shunt optimization option. For Shunt optimization, the number of Capacitors moved is 4 (2 on each capacitor bank). For Transformer Tap Optimization 2 Transformer Taps (steps) (1.05 PU) are moved from its nominal value of (1.0 PU).

5.6 Performance Index

Volt Var Optimization is computed using 55 Bus Data [1] with the steps explained in Sects. 4.1, 4.2 and 4.3. The Performance Index is calculated as per the formulae explained in Sect. 2.6 once the VVO problem is solved. Weighting Factor of 100 is considered while computing the Violation penalty explained in Sect. 2.5. The Performance Indices are tabulated in Table 6. Highlighted Green color indicates the optimal Performance index for various configurations which was obtained after the VVO optimization was solved.

Table 5 Controls summary

Phase	Shunt off transformer tap on	Shunt on transformer tap off	Shunt on transformer tap on
A	TAP = 1.05 PU (2 steps)	4 shunts (2 shunts of each capacitor bank)	4 shunts (2 shunts of each capacitor bank) TAP = 1.025 PU (1 step)
B	TAP = 1.05 PU (2 STEPS)	4 shunts (2 shunts of each capacitor bank)	4 shunts (2 shunts of each capacitor bank) TAP = 1.025 PU (1 step)
C	TAP = 1.05 PU (2 steps)	4 shunts (2 shunts of each capacitor bank)	4 shunts (2 shunts of each capacitor bank) TAP = 1.025 PU (1 step)

Table 6 Performance index

Shunt off transformer tap on		Shunt on transformer tap off		Shunt on transformer tap on	
Initial value	661.57	Initial value	661.57	Initial value	661.57
T/F Tap (1.025)	84.383	1 shunt	638.37	1 shunt	638.37
T/F Tap (1.05)	0.07637	2 shunts	566.75	2 shunts	566.75
T/F Tap (1.075)	124.71	3 shunts	494.23	3 shunts	494.23
–		4 shunts	430.52	4 shunts	430.52
		–		1.025 Tap	0.0668
				1.05 Tap	0.0668

5.7 Bus Voltages

Volt Var Optimization is computed using 55 Bus Data [1] with the steps explained in Sects. 4.1, 4.2 and 4.3. Voltages at each bus are improved due to VVO as compared to the base Power Flow Solution. Transformer Tap optimization and Transformer Tap with Shunt optimization option have a greater impact on the voltage improvement as compared to Shunt Optimization option since Shunts typically supply reactive Power whereas Transformers help in increasing the voltage at the feeder input. Figure 6 shows the line graphs of bus voltages for the 55 bus system after Power Flows are executed. Figure 7 shows the line graphs of bus voltages with Shunt "ON" and Transformer Tap "OFF" options. Figure 8 shows the line graphs of bus voltages with Shunt "OFF" and Transformer Tap "ON" options. Figure 9 shows the line graphs of bus voltages with Shunt "ON" and Transformer Tap "ON" option.

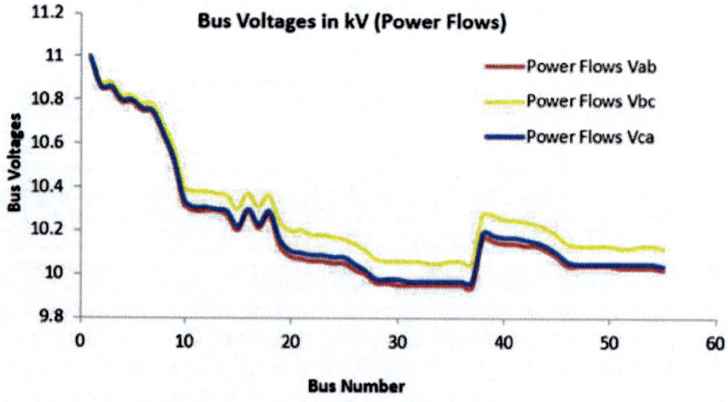

Fig. 6 Bus voltages in kV with power flows (line graph)

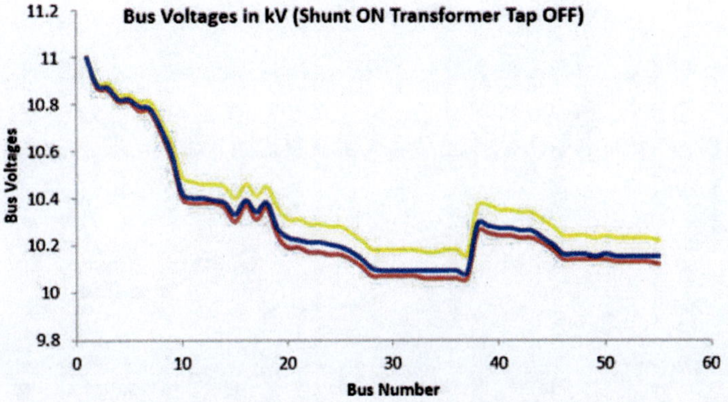

Fig. 7 Bus voltages in kV with shunt ON transformer tap OFF (line graph)

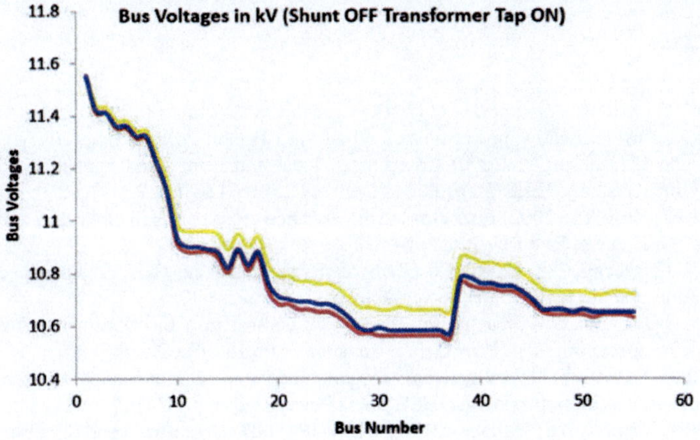

Fig. 8 Bus voltages in kV with shunt OFF transformer tap ON (line graph)

6 Conclusions

This paper presents VVO optimization approach used to find the optimal dispatch of Tap Changing Transformers and Capacitor banks, in a distribution system. The objective function used in this thesis is the minimization of the line losses. 55 bus Unbalance Radial system is used to demonstrate the effectiveness of the proposed method. Results illustrate how VVO algorithm with Loss Minimization objective function can be used to improve bus voltage profiles and minimize the losses. The proposed method is compared with an existing deterministic algorithm. The results signify the merit of the proposed algorithm in terms of execution time without compromising the accuracy of the results.

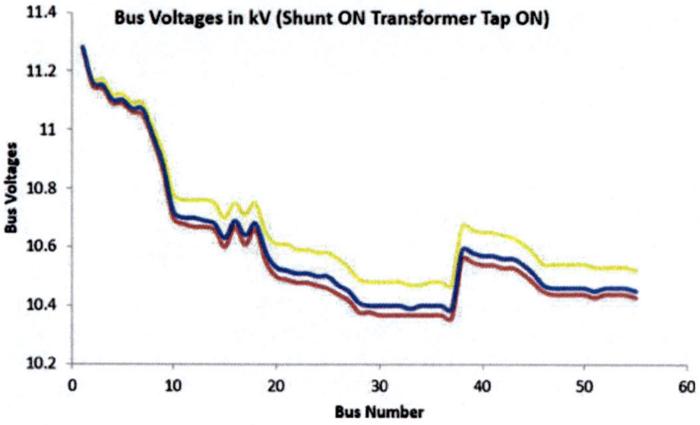

Fig. 9 Bus voltages in kV with shunt ON transformer tap ON (line graph)

References

1. Sarada Devi MSNG, Yesuratnam G (2019) Demand re-allocation in conjunction with feeder input power using iterative power flows method. Int J Recent Technol Eng (IJRTE) 8(4)
2. Roytelman I, Ganesan V (2000) Coordinated local and centralized control in distribution management systems. IEEE Trans Power Delivery 15(2):718–724
3. Liang RH, Wang YS (2002) Fuzzy-based reactive power and voltage control in a distribution system. IEEE Power Eng Rev 22(12):64–64
4. Liang R-H, Cheng C-K (2001) Dispatch of main transformer ULTC and capacitors in a distribution system. IEEE Trans Power Delivery 16(4):625–630
5. Lu FC, Hsu YY (1995) Reactive power/voltage control in a distribution substation using dynamic programming. IEE Proc Gener Transm Distrib 142(6):639–645
6. Gu Z, Rizy DT (1996) Neural networks for combined control of capacitor banks and voltage regulators in distribution systems. IEEE Trans Power Delivery 11(4):1921–1928
7. Vovos PN, Kiprakis AE, Wallace AR, Harrison GP (2007) Centralized and distributed voltage control: impact on distributed generation penetration. IEEE Trans Power Syst 22(1):476–483
8. Sheng W, Member S, Liu K, Cheng S, Meng X, Dai W (2016) A trust region SQP method for coordinated voltage control in smart distribution grid, vol 7, no 1, pp 381–391
9. Ranamuka D, Agalgaonkar AP, Muttaqi KM (2014) Online voltage control in distribution systems with multiple voltage regulating devices. IEEE Trans Sustain Energy 5(2):617–628
10. Madureira AG, Lopes JP (2009) Coordinated voltage support in distribution networks with distributed generation and microgrids. IET Renew Power Gener 3(4):439–454
11. Oshiro M, Tanaka K, Senjyu T, Toma S, Yona A, Saber AY, Funabashi T, Kim CH (2011) Optimal voltage control in distribution systems using PV generators. Int J Electr Power Energy Syst 33(3):485–492
12. Cheng L, Chang Y, Huang R (2015) Mitigating voltage problem in distribution system with distributed solar generation using electric vehicles. IEEE Trans Sustain Energy 6(4):1475–1484
13. Zhang C et al (2015) Model-based Volt-Var optimization using advanced metering infrastructure in distribution networks. In: 2015 IEEE Eindhoven power tech, pp 1–5. https://doi.org/10.1109/PTC.2015.7232281

Reactive Power Procurement as Ancillary Service for Dispersed Electric Vehicles in Radial Distribution System

Dhruv Rajput, Nitin Kumar Saxena, Abhinav Srivastava, Dev Verma, and Praveen Kumar Tyagi

Abstract Reactive power management which is termed a local area phenomenon has a serious concern with new era distribution system where an enormous burden has been put on to existing system due to the insertion of Electric Vehicles. These EVs as load may disturb the load flow profile of Radial Distribution System (RDS) if not attended properly. Voltage drop can be adjusted by the existing generators in system as they can expand their reactive power capabilities but this can make them overburden and the system may suffer from real power crises. According to utility, reactive power is designated as one of the most important ancillary services that can be provided at a local level through the Distribution Generator (DG) owner or others in lieu of additional charges from the customers. This paper presents an iteration-based reactive power procurement for attending the voltage profile which was being disturbed through the insertion of dispersed electric vehicles in the radial distribution system. The method is framed for IEEE 14 Bus RDS to demonstrate how the customers at EV charging station has to pay extra for reactive power procurement as an ancillary service.

Keywords Fast voltage stability index (FVSI) · Distribution station · Reactive power compensation · Ancillary services · Electric vehicle

---

D. Rajput (✉) · N. K. Saxena · A. Srivastava · D. Verma · P. K. Tyagi
KIET Group of Institutions, Delhi-NCR, Ghaziabad, India
e-mail: dhruvrajput0408@gmail.com

N. K. Saxena
e-mail: nitinsaxena.iitd@gmail.com

A. Srivastava
e-mail: abhinav.2024en1050@kiet.edu

D. Verma
e-mail: dev.2024en1096@kiet.edu

P. K. Tyagi
e-mail: praveen.tyagi@kiet.edu

N. K. Saxena
Electrical and Computer Engineering Department, University of Denver, Denver, United States

A. Kumar et al. (eds.), *Decarbonisation and Digitization of the Energy System*, Lecture Notes in Electrical Engineering 1099, https://doi.org/10.1007/978-981-99-7630-0_12

1 Introduction

Electric vehicles (EV) are the most budding sector in the present electric power system. The concept of deregulated power market still has multidimensions to explore in new era. Electric vehicle has implanted enormous scope of research into it. Starting with environmental issues due to fossil fuel based transportation system, electrical vehicles (EVs) demand accelerates significantly to down the CO_2 level emissions [1]. Operating cost of traditional combustion-based engines also gives a chance for getting market to EVs [2]. Apart from environmental issues, EV challenges may be classified into market planning and charging facility. According to industries and government planning, EV market is expected to 974,102.5 US million Dollar by 2027 growing at a healthy compound annual growth rate in the forecast period of 2020 to 2027 (Electric vehicle, 2020). Electric vehicle charging station is also a big challenge for their acceptability because the EV market is increasing exponentially and power demand from existing system raising new challenges to the existing distribution network and to distribution network operators (DNO).

Technical structure of any distribution system focuses on several key parameters while integrating new loads or generations. These parameters are bus voltage, stability, power loss, total harmonic distortions, voltage variations etc. Electric vehicles have their unique terms with specific characteristics such as state of charge (SoC), state of discharge (SoD), battery capacity, charging time, discharging time, fast charging capability etc. [3, 4]. These parameters further complicate the bus and distributed network parameters when integrated into the system but a distribution system with accurate planning can resolve these issues too [5].

EV implementation in any radial distribution network can be done; however, planning and some modifications are required for a better support, steady-state and stable system operation. As the EVs are increasing in RDS so existing system must go with a more distributed generating system so that additional demand can be met. Renewable-based distributed generators (DGs) can help in meeting the additional power requirement for an increasing load. The power engineers have to protect the extending lines limit for supporting to the greater flow of power on it [6]. Depending on the geographical areas, these DGs may have operated through different input resources or even a hybrid approaches can also be used to develop such DG structures. DGs placement can be done with different techniques as suggested by many researchers in their respective works [7, 8]. As reported in ref. [9–11], Fast Voltage Security Index (FVSI), Line Quality Factor (LQF) and Voltage Stability Index (VSI) can be used to investigate suitable bus and line for DG placement.

One important aspect of today's power system scenario is to deregulate the power structure and tend it towards value-added services required for the customers. These value-added services are coined as Ancillary Services in power system. Three broad classified ancillary services are services required for routine operation, services needed to avoid blackout, and services needed to be restored to the system after blackout. However, Ancillary services, as identified by the North American Electric Reliability Council (NREC) along with the Electric Power Research Institute (EPRI), are 12 in number. Indian power market mainly focuses on providing three ancillary services to their customers [12]. These three services are frequency control ancillary service (FCAS), voltage control ancillary services (VCAS) and black start ancillary services (BSAS). Out of all these three services, voltage control ancillary services become more interesting for putting efforts into new research work because of the following motives. This voltage control needs reactive power support from the local area management. This reactive power management can be done with a wide range of compensating devices from the range of static to rotating to dynamic nature. The cost function of these all compensating devices opens financial market-based studies. Dynamic compensating devices have controlling requirements and power quality issues. This controlling requirement opens a wide area of optimization techniques for machine learning practitioner. Power quality issues mitigation opens a wide area for power engineers dealing with power electronics and derives. The DG allocation in the radial distribution system requires additional reactive power, and therefore, it must be procured for the system. EVs are the additional load to the system also require reactive power from the system to avoid voltage dip due to the EVs addition power system. More real power demand from the system needs more reactive power and if reactive power is not compensated to the system, the voltage level will drop and even cause voltage collapse [13]. Therefore, reactive power compensation is also an important component of such systems. Therefore, this paper deals with the VCAS problem statement for implanting EVs in a radial distribution network.

As stated for the ancillary service, customers must pay additionally for their specific need of quality services in addition to the cost of watt power consumption. Further, most of the DGs are commissioned by private–public participation (PPP) in the market and make cost-based studies more important [14].

Reactive power cost using a condenser with an alternator is presented in [15]. This study is done for IEEE 14 bus deregulated power market model. Reactive power estimation for maintaining the bus voltages is proposed using the least error iterative procedure with voltage as an optimizing constraint [11].

In this paper, EVs of different capacities are installed at several locations of IEEE 14 bus radial distribution system, and reactive power procurement is done as an important study of ancillary service. This is done by following procedural steps in sequence. In this paper, the most suitable bus is selected using voltage indices. Suitable size DG is identified and placed at this bus. In the next step, dispersed EVs are implanted in existing RDS, and finally estimation of reactive power and its cost estimation using the least error voltage constraint iterative procedure.

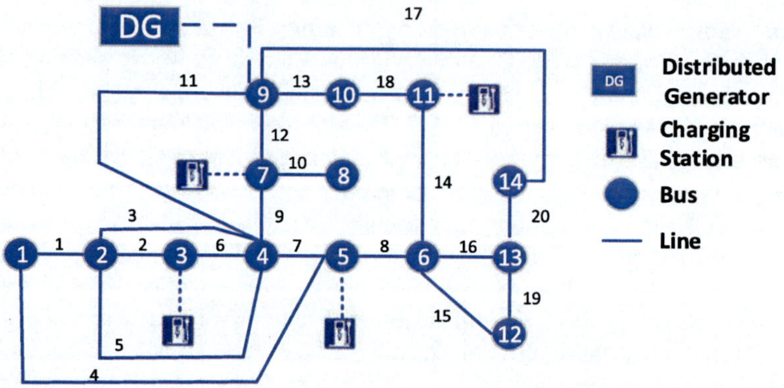

Fig. 1 Block diagram of the proposed IEEE 14 bus radial distributed system

2 Proposed Modified RDS Structure for Developing EVs Charging Station Structures

IEEE standard 14 bus radial distribution system is used for the proposed study. Modifications are done by clubbing a distributed generator at Bus 9 as in ref. [11]. Figure 1 presents a modified structure after including a distributed generator, dynamic compensator and proposed electric vehicle charging stations. The methodology adopted for the selection of either of them is sequenced in the sections below. However, the pointwise illustration is as follows:

- Selection of bus at which DG must be placed
- Selection of DG rating
- Selection of EVs as different proposed case studies
- Selection of reactive power requirement for regulating voltage level at all buses
- Formulation for reactive power remuneration concept

Please note that the first paragraph of a section or subsection is not indented. The first paragraph that follows a table, figure, equation etc. does not have an indent, either.

Subsequent paragraphs, however, are indented.

3 Selection of Bus at Which DG Must Be Placed

The bus selection for placing DG as well as compensator decides with the voltage limits on every bus as one of the important factors. The voltage range in the distribution system, suggested by IEC 60,038, is from 0.90 pu to 1.1 pu. The most sensitive bus can be identified in the distribution system by applying voltage indices-based studies. Some of the popular voltage indices-based studies are FVSI, LQF and VSI. The expressions for FVSI as in ref. [16, 17] is given in Eq. (1).

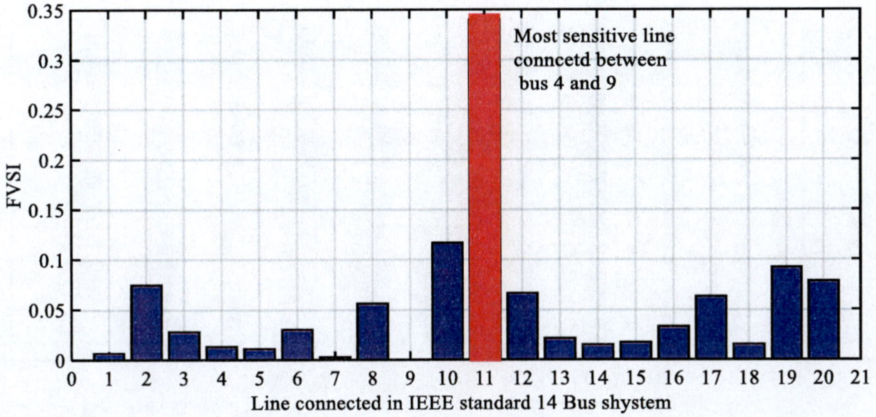

Fig. 2 FVSI for all the 14 lines in 14-bus RDS

Fast Voltage Stability index (FVSI) with the highest value can be standardized for the selection of line in RDS.

$$FVSI_{ij} = \frac{4Z_{ij}^2 Q_j}{V_i^2 X_{ij}} \tag{1}$$

The proposed study is done with steady-state load flow analysis using Newton-Raphson load flow (NRLF) method. The most suitable location for placing DG and compensator is found at line 11 connected with Bus 4 and 9 because of the highest value of FVSI at this line as shown in Fig. 2. Further, the selection of either 4th or 9th bus is a challenge which is countered by piecewise increase in the generation on both bus and corresponding sensitivity is checked again. Finally, it is concluded with the help of Fig. 3 that Bus 9 can be chosen as the most sensitive bus. Therefore, the most suitable Bus for placing DG in the proposed IEEE 14th bus RDS is ninth in number.

4 DG Selection Method

After selecting the 9th bus for placing DG, the next challenge is to find a suitable rating for it. This çan be done by adopting an iteration procedure with the consideration of voltage magnitude at all the buses which must remain in ±0.1 pu.

IEEE standard 14 Bus data, obtained from NRLF iterative procedure, is tabulated in Table 1. Slack Bus shows that 0.2329372 GW is supplied to the RDS. Additional DG is required to meet the additional load requirement and must be placed at the appropriate bus. The appropriate bus selection is done with the optimization problem considering the following considerations,

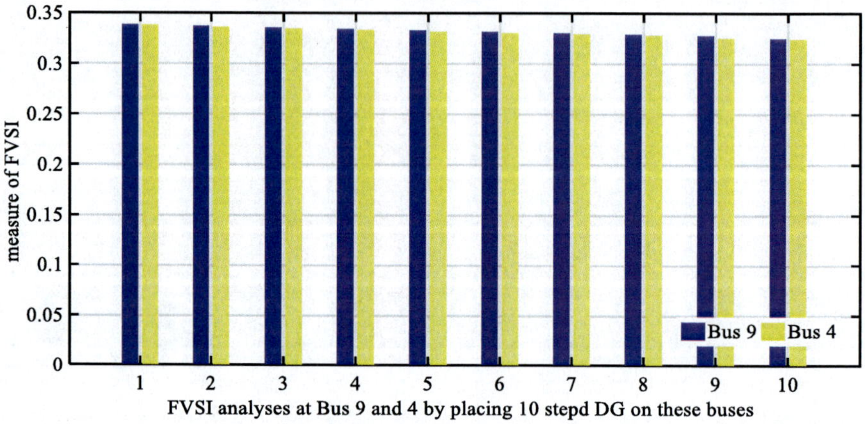

Fig. 3 Sensitivity comparison along Line 11 connected with bus 9 and bus 4 in IEEE 14 bus RDS

Table 1 Analysis for standard and modified IEEE 14 Bus

Bus no.	Standard IEEE 14 bus without DG				Modified IEEE 14 bus with DG [58 MW UPF] at bus 9			
	P (MW)	Q (MVAR)	$\|V\|$ (pu)	θ (°)	P (MW)	Q (MVAR)	$\|V\|$ (pu)	θ (°)
1	232.9372	−31.2106	1.0600	0.0000	170.1842	−24.6956	1.0600	0.0000
2	18.3000	−2.9498	1.0450	−4.9239	18.3000	−26.2436	1.0450	−3.5618
3	−94.2000	−9.7129	1.0100	−12.4936	−94.2000	−17.7701	1.0100	−10.1623
4	−47.7999	3.9001	1.0423	−10.5174	−47.8000	3.9000	1.0566	−7.6372
5	−7.5999	−1.5999	1.0516	−9.1964	−7.6000	−1.6000	1.0641	−6.7065
6	−11.2000	6.1199	1.0761	−14.7492	−11.2000	6.1200	1.0860	−10.1883
7	0.0000	0.0001	1.0505	−13.3727	0.0000	0.0001	1.0708	−7.6872
8	0.0000	18.2400	1.0803	−13.3727	0.0000	18.2400	1.1000	−7.6872
9	−29.5001	−16.6000	1.0250	−14.9340	28.5001	−16.6000	1.0476	−7.7144
10	−9.0000	−5.8000	1.0264	−15.1790	−9.0000	−5.8000	1.0472	−8.4296
11	−3.5000	−1.7999	1.0473	−15.0734	−3.5000	−1.8000	1.0632	−9.4123
12	−6.1000	−1.5999	1.0585	−15.5866	−6.1000	−1.5999	1.0690	−10.8217
13	−13.5000	−5.7999	1.0511	−15.5971	−13.5000	−5.7999	1.0634	−10.6525
14	−14.9000	−5.0000	1.0181	−16.2539	−14.9000	−5.0000	1.0364	−9.9981

The total real power loss (TPL) should be minimal,

$$TPL_k = \min_k\left(\sum_{j=1}^{N}(P_{G,j} - P_{L,j})\right) \quad (2)$$

where, $k \in DG\ sample\ \&\ j = No.\ of\ Bus$

The total reactive power loss (TQL) should be minimal.

$$TQL_k = \min_k\left(\sum_{j=1}^{N}(Q_{G,j} - Q_{L,j})\right) \tag{3}$$

where, $k \in DG\ sample\ \&\ j = No.\ of\ Bus$

Voltage for all the buses;

$$V_i \in [0.9, 1.1], \forall i \in [1, 2, 3, \ldots, N] \tag{4}$$

And, the FVSI value in each sample of DG rating.

Slack bus has a real power of 230 MW; so DG size can be iterated by analysing the performance of different size DGs between 1 and 230 MW at max. Each iteration gives the values corresponding to the above explained constraints and the best sample can be chosen for the same. The voltage is chosen as the most sensitive parameter for this analysis along with parameters such as TPL, TQL and FVSI. Therefore, from the results obtained for IEEE 14 Bus, 58 MW unity power factor (UPF) DG is proposed to be placed at the 9th Bus. The algorithm to decide the size of DG is also elaborated with the help of a flow chart as shown in Fig. 4.

Voltage study after Data obtained from this study is summarized in Table 1 for more understanding. Parameters P, Q V and δ are tabulated before and after placing DG of 58 MW UPF. It is found that voltage is in the specified range in both cases. Therefore, the system may accept it without any technical challenge.

5 Implanting Electric Vehicles in RDS

As in Table 1, it has been checked how system voltage remains within the limit even after placing DG of 58 MW UPF at Bus 9 for IEEE 14 bus RDS. Now to analyse the effect of electric vehicles in terms of voltage fluctuations, four EVS of rating below are chosen at different buses. The voltage profile in case of modified IEEE bus as in Table 1, voltage level after placing EVs and voltage level regulation with the help of reactive power solution is evaluated. The procedure till corresponding reactive power requirement to maintain the voltage level again is summarized in this section.

To clarify on the logic applied for getting adequate reactive power, so that the voltage profile can remain regulated in the system, an algorithm is explained with the help of Fig. 5. A new iterative procedure is adopted which gives the most suitable reactive power at DG end so that overall profile of the system can be maintained. The iterative procedure requires fitness function and corresponding constraints which are also highlighted. Modified RDS with 58 MW UPF DG at Bus 9 is considered as a reference case. Corresponding voltages at all the buses as in Table 1 are marked as reference voltage. As the load is added in terms of EVs as mentioned in Table 2, bus voltages drop due to this load increment. The voltage has to reach near its

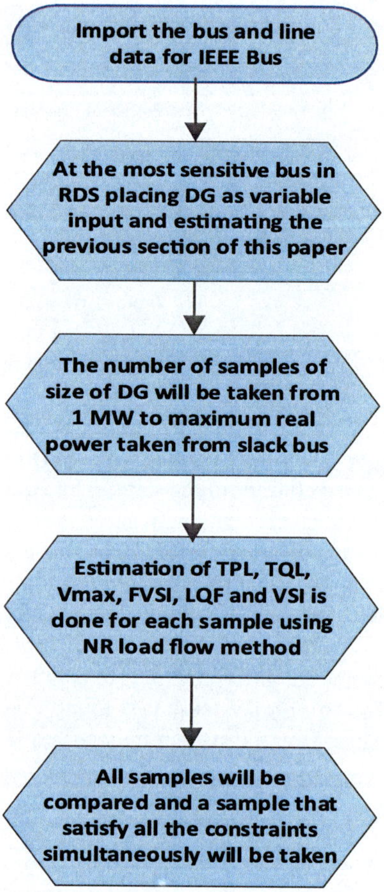

Fig. 4 Flow chart for DG selection using multi constraints optimization procedure

original value so that the effect of EV inclusion in RDS can be mitigated. This can be done by adding some reactive power compensator but the selection of this additional reactive power requirement must be remunerated from the EV owner and so has to be precisely known. The correctly procured reactive power will control the system voltage only. Therefore, the instantaneous load change must be attended with instantaneous reactive power through optimization technique. For this, a least error iterative procedure for examining the systems' reactive power requirement is presented in this paper.

Reactive power increment in each iteration can be supplied through the formula;

$$Q_{com}^{k} = Q_{com}^{min} + k\frac{Q_{com}^{max} - Q_{com}^{min}}{no.\ of\ samples} \tag{5}$$

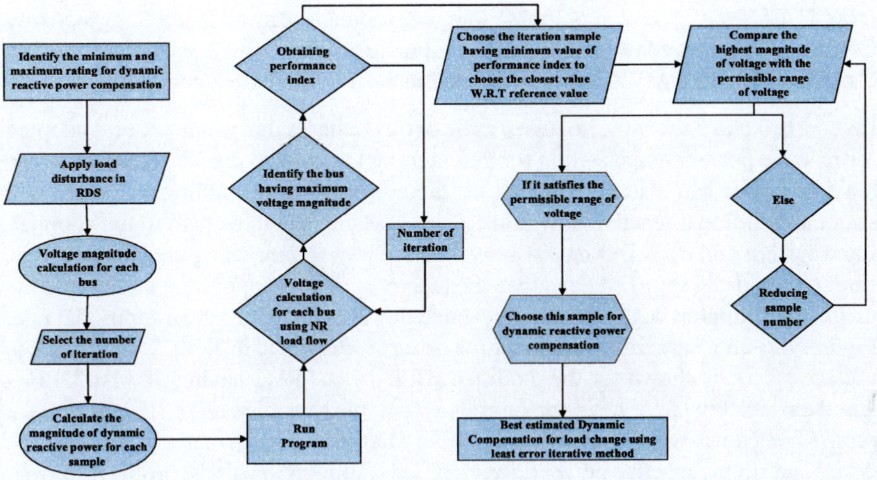

Fig. 5 Flow chart of steps for estimating the dynamic compensation for load change using the least error iterative method

Table 2 Proposed EVs with their location

S. no.	Bus no.	Size of EVs (MW)
1	3	7.2
2	5	3.6
3	7	14.4
4	11	10.8

For every sample of Q_{com}, load flow procedure estimates voltage for all the system buses V_j^k. The voltage deviation is collected for all buses and summed together to find an objective function as presented in Eq. (6).

$$\eta^k = Min \sum_{j=1}^{N} \left| V_j^{ref} - V_j^k \right| \tag{6}$$

Compensation's optimal value will come with the optimum value of Eq. (6) keeping voltage magnitude in range for all the system buses. The optimization procedure followed in this paper is explained in detail and presented in Fig. 5.

Now the three cases are drawn to see the reactive power requirement and voltage patterns at all the buses. These three cases are developed with the help of EVs implantation in RDS as shown in Table 2. The reactive power requirement and system self-healing for voltage control can be understood with the help of the results and figures described below.

Case 1 EVs of 7.2 MW and 3.6 MW are connected in Bus no 3 and 5 respectively
Case 2 eV of 14.4 MW further added at Bus no 7 with case 1 condition
Case 3 eV of 10.8 MW further added at Bus no 11 with case 2 condition

Figures 6, 7 and 8 show the voltage profile of all the buses during the process adopted for reactive power compensation for maintaining the voltage profile at each bus. The reactive power can also be supplied by the system and this additional power will count as additional reactive power losses in RDS and has to be paid to the supplier. Each system and even DG has its own reactive power generating capability and so cannot provide after its limit and then the reactive power compensator comes into the picture and supplies such additional reactive power and charges at an additional rate. Figures 6, 7 and 8 are drawn for the three cases as illustrated in Table 2. In Fig. 6, the voltage profile is shown for the modified IEEE 14 bus RDS having 58 MW UPF at Bus 9 through blue line. After implanting EVs at two buses as said in Table 2, voltage profile goes down as shown in the pink line. The system participates to mitigate the issues and supply reactive power through DG and support to voltage profile as further shown by red line. This balancing is mathematically estimated using the iterative procedure and dots representation in the figure showing the voltage magnitude with different reactive power quantity and finally the most optimum value in terms of violate deviation from reference case (blue line) is chosen. Table 3 explains how the system will self-sustain with this additional EV load without asking for the reactive power from the compensator and hence reactive power requirement from compensator for case I is zero.

The same explanations are true for case II and case III where the EVs load is further increased and the role of reactive power through compensator comes into the picture as mentioned in Table 3.

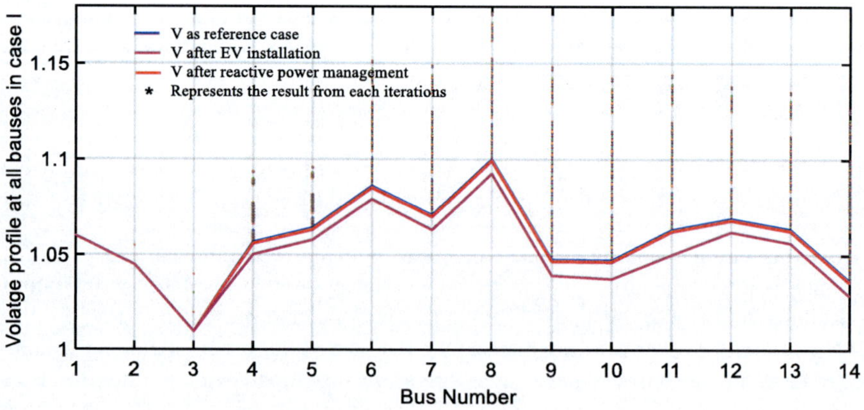

Fig. 6 Voltage profile comparison for placing EVs as in case I

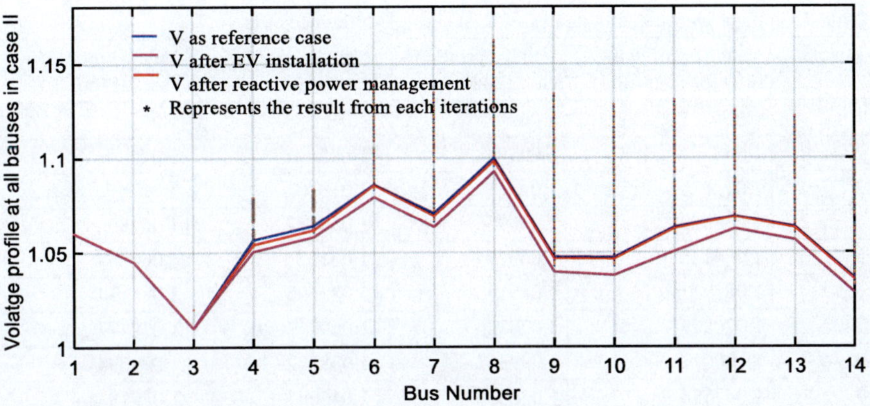

Fig. 7 Voltage profile comparison for placing EVs as in case II

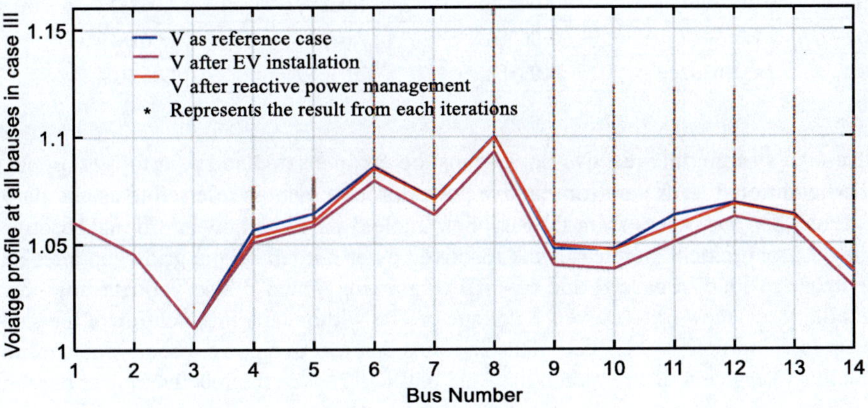

Fig. 8 Voltage profile comparison for placing EVs as in case III

Table 3 Summary of the reactive power requirement and system power losses

Case no.	Additional reactive power requirement from compensators (MVAR)	Real power losses in system (MW)	Reactive power supplied without compensator in the system (MVAR)
1	0	10.21425	−74.8997
2	1.995605	11.28771	−70.2216
3	4.989011	12.29846	−65.5592

Table 4 highlights the voltage profile in all three cases. It can also be verified that voltage remains in the predefined range of ±10% always which further validates the results from the proposed study. As additional reactive power is required

Table 4 Voltage profile in all three cases comparing to reference case

Bus no.	Voltage profile in reference case (pu)	Voltage profile in case I (pu)	Voltage profile in case II (pu)	Voltage profile in case III (pu)
1	1.06	1.06	1.06	1.06
2	1.045	1.045	1.045	1.045
3	1.01	1.01	1.01	1.01
4	1.056577	1.055956	1.053917	1.052896
5	1.064125	1.063275	1.06162	1.060254
6	1.085993	1.085193	1.085595	1.085134
7	1.070819	1.070126	1.069263	1.070477
8	1.100027	1.099352	1.098512	1.099694
9	1.047566	1.046832	1.046697	1.04918
10	1.047241	1.046489	1.046416	1.046787
11	1.063201	1.062422	1.062554	1.057919
12	1.069011	1.068208	1.068677	1.068411
13	1.063412	1.062607	1.0629	1.062879
14	1.036362	1.035585	1.035694	1.037104

from the system, this reactive power must be compensated and paid to its supplier. This additional tariff paid for reactive power procurement is referred to as ancillary services for customers from the supplier. Table 3 presents how additional reactive power procurement suppresses the reactive power burden on the grid even after an increase in load in case II and case III. Figures 6, 7 and 8 also validate how this ancillary service will improve the voltage profile without any intervention of the grid after load increment at EVCSs. This signifies that reactive power procurement as an ancillary service will strengthen the RDS and will reduce the reactive power burden on the power system.

6 Conclusion

Electric Vehicles involvement in modern distribution system is exponentially increasing due to which existing system has to face new challenges. All the features cannot be addressed uniformly to all customers because of the different levels of requirements such as location size time of operation etc. For such non-deterministic conditions, ancillary service concept is very overwhelming because customers can be asked to pay additionally for their value-added services. Reactive power compensation after a certain amount cannot be given by the existing system and so, special reactive power compensator must be installed as DG owner or separate utility. These utilities can charge through the concept of reactive power compensator as ancillary services from the customer who wishes to join RDS at different locations as per their convenience and requirement. This study explains how standard IEEE 14 bus RDS can be replaced with proposed RDS having additional DG to overcome the real power load demand of the existing system and then how a reactive power compensator can be placed with this DG and supply for additional reactive power in the system so that overall voltage may remain in predefined limit. The study can be expanded for a statistical approach of inserting EVs in other IEEE buses. EVs can also be classified in several sectors for which the remuneration models may be changed according to load size, time of operation, charging time etc.

References

1. Parker N, Breetz HL, Salon D, Conway MW, Williams J, Patterson M (2021) Who saves money buying electric vehicles? Heterogeneity in total cost of ownership. Transp Res D 96(May):102893
2. Zhou M, Long P, Kong N, Zhao L, Jia F, Campy KS (2021) Characterizing the motivational mechanism behind taxi driver's adoption of electric vehicles for living: Insights from China. Transp Res A 144(2):134–152
3. Zhao G, Wang X, Negnevitsky M (2022) Connecting battery technologies for electric vehicles from battery materials to management, vol 25, no 2. https://doi.org/10.1016/j.isci.2022.103744
4. Zeb MZ, Imran K, Khattak A, Janjua AK, Pal A, Nadeem M, Zhang J, Khan S (2020) Optimal placement of electric vehicle charging stations in the active distribution network. IEEE Access 8:68124–68134
5. Steen D, Tuan LA (2017) Impacts of fast charging of electric buses on electrical distribution systems. CIRED Open Access Proc J 2017(1):2350–2353
6. Ackermann T, Andersson G, Soder L (2001) Distributed generation: a definition. Electr Power Syst Res 57:195–204
7. Ghosh S, Ghoshal SP, Saradindu G (2010) Optimal sizing and placement of distributed generation in a network system. Electr Power Energy Syst 32:849–856
8. Kumar A, Babu PV, Murty VVSN (2017) Distributed generators allocation in radial distribution systems with load growth using loss sensitivity approach. J Inst Eng India Ser B 98(3):275–287
9. Chakrabarti A, Kothari DP, Mukhopadhyay AK et al (2010) Reactive power control and voltage stability in power transmission systems. PHI Press, New Delhi
10. Murty VVSN, Kumar A (2015) Optimal placement of DG in radial distribution systems based on new voltage stability index under load growth. Electr Power Energy Syst 69:246–256
11. Saxena NK, Kumar A (2021) Estimation of dynamic compensation for renewable-based hybrid DG in radial distribution system using least error iterative method. Iran J Sci Technol Trans Electr Eng 45:15–28. https://doi.org/10.1007/s40998-020-00345-1
12. Saxena NK, Mekhilef S, Kumar A, Gao DW (2022) Marginal cost-based reactive power reinforcement using dynamic and static compensators. IEEE J Emerg Sel Topics Power Electron 10(4):4001–4013. https://doi.org/10.1109/JESTPE.2022.3145871
13. Calderaroa V, Coniob G, Galdia V et al (2012) Reactive power control for improving voltage profiles: a comparison between two decentralized approaches. Electr Power Syst Res 83:247–254
14. Saxena NK, Kumar A (2018) Dynamic reactive power compensation and cost analysis for isolated hybrid power system. Electr Power Compon Syst 45(18):2034–2049
15. Kumar A, Kumar R (2015) Reactive power cost analysis with generators and condensers in deregulated power system. In: Proceedings of the 4th international conference on informatics, environment, energy and applications, India

16. Bhadoriya JS, Gupta AR, Zellagui M, Saxena NK, Arya AK, Bohre AK (2022) Optimal allocation of electric vehicles charging station in distribution network beside DG using TSO. In: Bohre AK, Chaturvedi P, Kolhe ML, Singh SN (eds) Planning of hybrid renewable energy systems, electric vehicles and microgrid. Energy Systems in Electrical Engineering. Springer, Singapore. https://doi.org/10.1007/978-981-19-0979-5_29
17. Panda SB, Mohanty S (2022) Assessment of power system security for different bus systems through FVSI/RFVSI and fuzzy logic approaches. IETE Tech Rev 39(6):1485–1500. https://doi.org/10.1080/02564602.2022.2028587

Radial Unbalance Distribution System Analysis with DG and D-STATCOM Allocation Using Arithmetic Optimization Algorithm

Anchal Maurya and Ashwani Kumar

Abstract Distribution system is the key link of the utility with the consumers. It plays a key role in the distribution of power to the residential, commercial, industrial and agricultural consumers. The distribution system is not perfectly balanced and can be unbalanced due to the traction load or unequal loadings on the different phases. Higher unbalance can cause more losses in the system and mal operation of relays. Therefore, the distribution system operator must analyze the unbalanced distribution system to reduce the unbalances. Distributed generation and devices can play a major role in the correction of unbalances meeting the active and reactive power load requirements. In this paper, the authors present the analysis of the radial unbalance (UB) distribution system with optimal placement of distributed generation (DG) and D-STATCOM. The main goals of the study are optimal DG location and sizing, optimal D-STATCOM location and sizing, losses reduction in the system, and voltage profile improvement using an Arithmetic Optimization Algorithm (AOA). The analysis is also carried out with DG and D-STATCOM's effect on overall cost reduction due to the decrease in energy losses. In addition, results are also obtained with different loading scenarios and load growth which can help distribution network operators for better power distribution system planning. The results are determined for a conventional 25 bus radial UB distribution system using MATLAB 2021a.

Keywords Distributed generation (DG) · Distribution static compensator (D-STATCOM) · Unbalance (UB) · Distributed network operator (DNO) · Arithmetic optimization algorithm (AOA)

A. Maurya (✉) · A. Kumar
Department of Electrical Engineering, NIT Kurukshetra, Kurukshetra, Haryana, India
e-mail: anchal_32114319@nitkkr.ac.in

A. Kumar
e-mail: ashwani.k.sharma@nitkkr.ac.in

A. Kumar et al. (eds.), *Decarbonisation and Digitization of the Energy System*, Lecture Notes in Electrical Engineering 1099, https://doi.org/10.1007/978-981-99-7630-0_13

1 Introduction

In the competitive electricity markets, the distribution system operator (DSO) shall design efficient, reliable, and cost-effective distribution networks that will defer the cost due to the addition of lines and generation sources. The distribution systems shall be planned to accommodate the growing electricity demands of residential, commercial, and industrial. Distribution supply system quality is judged with reduced losses, better voltage profile, and higher efficiency. Recommended practices to reduce power losses and voltage drop include reconfiguration, reactive power compensation, automatic power factor correction, distribution generation, energy storage facilities, demand side management, voltage regulators, load balancing, optimal cable sizing, usage of induction motors and energy efficient transformers which call for radical change in the distribution system operation. The present power system networks are weak and highly loaded and being operated at the verge of voltage stability limits due to economic and environmental constraints. Therefore, for system operator, maintaining voltage stability in the present power systems is a challenging task which is crucial for better planning and operation of power systems. Distribution side loads in a power system demand reactive power as they are mostly lagging in nature. Because of high reactive power demand, distribution network voltage profiles become undesirable and lead to high losses.

DG placement, capacitor placement, Flexible AC transmission systems (FACTS) placement, network reconfiguration and load management are various approaches that can be used for loss reduction. Due to improvements in technology, the addition of electric vehicles, population growth and power demand has been increasing day by day. To meet present power demand renewable energy sources are being integrated with existing power systems and also considering the environmental concerns. Different types of loads are connected to the electrical power distribution system because of which power loss and voltage fluctuation problems arise and lead to poor performance of the system. To overcome these problems, distributed generators (DG) and D-STATCOM can be used in electrical power distribution networks. DG helps in regulating real power in radial distribution systems and also it is possible for consumer to generate electricity on their own with DG but it cannot regulate reactive power because of which voltage at certain buses gets hampered. To overcome this problem, D-STATCOM is also allocated along with DG. D-STATCOM regulates reactive power in radial distribution system. Allocating both D-STATCOM and DG helps in reducing total power loss, improving power factor and quality, voltage profile and stability.

To allocate D-STATCOM and DG optimally in radial UB distribution system different optimization techniques have been used. In [1], ASO algorithm has been used to allocate D-STATCOM and DG in UB distribution system. In [2], Novel techniques have been used to find the optimal size and location of DG in three-phase UB distribution system while improving voltage profile and minimizing power loss. In [3], for allocating DG units optimally in UB distribution system, novel sensitivity analysis is used. In [4], the Harmony search algorithm has been used for optimization of radial UB distribution system network with DG units by network configuration. In [5], loss sensitivity approach has been used to allocate DG in RDS considering load growth. In [6], Hybrid gray wolf optimization is used for allocating DG optimally in three-phase UB distribution system. In [7], DG is allocated optimally in UB distribution system using Jaya algorithm, also actual distribution system load and configurations have been considered. In [8], a capacitor is allocated in a radial UB distribution system considering loading and unbalancing conditions. In [9], Sine Cosine algorithm has been used for the optimal allocation of DG and D-STATCOM and also different unbalanced conditions are taken into consideration. In [10], EV charging stations are allocated optimally along with DG in the distribution network.

The literature survey shows that DG has been placed optimally in UB distribution system by authors for power loss minimization and voltage profile improvement. But allocation of both DG and D-STATCOM considering load growth can be considered for the supply of both active and reactive power optimally in the network. In this paper, DG and D-STATCOM are placed optimally in the radial UB distribution system and also load growth is considered. Percentage of energy saving annually is also determined.

2 Arithmetic Optimization Algorithm

The advantage of distribution behavior of the four fundamental arithmetic operators in mathematics—addition (A), subtraction (S), multiplication (M) and division (D)—is taken by an algorithm known as the Arithmetic Optimization Algorithm (AOA). For the purpose of performing the optimization procedures in a variety of search areas, AOA is mathematically modeled and implemented. The meta-heuristic optimization algorithm uses the two crucial search techniques of exploration and exploitation. Exploration searches the whole search space, avoiding local optima. Exploitation, on the other hand, focuses on finding the best local solutions to raise quality locally. Initialization, exploitation, exploration and termination are the four steps of AOA [11].

$$MOA(C\_Iter) = Min + C\_Iter\left(\frac{Max - Min}{M_{Iter}}\right) \tag{1}$$

$$X_{i,j}(\mathrm{C\_Iter} + 1) = \begin{cases} best(X_j)/(MOP + \in) \times \left((UB_j - LB_j) \times \mu + LB_j\right), r2 < 0.5 \\ best(X_j) \times MOP \times \left((UB_j - LB_j) \times \mu + LB_j\right), otherwise \end{cases} \tag{2}$$

$$MOP(C\_Iter) = 1 - \frac{C\_Iter^{\frac{1}{\alpha}}}{M\_Iter^{\frac{1}{\alpha}}} \tag{3}$$

$$X_{i,j}(\mathrm{C\_Iter}+1) = \begin{cases} best(X_j) - MOP \times \left((UB_j - LB_j) \times \mu + LB_j\right), r3 < 0.5 \\ best(X_j) + MOP \times \left((UB_j - LB_j) \times \mu + LB_j\right), otherwise \end{cases} \quad (4)$$

Algorithm: Arithmetic Optimization Algorithm.

Input: Load and Line data, DG & D-STATCOM upper bound and lower bound limits.

Output: *Xi* solutions-optimal sizes and location of DG & D-STATCOM.

Step 1	Use the DLF method to calculate TRPL for the base case.
	Initialize:
Step 2	Population size, number of iterations, solution position randomly, Loop process:
Step 3	While (C_iter < M_iter) do
Step 4	Calculate the fitness function then determine the best solution
Step 5	Calculate MOA and MOP using Eqs. (1) and (3)
Step 6	if r_1>MOA (exploration phase)
Step 7	if r_2>0.5 then use the first rule of Eq. (2) to update the solution otherwise use the second rule
Step 8	end if
Step 9	else (exploitation phase)
Step 10	if r_3>0.5 then use the first rule of Eq. (4) to update the solution otherwise use the second rule
Step 11	end if
Step 12	end if
Step 13	C_iter = C_iter +1
Step 14	end while

Return: Size and location

3 Problem Formulation

3.1 Power Loss

Because of mutual resistance between phases, the straightforward formula that is I^2R will not work to compute losses in the radial UB distribution network. Unbalanced power loss formula is proposed in [12].

$$L_a = [\left(V_{sa} * I_a{}^*\right) - \left(V_{ra} * I_a{}^*\right)] \quad (5)$$

$$L_b = [(V_{sb} * I_b{}^*) - (V_{rb} * I_b{}^*)] \tag{6}$$

$$L_c = [(V_{sc} * I_c{}^*) - (V_{rc} * I_c{}^*)] \tag{7}$$

where V_{sa} = Phase a sending end voltage, V_{ra} = Phase a receiving end voltage, I_a = Current between V_{sa} and V_{ra}

Active power losses for phases A, B and C are given by the real part of Eqs. (5)–(7) and reactive power loss by imaginary part. BIBC and BCBV load flow techniques are used in this paper.

3.2 Constraints

Voltage magnitude constraints

$$Voltage_{imin} \leq Voltage_i \leq Voltage_{imax} \tag{8}$$

Thermal constraint

$$I_{ik} \leq I_{ikmax} \tag{9}$$

Active Power Compensation

$$P_{min,DG} \leq P_{DG} \leq P_{max,DG} \tag{10}$$

Reactive Power Compensation

$$Q_{min,DS} \leq Q_{DS} \leq Q_{max,DS} \tag{11}$$

Equality Constraints

$$P_D + P_{TL} = P_{DG} + P_{DS} + P_G \tag{12}$$

$$Q_D + Q_{TL} = Q_G + Q_{DS} \tag{13}$$

3.3 Load Growth

Understanding load growth is essential for future distribution system design and expansion.

$$Load_i = Load * (1 + rate)^m \tag{14}$$

where m = planning period that has been taken into account by DNO and r = annual rate of load growth. In this study, m = 1 and r = 2.5%

3.4 % Energy Saving Annually

$$\%ESA = \left(\frac{(TPL_{Before} - TPL_{After})}{TPL_{Before}}\right) * 8760 \tag{15}$$

TPL_{Before} is the total power loss that occurred before DG and D-STATCOM placement.

TPL_{After} is the total power loss that occurred after DG and D-STATCOM placement.

3.5 Load Unbalancing: Type A and Type B

Type A

In Type A unbalancing, total load on network remains constant. Phase B load is reduced by a certain percentage and in Phase C load is increased by the same percentage.

$$S_a = S_a \tag{16}$$

$$S_b = S_b * (1 - \%unbalance) \tag{17}$$

$$S_b = S_b * (1 - \%unbalance) \tag{18}$$

Type B

In Type B unbalancing, the total load on network reduces. Load is constant on Phase A. On Phase B, a certain percentage of load is reduced and in Phase C reduced percentage of load gets doubled.

$$S_a = S_a \tag{19}$$

$$S_b = S_b * (1 - \%unbalance) \tag{20}$$

$$S_c = S_c * (1 - (2 * \%unbalance) \tag{21}$$

%unbalance taken in this paper is 20%.

4 Simulation and Results

Analysis for 25 bus radial UB distribution system is carried out in this paper considering Type A and Type B load UB. The case studies are considered without and with load growth. Results for the base case of each case considered in the paper match with the base case of reference [1], hence it validates the algorithm. The load and line data for 25 bus system is taken from [13]. 4.17 kV is taken as base kVA and 30MVA is taken as base MVA. Different cases have been considered to obtain the result (Figs. 1, 2, 3 and 4).

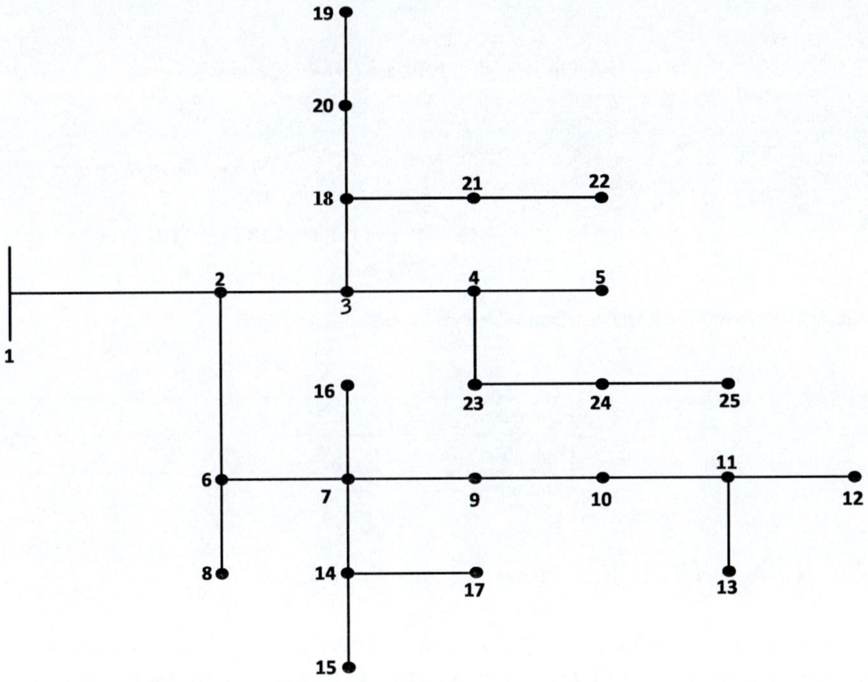

Fig. 1 IEEE 25 bus test system single line diagram

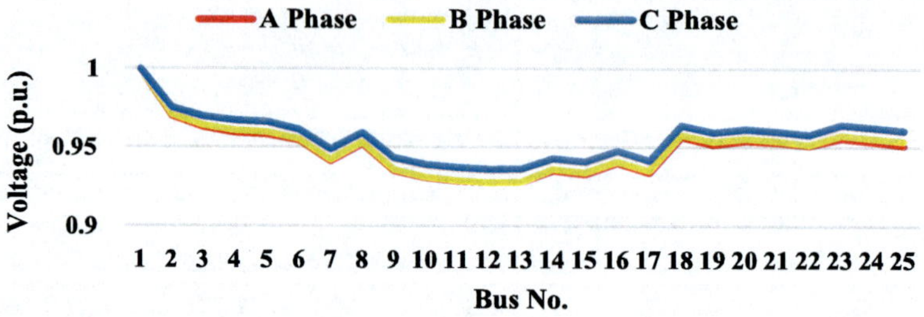

Fig. 2 Voltage profile of UB distribution system for case a

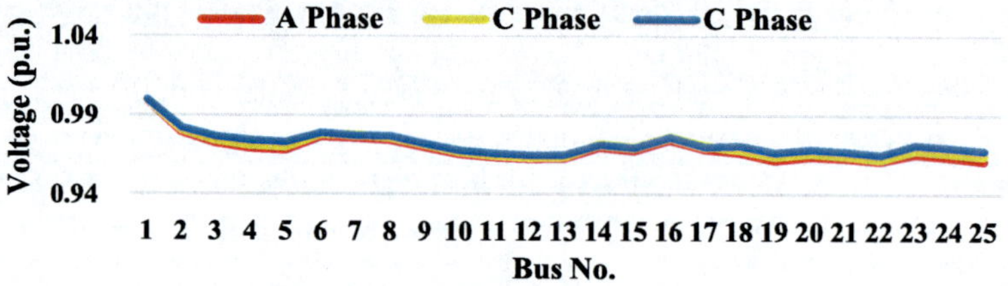

Fig. 3 Voltage profile of UB distribution system for case b

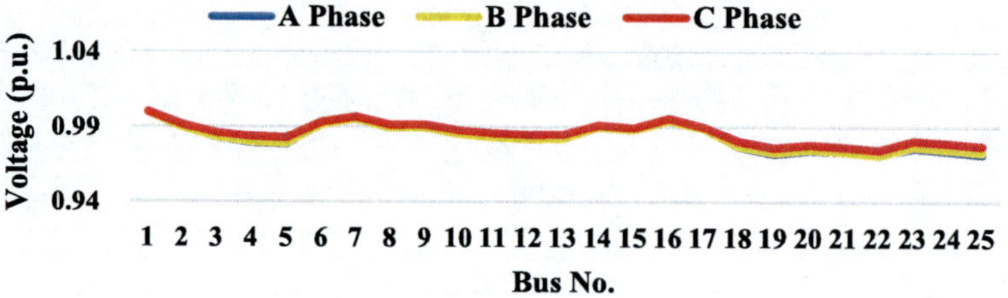

Fig. 4 Voltage profile of UB distribution system for case c

Table 1 AOA parameters

Parameter	Value
Population size	50
Iterations	100

4.1 Case 1. UB Distribution System Base Case

Case a: Base case-without DG and D-STATCOM allocation: Each phase loss can be found in Fig. 5. 150.13 kW is TRPL and 167.28 kVAR is TQPL for this scenario. Figure 2 shows the voltage profile of each phase.

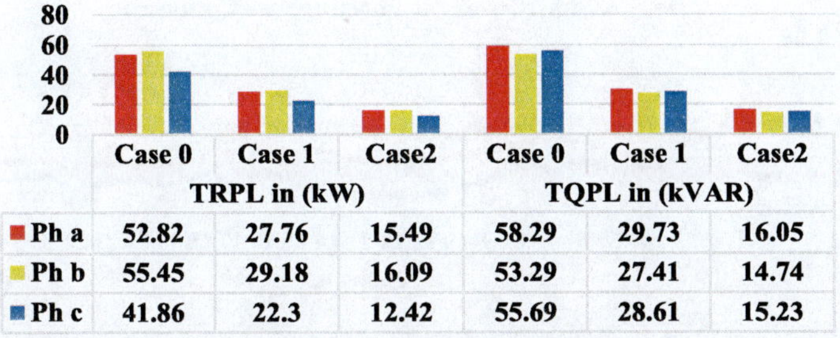

	TRPL in (kW)			TQPL in (kVAR)		
	Case 0	Case 1	Case2	Case 0	Case 1	Case2
Ph a	52.82	27.76	15.49	58.29	29.73	16.05
Ph b	55.45	29.18	16.09	53.29	27.41	14.74
Ph c	41.86	22.3	12.42	55.69	28.61	15.23

Fig. 5 Effects of D-STATCOM and DG in various UB distribution system

Case b: Single DG: Each phase loss can be found in Fig. 5. 79.25 kW is TRPL and 85.75 kVAR is TQPL for this scenario. Figure 3 shows the voltage profile of each phase.

Case c: Single DG and single D-STATCOM: Each phase loss can be found in Fig. 5. 44.00 kW is TRPL and 46.02 kVAR is TQPL for this scenario. Figure 4 shows the voltage profile of each phase.

4.2 *Case 2: Type A UB Distribution System*

Case a: Base case-without DG and D-STATCOM allocation: Each phase loss can be found in Fig. 9. 155.83 kW is TRPL and 176.60 kVAR TQPL for this scenario. Figure 6 shows the voltage profile of each phase (Figs. 7 and 8).

Case b: Single DG: Each phase loss can be found in Fig. 9. 84.22 kW is TRPL and 94.78 kVAR is TQPL for this scenario. Figure 7 shows a voltage profile of each phase.

Case c: Single DG and single D-STATCOM: Each phase loss can be found in Fig. 9. 53.96 kW is TRPL and 58.26 kVAR is TQPL for this scenario. Figure 8 shows the voltage profile of each phase.

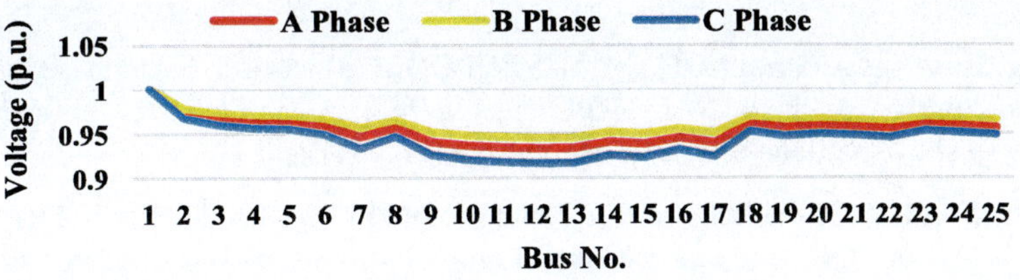

Fig. 6 Voltage profile of Type A UB distribution system for case a

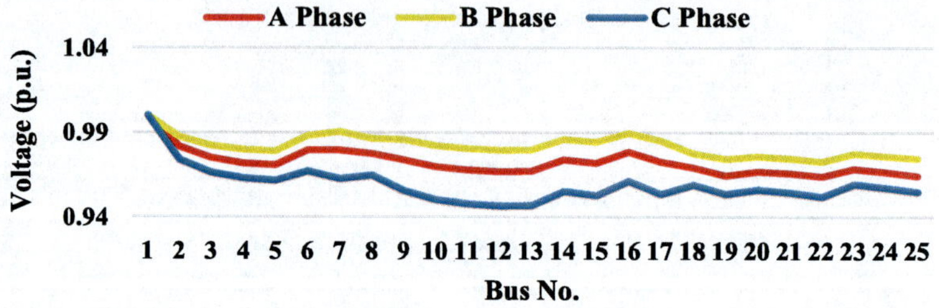

Fig. 7 Voltage profile of Type A UB distribution system for case b

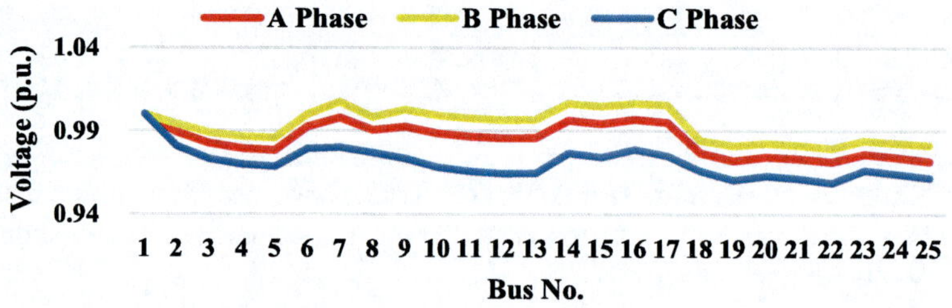

Fig. 8 Voltage profile of Type A UB distribution system for case c

	Case 0	Case 1	Case 2	Case 0	Case 1	Case 2
Ph a	47.68	24.83	15.73	59.02	28.57	17.19
Ph b	37.92	23.05	15.31	28.02	16.56	10.8
Ph c	70.23	36.34	22.92	89.56	49.65	30.26

Fig. 9 Effects of D-STATCOM and DG in various Type A UB distribution system

4.3 Case 3: Type B UB Distribution System

Case a: Base case-without DG and D-STATCOM allocation: Each phase loss can be found in Fig. 13. 98.34 kW is TRPL and 112.74 kVAR is TQPL for this scenario. Figure 10 shows voltage profile of each phase (Figs. 11 and 12).

Case b: Single DG: Each phase loss can be found in Fig. 13. 54.02 kW TRPL and 60.42 kVAR is TQPL for this scenario. Figure 11 shows a voltage profile of each phase.

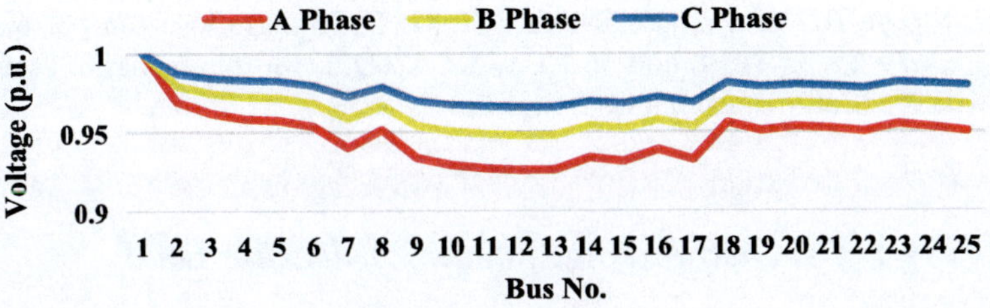

Fig. 10 Voltage profile of Type B UB distribution system for case a

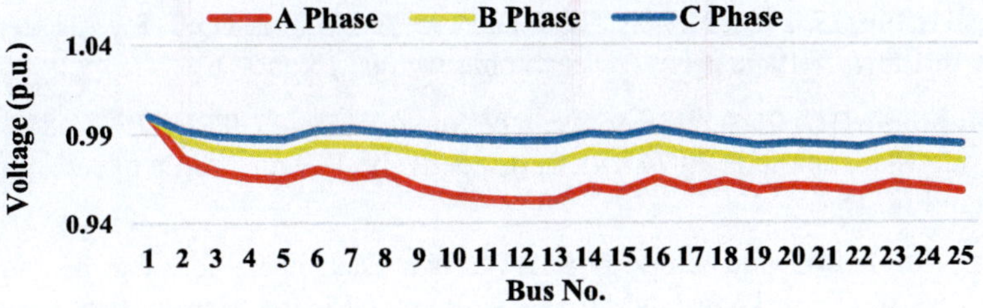

Fig. 11 Voltage profile of Type B UB distribution system for case b

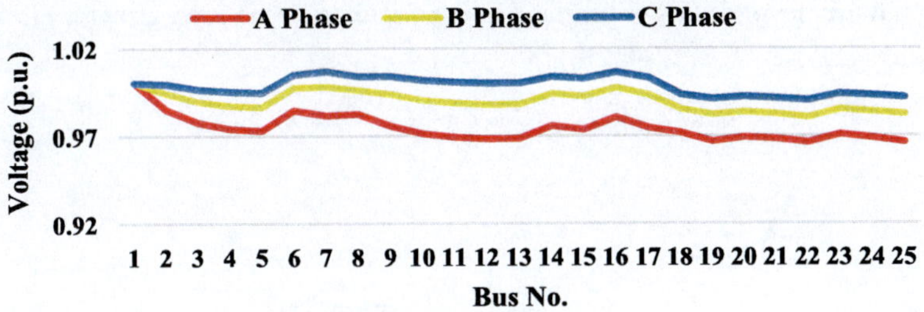

Fig. 12 Voltage profile of Type B UB distribution system for case c

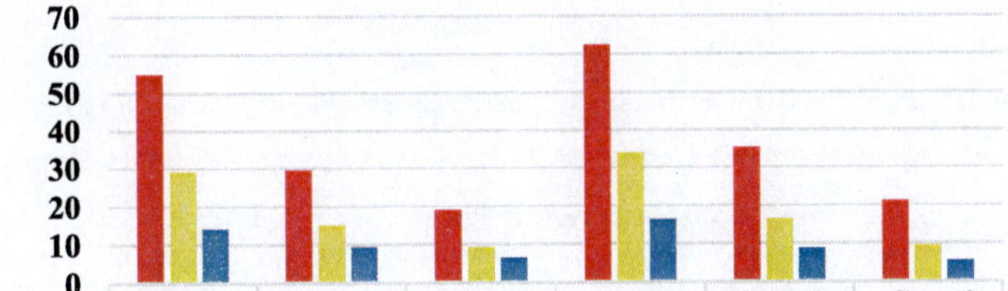

	TRPL in (kW)			TQPL in (kVAR)		
	Case 0	Case 1	Case 2	Case 0	Case 1	Case 2
Ph a	54.85	29.52	18.96	62.29	35.14	21.05
Ph b	29.31	15.17	9.42	34.01	16.49	9.58
Ph c	14.21	9.33	6.51	16.44	8.77	5.46

Fig. 13 Effects of D-STATCOM and DG in various Type B UB distribution System

Case c: Single DG and single D-STATCOM: Each phase loss can be found in Fig. 13. 34.89 kW is TRPL and 36.10 kVAR is TQPL for this scenario. Figure 12 shows voltage profile of each phase.

4.4 *Case 4: UB Distribution System Considering Load Growth*

Case a: Base case-without DG and D-STATCOM allocation: Each phase loss can be found in Fig. 17, 175.79 kW is TRPL and 195.82 kVAR is TQPL for this scenario. Figure 14 shows voltage profile of each phase (Figs. 15 and 16).

Case b: Single DG: Each phase loss can be found in Fig. 17, TRPL is 92.48 kW and TQPL for this scenario is 99.74 kVAR, respectively. Voltage profile of each phase is shown in Fig. 15.

Case c: Single DG and single D-STATCOM: Each phase loss can be found in Fig. 17, 53.26 kW is TRPL and 56.68 kVAR is TQPL for this scenario. Figure 16 shows voltage profile of each phase.

It can be noted that in case 1-UB distribution system, certain buses have voltage levels that are beyond the set limitations even if there is no load growth. However,

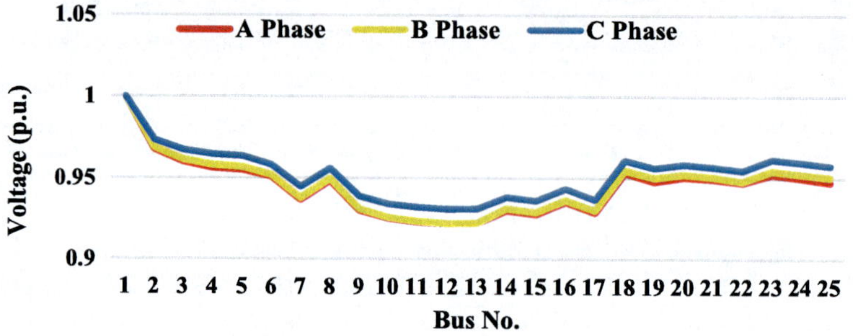

Fig. 14 Voltage profile of UB distribution system considering load growth for case a

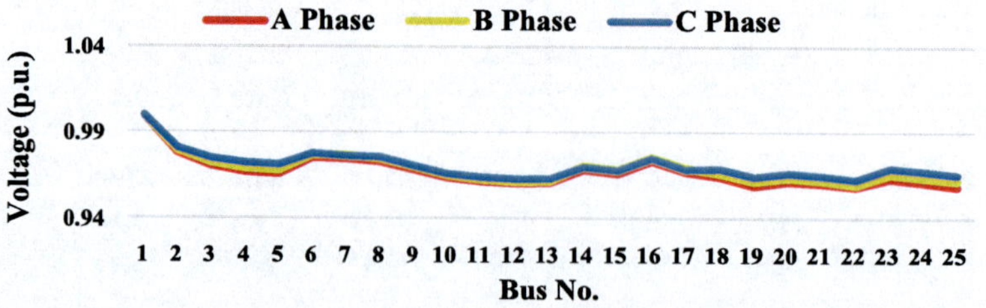

Fig. 15 Voltage profile of UB system considering load growth for case b

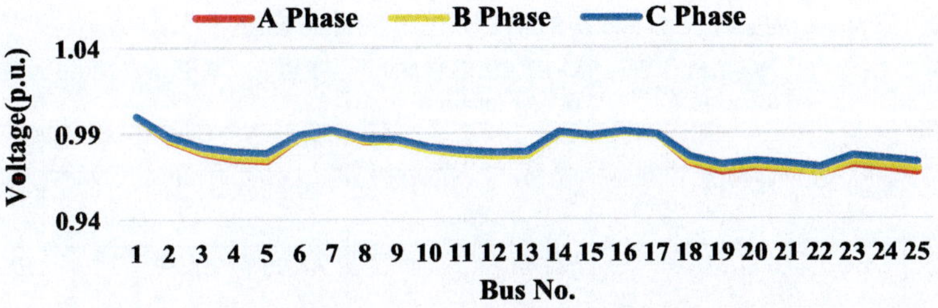

Fig. 16 Voltage profile of UB distribution system considering load growth for case c

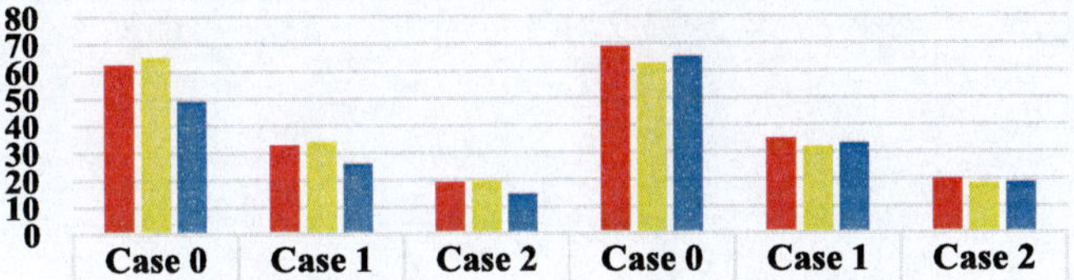

	TRPL in (kW)			TQPL in (kVAR)		
	Case 0	Case 1	Case 2	Case 0	Case 1	Case 2
Ph a	61.88	32.41	18.77	68.29	34.6	19.76
Ph b	64.92	34.03	19.59	62.41	31.88	18.23
Ph c	48.98	26.04	14.89	65.12	33.26	18.68

Fig. 17 Effects of D-STATCOM and DG in various UB distribution system considering load growth

with an optimal DG allocation of 707.33 kW per phase, the voltage is increased while cost savings of 41.36% are realized. However, when DG and D-STATCOM of 604.31 kW and 571.61kVAR are allocated simultaneously, improvement in voltage profile to values close to unity is observed from Figs. 2, 3 and 4 in each case and with significant savings of 61.93%. From Fig. 5, it is observed that losses are least in each phase after allocation of both D-STATCOM and DG.

It can be noted that in case 2-Type A UB distribution system because of load shifting voltage profile improves in phase b and increase in phase c loss can be observed in contrast with phases a and b with DG allocation of size 729.79 kW and allocating DG, D-STATCOM simultaneously of size 826.97 kW and 322.47 kVAR. From Figs. 6, 7 and 8 shows voltage profile for different cases and it is observed that because of load shifting to phase c there is less improvement in its voltage profile in each case in comparison to other two phases. From Fig. 9, it can be observed that allocation of only DG reduces loss but allocation of both D-STATCOM and DG gives less loss in each phase as compared to other cases.

It can be noted that in case 3-Type B UB distribution system more improvement in the voltage profile of phase c can be seen in each case from Figs. 10, 11 and 12 in comparison to other two phases because phase c has more load reduction. Phase c has much less TRPL and TQRL in comparison to the other two phases from Fig. 13 it can be seen, also allocation of both D-STATCOM and DG has reduced losses in

Table 2 Per phase values of DG and D-STATCOM for different cases

Cases		DG-size (kW) location	D-STATCOM size (kVA) location	TRPL	% Power loss	%ESA
Case 1	Case a	–	–	150.13	–	–
	Case b	707.33 7	–	79.25	47.21	41.36
	Case c	604.31 7	571.61 7	44.00	70.69	61.93
Case 2	Case a	–	–	155.83	–	–
	Case b	729.79 7	–	84.22	45.95	40.25
	Case c	826.97 7	322.47 14	53.96	65.37	57.27
Case 3	Case a	–	–	98.34	–	–
	Case b	534.19 7	–	54.02	45.06	39.48
	Case c	598.58 7	395.59 6	34.89	64.45	56.52
Case 4	Case a	–	–	175.79	–	–
	Case b	776.28 7	–	92.48	47.37	41.51
	Case c	743.10 7	411.20 14	53.26	69.70	61.06

each phase in comparison to other cases. 534.19 kW is the size of DG required and for DG and D-STATCOM simultaneous allocation size required are 598.58 kW and 395.59 kVAR as seen in Table 2.

It can be noted that in Case 4- UB distribution system considering load growth because of load growth total power taken from substation increases. Due to the increment in load DG and D-STATCOM size increases in comparison to all other cases as seen in Table 2, however, voltage profile is similar to case 1. From Fig. 17, it can be seen that in this case also the losses are least in each phase after the allocation of both D-STATCOM and DG.

5 Conclusion

This study has analyzed the radial UB distribution system taking into account the various radial UB distribution system situations and growth in load for the forthcoming year with a 2.5% load increase by using AOA. AOA uses simple mathematical operators so it is easy to implement in comparison to other algorithm. From the results it is observed that by allocating both DG and D-STATCOM, the loss is minimized and the voltage profile is improved in each case. It is inferred that as energy demand increases, TRPL and TQPL also increase. This study could be useful to DNO for planning distribution system in unbalance cases while considering the effect of various load unbalancing impact on the voltages and losses taking the load increase scenarios. The D-STATCOM and DG sizing can play an important role in balancing thc distribution network and improving the voltage profile under different unbalance scenarios. Therefore, DNO can plan a distribution network for better management and control of voltage. In the future EV charging station can also be incorporated with DG and D-STACOM.

References

1. Tahiliani G, Gupta AR (2022) Electrical distribution system analysis with atom search optimization based DG and DSTACOM allocation. In: 2022 IEEE Delhi section conference (DELCON), New Delhi, India, 2022, pp 1–6. https://doi.org/10.1109/DELCON54057.2022.9753420
2. Ponnam V, Swarnasri K (2020) Novel technique for optimal placement and sizing of DG in 3-phase unbalanced radial secondary distribution system, pp 2278–3075. https://doi.org/10.35940/ijitecD1665.029420
3. Murthy G, Suresh C, Sowjankumar K, Hanumantharao B (2019) Impact Of DG On unbalanced RDS. Int J Sci Technol Res 8:539
4. Roosta A, Eskandari HR, Khooban MH (2019) Optimization of radial unbalanced distribution networks in the presence of distribution generation units by network reconfiguration using harmony search algorithm. Neural Comput Appl 31:7095–7109. https://doi.org/10.1007/s00521-018-3507-0
5. Ashwani Kumar, Vijay Babul P, Murty VVSN (2017) Distributed generators allocation in radial distribution systems with load growth using loss sensitivity approach. J Inst Eng India Ser B 3(98):275–287. https://doi.org/10.1007/s40031-016-0242-8
6. Sanjay R, Jayabarathi T, Raghunathan T, Ramesh V, Nadarajah M (2017) Optimal allocation of DG using hybrid grey wolf optimizer. IEEE Access 1–1. https://doi.org/10.1109/ACCESS.2017.2726586
7. Kumawat M, Gupta N, Jain N, Bansal RC (2017) Optimally allocation of distributed generators in three-phase unbalanced distribution network. Energy Proc 142:749–754. ISSN 1876-6102. https://doi.org/10.1016/j.egypro.2017.12.122
8. Murty VVSN, Kumar A (2014) Capacitor allocation in unbalanced distribution system under unbalances and loading conditions. Energy Proc 54:47–74. https://doi.org/10.1016/j.egypro.2014.07.248
9. Ray R, Gupta AR (2022) Analysis of radial unbalanced distribution network for different loading conditions with DG and D-STATCOM placement. In: 2022 IEEE Delhi section conference (DELCON), New Delhi, India, pp 1–8. https://doi.org/10.1109/DELCON54057.2022.9752872
10. Bhadoriya J, Gupta A, Zellagui M, Saxena N, Arya A, Bohre A (2022) Optimal allocation of electric vehicles charging station in distribution network beside DG using TSO. https://doi.org/10.1007/978-981-19-0979-5_29
11. Abualigah L, Diabat A, Mirjalili S, Abdelaziz M, Gandomi AH (2021) The arithmetic optimization algorithm. Comput Meth Appl Mech Eng 376:113609
12. Mishra S, Das D (2009) Load flow analysis for unbalanced radial distribution systems: a comparison of different practical methods. ICFAI Univ J Electr Electron Eng II(3)
13. Ganesh V, Sivanagaraju S, Ramana T. Feeder reconfiguration for loss reduction in unbalanced distribution system using. Int J Electr Electron Eng 200:754–762

A Construction of Secure and Efficient Lightweight Authenticated Key Agreement Based on Elliptic Curve Cryptography for Smart Grid

Dharminder Chaudhary, M.S.P. Durgarao, Pasupuleti Gnaneshwar Parikshith, and Yarragoti Ravi Theja

Abstract Smart grids are important in modern smart cities for setting and managing the response demands, for this, they use Information and Communication Technologies. Many kinds of sensors are used in Smart Grid Environment for the surveillance purposes in an area. Their response time is less which is the reason these are deployed in high-tension areas such as power supply lines so that they can pass and share important information to the control center about any flaw, damage, or any emergency message. Usually, these grids face cyberattacks along with physical attacks and damages when they are deployed to high-tension open environment. Therefore, the objective is the requirement of prevention of such attacks so that the smart grids environment can work efficiently for surveillance. The proposed protocol helps to establish an authenticated key exchange. This is useful to communicate over public channel with grids authentication.

Keywords Authentication · Cyber system · Smart grids · Surveillance

1 Introduction

The widespread adoption of smart grids has necessitated the use of Information and Communication Technology (ICT). Smart grids are used in various fields, such as response management, creating smart energy environments with responsible detection of faults, and recovering any entity in the grid system. These grids offer benefits like the capacity to expand and manage renewable energy sources. Recent advancements in ICT have resolved both security and privacy issues associated with smart

D. Chaudhary (✉) · M.S.P. Durgarao · P. G. Parikshith · Y. R. Theja
Department of Computer Science and Engineering, Amrita School of Computing, Amrita Vishwa Vidyapeetham, Chennai, India
e-mail: manndharminder999@gmail.com

A. Kumar et al. (eds.), *Decarbonisation and Digitization of the Energy System*, Lecture Notes in Electrical Engineering 1099, https://doi.org/10.1007/978-981-99-7630-0_14

grid technologies. The smart grid system involves numerous parameters that require consideration, including smart meters, internet communication, and remote power control [5, 6]. However, due to the critical nature of the smart grid framework, it is vulnerable to cyberattacks, physical attacks, and natural disasters, among other threats. In this regard, extensive research has been conducted to guarantee the security of the smart grids system. Both security and privacy concerns surrounding smart grid technology have been addressed in recent years through the rapid development of information and communication technology [7]. The smart grid system encompasses numerous parameters and considerations, including smart meters, internet communication, and remote power control, in order to establish an advanced and intelligent platform [5]. This platform facilitates communication between clients and service providers [19], but it may also be vulnerable to cyberattacks due to its critical nature [12]. The deployment of smart grids poses significant risks, such as physical attacks, cyberattacks, and natural disasters, which may lead to infrastructure failures, energy crisis, breaching consumer's data, and employee security concerns. As a result, extensive studies have been conducted to guarantee the security of the smart grid [8, 9, 13]. Elegant and effective solutions are needed to withstand cyberattacks on the smart grid system. Security strategies are crucial for establishing cyber-attack-resistant solutions for smart grid applications [4, 17]. Several authentication protocols have been proposed to secure the smart grid system. Braeken et al. [2] introduced a communication model for smart grid systems, while Moghadam et al. [16], Mostafa Farhadi, and others proposed a protocol that guarantees authenticated key agreement and effective communication from station to data center and vice versa. However, this protocol requires significant hardware and software to match the specifications, and it suffers from significant communication overhead. Li et al. [14] proposed an advanced silent verification technique for smart grid architecture, which does not ensure data confidentiality and anonymity, and the enrollment process is more tedious among other security needs. Khan et al. [11] proposed a more efficient protocol than Li et al. [14], which ensures verification and interaction between two smart grids. Recently, Chaudhry et al. [3]. proposed a secure and efficient authenticated key exchange and identified vulnerabilities in Khan et al. [10, 11]'s protocol. This article builds on Chaudhry et al. [3] work and offers a secure and efficient protocol that can withstand a range of attacks.

2 Motivation and Contribution

Ensuring the security of smart grids is crucial due to the increasing prominence of authentication threats in the modern digital landscape. Despite the existence of various authentication protocols, none of them provide optimal security functionalities and features that take into account smart grid communication [15]. To address this gap, this paper proposes a novel method that aims to mitigate the current shortcomings of the smart grid framework. This study is expected to contribute to the improvement of smart grid security and alleviate researchers' concerns. This paper proposes a novel, secure, and efficient lightweight authenticated key agreement for smart grids. The proposed model utilizes efficient cryptographic one-way hash functions [18] and maintains all well-known security features while ensuring anonymity and privacy. The suggested system is compared to relevant existing schemes [1, 2, 11, 14, 15], demonstrating a better balance between security and functionality aspects, as well as communication and computation overheads. Finally, the suggested technique has been proved to be robust to a variety of known assaults, which are discussed in Sect. 5. The proposed registration technique resists password guessing in two-factor authentication protocols.

3 Threat Model

The most widely used threat model, for finding the security of secure lightweight authentication protocol threat model. The actions which can be performed by the malicious attacker or adversary (MA) are discussed here. The Dolev and Yao model states that attackers can compromise, delete, eavesdrop, inject some codes, and modify some of the data shared through the public medium. A malicious attacker can also perform a powerful analysis attack on smartphones by acting as a legal user. This leads to the compromising of sensitive information present in the smartphone. Apart from this, the attacker can capture the physical device. After this, the login details can be extracted and the attacker can behave like a legal user. The attacks performed by the attacker after getting the login details are forgery and impersonation attacks. Another scheme that is more secure and efficient than Dolev and Yao's threat model is Canetti and Krawczyk, model. It is also known as the CK threat model. The standard CK model is de facto for all AKA schemes. The Canetti and Krawczyk threat model states that the attacker (MA) has all the capabilities mentioned in the DY model. But in addition to it, the adversary can also compromise sensitive information by performing attacks based on session hijacking.

4 Elliptical Curver Cryptography Fundamentals

An elliptic curve over Z_p (where $p > 3$) is defined as a set of all pairs (x, y) that belong to Z_p and meet the equation $y^2 \equiv x^3 + a.x + b(modp)$ along with an imaginary point referred to as infinity ($\mathcal{O}$). The values a and b belong to Z_p, and the condition $4.a^3 + 27.b^2 \neq 0modp$ must be satisfied. The term "elliptic curve" implies that the curve has no singularities. In terms of geometry, this means that the plot does not cross over itself or have any vertices, which is guaranteed when the discriminant of the curve, $-16(4a^3 + 27b^2)$, is not equal to zero. It is essential to examine the curve in a prime field for cryptographic purposes, as defined. However, when we plot the curve over Z_p, the result does not look like a curve. Nevertheless, there is no restriction on using an elliptic curve equation and plotting it over real numbers. The elliptic curve groups, denoted as G, operate using addition and scalar multiplication, with point addition defined as (x3, y3) = $(\lambda^2 - x_1 - x_2(modq), (\lambda(x_1 - x_2) - y_1)(modq))$.

$$P + Q = (x_3, y_3) \tag{1}$$

$$(x3, y3) = (\lambda^2 - x_1 - x_2(modq), (\lambda(x_1 - x_2) - y_1)(modq)) \tag{2}$$

$$\lambda = \frac{y2 - y1}{x2 - x1}(mod\ q)\ if\ P \neq Q \tag{3}$$

$$\lambda = \frac{3x_1^2 + a}{2y_1}(mod\ q)\ if\ P = Q \tag{4}$$

4.1 Discrete Logarithm Problem with Elliptic Curves

The Elliptic Curve Discrete Logarithm Problem (ECDLP) involves a given elliptic curve E. The problem is to find an integer d, where $(1 \leq d \leq \#E)$ and # E is number of points on the curve, such that the primitive P added to itself d times equals T. Where T is another element on the curve.

$$T = dP = P + P + \cdots + P\ (d\ times) \tag{5}$$

In cryptography, d is viewed as a private integer key, and public key T is represented as a point on the curve with coordinates (x_T, y_T). The ECDLP differs from the discrete logarithm problem for Z_p^*, where both keys are integers. The expression $T = dP$, called point multiplication, is just a notation for repeatedly applying the group operation $(y^2 \equiv x^3 + a.x + b(modp))$, not the actual multiplication of an integer d and a curve point P.

4.2 Diffie-Hellman Key Exchange with Elliptic Curves

Given the generator g, values a, and b, it is challenging to compute abg for group G in the Elliptic Curve Diffie-Hellman Problem (ECDHP). In terms of key sizes, Elliptic Curve Cryptography (ECC) is more effective than other cryptographic protocols like RSA, DSA, and Diffie-Hellman, which impacts the performance and bandwidth of the systems.

4.3 Hash Function Characteristics

A hash function converts an input into an encrypted output known as a hash. The following properties are what a good hash function should possess:

- Arbitrary message size: Hash function, h(x), is flexible about message size; it may be used with messages of any size, x.
- Fixed output length: Output with a definite length is produced by the hash function, $h(x) = z$.
- Simple computation: The hash function $h(x)$ is efficient.
- One-way resistance: It is impossible to identify any input x, such that $h(x) = z$ for given z that means $h(x)$ is a one-way function.
- Second preimage resistance: It is computationally impossible to identify any alternative input, x_2, that would result in the same hash value, $h(x_1) = h(x_2)$, for given an input x_1, and its associated hash values $h(x_1)$, and x_2.
- Collision resistance: Finding any two inputs, x_1 and x_2 such that $h(x_1) = h(x_2)$ is computationally impossible.

5 Proposed Secure and Efficient Lightweight Authenticated Key Agreement for Smart Grid

In this paper, we have proposed a better dynamic and lightweight protocol. There are three phases include login, registration, and authentication.

5.1 Registration Phase

- Two parameters Id_A and Pw_A are selected by user U_A. Then, he chooses a random number a_A belongs to the set of positive integers modulo prime q and computes N_A = h(Id_A‖h(Pw_A‖a_A)). He sends Id_A and N_A using a private channel to Trusted Authority(TA).
- Once Id_A and N_A are received by TA, then TA computes five parameters known as A_A, C_A, , , as = , = + , = h(), = + , $SK_A = N_A + ms_A$. These five parameters are transferred to U_A through a private channel.
- After receiving the required parameters from TA, U_A computes A_A and PK_A as $A_A = N_A.P$, $PK_A = SK_A.P$. If $PK_A = C_A + N_A.PK_T$, then only the verification is successful. Finally, the values $\varrho_A = a_A \oplus h((h(id_A \| pw_A) \| h(pw_A))$ $mod(q))$ and A_A are stored in the database by U_A.

5.2 Login and Authentication

In order to establish a secure communication following procedures need to be executed between the users U_A and U_B. U_A = Initial Entity, U_B = Responder Entity

- Id_A^* is the identity of user U_A and Pw_A^* is the password. He computes $a_A = \varrho_A \oplus h((h(id_A \| pw_A) \| h(pw_A)) mod(q))$, N_A^* = $h(Id_A^* \| Pw_A \| a_A)$, $A_A^* = N_A^*.P$. If $A_A = A_A^*$, then verification is successful. U_A sends a timestamp T_1, and parameters A_A, N_A, T_1 to U_B.
- U_B receives A_A, N_A, T_1, and chooses a new timestamp T_2, then the timestamp freshness is checked through following relation $|T_2 - T_1| \leq \Delta T$, then U_B selects random number g_B, q_B and computes $G_B = g_B.p$, $K_B = g_B.(A_A + PK_T + N_A.PK_T)$, $SK_{BA} = h(g_B.A_A \| T_2 \| q_B)$, $Aut_B = h(SK_{BA} \| q_B)$, $E_B = ENC_{KB}(q_B \| T_2)$. The parameter E_B is encrypted using key K_B by U_B. Then the parameters T_2, G_B, E_B, Aut_B are sent to U_A by U_B
- A new timestamp T_3 is set by the user U_A by checking $|T_3 - T_2| \leq \Delta T$. The equation is used to check the timestamp freshness. Further it computes $K_A = SK_A.G_B$, $DEC(E_B)_{KA} = (q_B, T_2), SK_{AB} = h(N_A.g_B \| T_2 \| q_B), Aut_A = h(SK_{AB} \| q_B)$. The parameters SK_{BA} and Aut_A are computed by U_A. If $Aut_A = ?Aut_B$, then verification is successful, and it chooses a random number X_A. He computes parameters $F_A and E_A$ through key K_A as $F_A = h(Aut_A \| X_A)$, $E_A = ENC_{KA}(F_A, X_A)$. These parameters E_A, $F_A and T_3$ are sent to U_B through a public channel by U_A
- A new timestamp T_4 is set by U_B after receiving all the parameters E_A, F_A, T_3 from U_A. The freshness of timestamp is checked by the relation $|T_4 - T_3| \leq \Delta T$, and decrypts $DEC(EA)_{KB} = (Aut_A, X_A)$ and it computes $F_B = h(Aut_A \| X_A)$. If $F_B = ?F_A$, then the session and authentication key are successfully established.

5.3 Password Change Phase

Id_A and Pw_A are entered by the user of his own choice at the registration phase. Lets consider user = U_A, Id = Id_A^*, Pw= PW_A^*, then U_A computes $N_A^* = h(Id_A^* \| PW_A^* \| a_A)$, $A_A^* = N_A^*.P$, if $A_A^* = ?A_A$ is successful, a new password can be entered by user which is $(PW_A^*)^*$ and performs $(N_A^*)^* = h(Id_A^* \| (PW_A^*)^* \| a_A)$, $(A_A^*)^* = (N_A^*)^*.P$, and A_A is replaced by $(A_A^*)^*$ in the database.

6 Informal Verification

Through logic and argumentation, it will be demonstrated that the suggested protocol satisfies several security requirements and can fend off known assaults.

1. **Perfect Forward Secrecy**: According to this security criteria, an attacker should not be able to use fixed public/private parameters, such as the parties' private and public keys, along with the parameters shared over the public channel, to determine the session key. Despite this flaw, even if the attacker managed to gain every private key and public key used in the protocol, they would still be unable to access the session key since it depends on the parameters $g_B.A_A$, $N_A.G_B$ and the session key and the two theorems of ECDLP and ECDHP.
2. **User Anonymity**: The attacker cannot obtain the identities of the entities engaged in the protocol using the parameters sent on the communication channel during the login and authentication phases. The proposed protocol satisfies this condition since it prevents access to the identities Id_A and Id_B even if the adversary possesses access to protocol run and possess all parameters exchanged on the channel.
3. **Password Guessing Attack**: In this kind of attack, the adversary eavesdrops on the messages sent during different stages such as authentication and key agreement in an effort to learn the passwords of the organizations engaged in the protocol. The suggested protocol avoids this by making sure the password cannot be derived using the relation $\varrho_A = a_A \oplus h(h(id_A \| pw_A) \| h(pw_A))$.
4. **Replay Attacks**: In this technique, the attacker uses previously recorded authentication and key agreement stage settings in subsequent phases. To combat this, the suggested protocol employs a timestamp to validate the validity of messages and identify duplicates.
5. **Stolen-verifier Attack**: The suggested protocol prevents this attack by making the session key dependent on certain parameters ($G_B.A_A$ and $N_A.G_B$) in each session. This attack entails the attacker obtaining the verification parameters stored in memory. As a result, even if the attacker has access to the associated entities' memory, they are unable to obtain the session key SK_AB = h $(N_A.G_B||T_1|q_B)$ or SK_BA = h $(g_B.A_A||T_2||q_B)$.
6. **Denial-of-service Attack**: A huge number of proxy or duplicate requests are sent to an object in this kind of attack, overwhelming its processing power. The suggested protocol employs timestamps to determine if a message is fresh and breaks the connection if it is not. As a result, the attacker finds it challenging to execute a denial-of-service assault.

7 Performance Comparisons

The assessment compares the communication, calculation, and overall execution times of the [3] with along protocol [11]. The registration stage was kept untouched and is not included in the comparison since the original [11] protocol's design defects were eliminated to enhance it. 4 T_{ESED} + 6 T_{PM} + 7 T_{HO} procedures are used to complete the authentication process in [3]. T_{ESED} stands for symmetric encryption/decryption time and refers to the authentication phase's use of two encryptions and two decryptions. When an integer is multiplied by a point on an elliptical curve,

Table 1 Comparison of the computation cost

Scheme	Operations	Computational Cost (ms)
[15]	$10T_{PM} + 4T_{PA} + 8T_{HO}$	22.3936
[14]	$7T_{ME} + 6T_{HO}$	26.9638
[1]	$9T_{PM} + 13T_{ESED} + 24T_{HO}$	20.149
[2]	$8T_{PM} + 3T_{PA} + 2T_{ESED} + 12T_{HO}$	17.9312
[11]	$4T_{ESED} + 8T_{PM} + 19T_{HO}$	17.8701
NEW	$4T_{ESED} + 6T_{PM} + 7T_{HO}$	13.3869

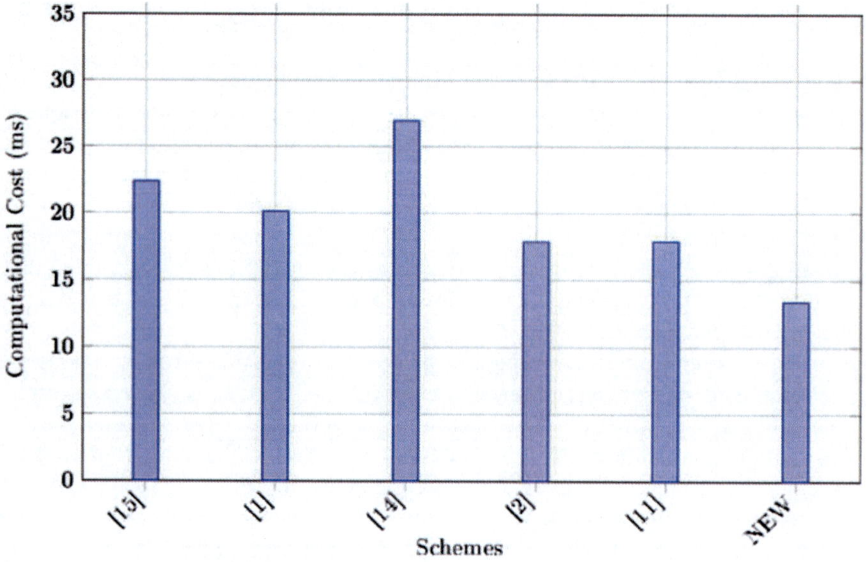

Fig. 1 Comparison of the computation cost

the result is the time for point multiplication or T_{PM}. Seven times throughout the protocol, the hash operation is referred to as T_{HO}. With T_{ESED} = 0.0046 milliseconds, T_{PM} = 2.226 milliseconds, and T_{HO} = 0.0023 milliseconds, the exact execution timings, as in the original paper proposing were utilized for simplicity, and compared to [11], which takes 17.8701 milliseconds, proposed takes 13.3869 milliseconds for authentication. The communication cost is the same for [11]. Two messages, M1 and M2, with a combined communication cost of 3136 bits, make up the original [11]. In comparison, LSPA needs three messages and has a communication cost of 1984 bits, which is 36.7% less expensive than [11]. In addition to the decrease in transmission costs, a 33.4% decrease in computing costs was also achieved (Table 1 and Fig. 1).

8 Conclusion

Smart grids are important in modern smart cities for setting and managing the response demands, for this, they use Information and Communication Technologies. The proposed protocol's response time is less and helps in secure communication over high-tension areas such as power supply lines so that they can pass and share the important information to the control center about any flaw, damage, or any emergency message. The protocol's objective is the requirement of prevention of such attacks so that the smart grids environment can work efficiently for surveillance. The proposed protocol helps to establish an authenticated key exchange. This is useful to communicate over public channel with grids authentication.

References

1. Abbasinezhad-Mood D, Nikooghadam M (2018) Design and extensive hardware performance analysis of an efficient pairwise key generation scheme for smart grid. Int J Commun Syst 31(5):e3507
2. Braeken A, Kumar P, Martin A (2018) Efficient and provably secure key agreement for modern smart metering communications. Energies 11(10):2662
3. Chaudhry SA (2021) Correcting "palk: password-based anonymous lightweight key agreement framework for smart grid." Int J Electr Power Energy Syst 125:106529
4. Ding D, Han Q-L, Xiang Y, Ge X, Zhang X-M (2018) A survey on security control and attack detection for industrial cyber-physical systems. Neurocomputing 275:1674–1683
5. Garg S, Kaur K, Kaddoum G, Rodrigues JJPC, Guizani M (2019) Secure and lightweight authentication scheme for smart metering infrastructure in smart grid. IEEE Trans Ind Inform 16(5):3548–3557
6. Gunduz MZ, Das R (2018) Analysis of cyber-attacks on smart grid applications. In: 2018 international conference on artificial intelligence and data processing (IDAP). IEEE, pp 1–5
7. Gunduz MZ, Das R (2020) Cyber-security on smart grid: threats and potential solutions. Comput Netw 169:107094
8. Hankerson D, Menezes AJ, Vanstone S (2006). Guide to elliptic curve cryptography. Springer Science & Business Media
9. Hwang S-J, Liao H-C (2005) Security of Tzeng-Hwang's authenticated encryption scheme based on elliptic curve discrete logarithm problems. Appl Math Comput 168(1):717–721
10. Khan AA, Kumar V, Ahmad M, Rana S (2021) LAKAF: lightweight authentication and key agreement framework for smart grid network. J Syst Arch 116:102053
11. Khan AA, Kumar V, Ahmad M, Rana S, Mishra D (2020) PALK: password-based anonymous lightweight key agreement framework for smart grid. Int J Electr Power Energy Syst 121:106121
12. Kimani K, Oduol V, Langat K (2019) Cyber security challenges for IoT-based smart grid networks. Int J CritAl Infrastruct Prot 25:36–49
13. Kumar V, Jangirala S, Ahmad M (2018) An efficient mutual authentication framework for healthcare system in cloud computing. J Med Syst 42:1–25
14. Li X, Wu F, Kumari S, Xu L, Sangaiah AK, Choo K-KR (2019) A provably secure and anonymous message authentication scheme for smart grids. J Parallel Distrib Comput 132:242–249
15. Mahmood K, Chaudhry SA, Naqvi H, Kumari S, Xiong L, Sangaiah AK (2018) An elliptic curve cryptography based lightweight authentication scheme for smart grid communication. Futur Gener Comput Syst 81:557–565

16. Moghadam MF, Nikooghadam M, Mohajerzadeh AH, Movali B (2020) A lightweight key management protocol for secure communication in smart grids. Electr Power Syst Res 178:106024
17. Shitharth S, Prince Winston D . A novel IDS technique to detect DDoS and sniffers in smart grid. In: 2016 world conference on futuristic trends in research and innovation for social welfare (Startup Conclave). IEEE, pp 1–6
18. Stinson DR (2006) Some observations on the theory of cryptographic hash functions. Des Codes Cryptogr 38:259–277
19. Yaghmaee MH, Leon-Garcia A, Moghaddassian M (2017) On the performance of distributed and cloud-based demand response in smart grid. IEEE Trans Smart Grid 9(5):5403–5417

Voltage Profile Improvement Using DSTATCOM in Three-Phase Unbalanced Radial Distribution System

Abhishek Gautam, Ashwani Kumar, and Sukriti Tiwari

Abstract The distribution system (DS) is a multi-phase system with a high R/X ratio and operates in a radially or weakly meshed topology. For such a complex system, a reliable and effective load flow algorithm is necessary to carry out real-time operations, including network optimization, VAr planning, voltage profile improvement, and state estimation. The incorporation of FACTS devices like DSTATCOM can actually aid in the voltage profile improvement of DS. Therefore, this study attempts to verify the precision of the load flow algorithm in DS, which uses a forward-backward sweep (FBS) approach after incorporating a DSTATCOM for voltage profile improvement. The proposed approach includes a sensitivity analysis to locate the most sensitive buses on the three-phase unbalanced IEEE-33 bus radial distribution system/network (RDS). Further, the probabilistic reliability models are used to calculate the reliability at different load locations. Once the sensitivity analysis is done, the voltage profile of the unbalanced IEEE-33 bus system is improved by incorporating DSTATCOM at an optimal location in reconfigured DS. The proposed approach also enables the estimation of voltage magnitudes and voltage angles at each bus and power losses in DS. All the results are validated on the modified IEEE-33 bus and indicate a significant improvement in the overall voltage profile of the DS, especially the most sensitive buses identified by sensitivity analysis.

Keywords Backward forward sweep · Distribution system · Data structure · Sensitivity factor · DSTATCOM

A. Gautam (✉) · A. Kumar · S. Tiwari
National Institute of Technology Hamirpur, H.P, India
e-mail: abhishekgautam1885@gmail.com

A. Kumar
e-mail: ashchandelin@nith.ac.in

S. Tiwari
e-mail: sukrititiwari@nith.ac.in

A. Kumar et al. (eds.), *Decarbonisation and Digitization of the Energy System*, Lecture Notes in Electrical Engineering 1099, https://doi.org/10.1007/978-981-99-7630-0_15

1 Introduction

The load flow analysis is crucial for power system operation and planning, contingency analysis, and state estimation. Due to the high R/X ratio in DS, traditional load flow methods like Gauss–Seidel (GS) and Newton–Raphson (NR) are unable to converge to an optimal solution [1]. Likewise, the DS is generally unbalanced due to single-phase, two-phase, and three-phase branches supplying unbalanced loads [2, 3]. The literature survey indicates that the flow of three-phase power systems in transmission networks has been widely investigated [4]. However, various studies like load flow studies, voltage profile improvement, and power compensation in unbalanced DS are emerging research areas [3]. In [5], the load flow study of DS used a phase component methodology based on the NR method to divide the power flow into each phase. This method, however, needs a considerable quantity of computer memory to create a simultaneous solution to an admittance matrix of order (6n $\times$ 6n) [6].

Identifying the node voltages is the primary step in the distribution system load flow analysis [7]. This is because the direct computation of current, power flows, system losses, and other steady-state variables is possible using these voltages [8]. Further applications such as VAr planning, network optimization, state estimation, and distribution automation require load flow studies to be performed again and again to determine the security status of DS [9]. The approach using the FBS notion is proved to be the most effective and quick for carrying out the load flow studies of RDS. This technique represents the DN as a tree, with the slack bus at the root [7]. In this method, during the backward sweep, the power or line currents flowing from the extremities to the root are predominantly added. Based on the estimated currents of the flows, the forward sweep computes the voltage drop and updates the voltage profile [7]. Once the voltage profile is computed, it will give a graphical representation of the nodal voltage presented in the power system [10]. However, there is a limitation on how much voltage can decrease at a particular node at the consumer end [10], which is generally $\pm 10\%$ of the nominal value. The voltage at the farthest end is kept under the permissible limit using the following devices:

- On-load Tap changing (OLTC) Transformer
- Bus step-voltage Regulator
- Voltage Booster or FACTS devices
- Feeder Voltage Regulators (Induction Regulator)
- Fixed and Switched Capacitor

Voltage sensitivity analysis is the pillar for evaluating a comprehensive range of power system optimization issues, including voltage control, loss reduction, network expansion planning [11], best locations for reactive sources, FACTS devices, generators, etc. Sensitivity analysis is so far utilized, specifically at the transmission level, to ascertain the voltage stability margin in a transmission network [8]. Subsequently, the present work utilizes sensitivity analysis to locate the most sensitive buses and, consequently, the voltage profile improvement of the unbalanced

IEEE-33 bus DS is achieved by incorporating a FACTS device called DSTATCOM at the optimal location in reconfigured DS.

This article is structured as follows: The system under study is explained in Sect. 2. Section 3 elaborates on the typical distribution system modelling and the proposed methodology for voltage profile improvement of DS is described. In Sect. 4, study outcomes are discussed. Section 5 concludes the work and provides the future scope.

2 System Configuration

The last segment of the power system, i.e., DS, is responsible for transferring electrical energy from the transmission system to specific users [3]. In this study, the IEEE 33-bus RDS is scrutinized for validating the proposed methodology for voltage profile improvement. The referred DS comprises 33 buses, 32 lines, 1 generator, a voltage source of 12.66 kV with a load capacity of 3.715 MW, and 2.3 MVAr [12].

The DS buses are linked to one another in a radial pattern, with one bus followed by another. In a similar fashion, buses (19, 20, 21, 22) are linked via bus 2, buses (23, 24, 25) are connected through bus 3, and buses (26, 27, 28, 29, 30, 31, 33) are connected through bus 6. The single-line diagram (SLD) of the IEEE-33 bus RDS is given in Fig. 1a to provide an idea of the DS topology. However, the actual network comprises several buses that are inherently unbalanced. To give an idea of an unbalanced system let us consider an example of a three-phase six-bus RDS as shown in Fig. 1b. In Fig. 2, buses 1, 2, and 3 have all three phases (a, b, and c) represented by (1a, 1b, and 1c), (2a, 2b, and 2c), and (3a, 3b, and 3c), respectively. Alternatively, bus 4 only has two phases, denoted by (4a, 4b), whereas bus 5 has one phase, i.e., (5b), and bus 6 has a c-phase, represented as bus (6c).

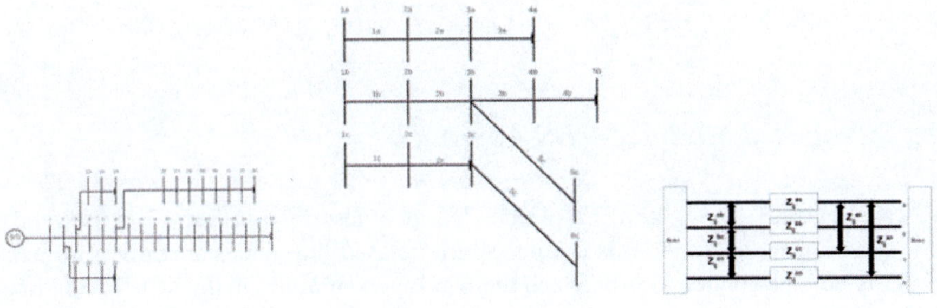

(a) SLD of IEEE-33 bus RDS [5]

(b) Three Phase Unbalanced IEEE-6 bus RDS [5]

(c) Distribution System Layout [12]

Fig. 1 IEEE-standard RDS representation

3 System Modeling

Figure 1c depicts a model of a three-phase, four-wire line segment between bus i and bus j of DS. The Carson and Lewis approach is used to determine the parameters of the line [12]. In Fig. 1a, the correlation between bus voltage and branch currents is given by the equation as follows:

$$\begin{bmatrix} V_i^a \\ V_i^b \\ V_i^c \end{bmatrix} = \begin{bmatrix} V_j^a \\ V_j^b \\ V_j^c \end{bmatrix} + \begin{bmatrix} Z_{ij}{}^{aa-n} & Z_{ij}{}^{ab-n} & Z_{ij}{}^{ac-n} \\ Z_{ij}{}^{ba-n} & Z_{ij}{}^{bb-n} & Z_{ij}{}^{bc-n} \\ Z_{ij}{}^{ca-n} & Z_{ij}{}^{cb-n} & Z_{ij}{}^{cc-n} \end{bmatrix} \begin{bmatrix} I_{ij}{}^a \\ I_{ij}{}^b \\ I_{ij}{}^c \end{bmatrix} \tag{1}$$

The two-phase line sections comprise of phases a–b, b–c, and a–c:

$$\begin{bmatrix} V_i^a \\ V_i^b \end{bmatrix} = \begin{bmatrix} V_j^a \\ V_j^b \end{bmatrix} + \begin{bmatrix} Z_{ij}{}^{aa-n} & Z_{ij}{}^{ab-n} \\ Z_{ij}{}^{ba-n} & Z_{ij}{}^{bb-n} \end{bmatrix} \begin{bmatrix} I_{ij}{}^a \\ I_{ij}{}^b \end{bmatrix} \tag{2}$$

$$\begin{bmatrix} V_i^b \\ V_i^c \end{bmatrix} = \begin{bmatrix} V_j^b \\ V_j^c \end{bmatrix} + \begin{bmatrix} Z_{ij}{}^{bb-n} & Z_{ij}{}^{bc-n} \\ Z_{ij}{}^{cb-n} & Z_{ij}{}^{cc-n} \end{bmatrix} \begin{bmatrix} I_{ij}{}^b \\ I_{ij}{}^c \end{bmatrix} \tag{3}$$

$$\begin{bmatrix} V_i^a \\ V_i^c \end{bmatrix} = \begin{bmatrix} V_j^a \\ V_j^c \end{bmatrix} + \begin{bmatrix} Z_{ij}{}^{aa-n} & Z_{ij}{}^{ac-n} \\ Z_{ij}{}^{ca-n} & Z_{ij}{}^{cc-n} \end{bmatrix} \begin{bmatrix} I_{ij}{}^a \\ 5 \end{bmatrix} \tag{4}$$

The line sections with a single phase are represented as

$$V_i^a = V_j^a + Z_{ij}^{aa-n} I_{ij}^a \tag{5}$$

$$V_i^b = V_j^b + Z_{ij}^{bb-n} I_{ij}^b \tag{6}$$

$$V_i^c = V_j^c + Z_{ij}^{cc-n} I_{ij}^c \tag{7}$$

3.1 Proposed Modified FBS Approach

To carry out this study, the IEEE-33 bus DS also described in Sect. 2 is used as a test system. The data model is constructed by classifying nodes according to their quantity and the connections between them as layers of nodes in the distribution line [13, 14].

3.2 Voltage Sensitivity Analysis

Let's look at the voltage at the i-th node, influenced by the net reactive currents and net active currents at all network nodes:

$$V_i = V_i(I_{p1}, I_{p2}, \ldots\ldots, I_{pn}, I_{q1}, I_{q2}, \ldots\ldots, I_{qn}) \tag{8}$$

The whole differential of function V_i may be expressed as follows:

$$dV_i = \sum_{j=1}^{n} \frac{\partial V_i}{\partial I_{pj}}.dI_{pj} + \sum_{j=1}^{n} \frac{\partial V_i}{\partial I_{qj}}.dI_{qj} \tag{9}$$

The derivatives $\frac{\partial V_i}{\partial I_{pj}}.dI_{pj}$ and $\frac{\partial V_i}{\partial I_{qj}}.dI_{qj}$ are denoted as voltage sensitivity coefficients with regard to the variation in node currents.

Taking into consideration the *n* equations that are provided by equation (15), we now have:

$$[dV] = [S][dI] \tag{10}$$

where [S] is the sensitivity matrix with the dimensions (n × n), and the components of this matrix represent the sensitivity coefficients:

$$[S] = \left[\begin{array}{ccc|ccc} \frac{\partial V_1}{\partial I_{p1}} & \cdots & \frac{\partial V_1}{\partial I_{pn}} & \frac{\partial V_1}{\partial I_{q1}} & \cdots & \frac{\partial V_1}{\partial I_{qn}} \\ \cdots & \cdots & \cdots & \cdots & \cdots & \cdots \\ \frac{\partial V_n}{\partial I_{p1}} & \cdots & \frac{\partial V_n}{\partial I_{pn}} & \frac{\partial V_n}{\partial I_{q1}} & \cdots & \frac{\partial V_n}{\partial I_{qn}} \end{array}\right] = [S_{Ip}|S_{Iq}] \tag{11}$$

After considering the factors called Constant Current Model(CCM) and Simplified Constant Current Model(SCCM) in this methodology that will be detailed in the upcoming sections, it is feasible to calculate the coefficients of voltage sensitivity.

3.3 CCM Voltage Sensitivity Coefficients

Let us consider the voltage dependency at the *i*th node from the perspective of CCM:

$$\begin{aligned}(S_{Ip})_{ij} = \left(\frac{-V_0}{V_i}\right).R_{ij} + \frac{1}{V_i}\Bigg[R_{ij}.\left(\sum_{k=1}^{n} R_{ik}.I_{pk} + \sum_{k=1}^{n} X_{ik}.I_{qk}\right) \\ +X_{ij}.\left(\sum_{k=1}^{n} X_{ik}.I_{pk} + \sum_{k=1}^{n} R_{ik}.I_{qk}\right)\Bigg]\end{aligned} \tag{12}$$

Similarly, it is also feasible to determine the analytical equation for using equivalent computations for $(S_{Iq})_{ij}$.

$$(S_{Iq})_{ij} = \left(\frac{-V_0}{V_i}\right).X_{ij} + \frac{1}{V_i}\left[X_{ij}.\left(\sum_{k=1}^{n} X_{ik}.I_{qk} + \sum_{k=1}^{n} R_{ik}.I_{pk}\right)\right.$$
$$\left. + R_{ij}.\left(\sum_{k=1}^{n} R_{ik}.I_{qk} + \sum_{k=1}^{n} X_{ik}.I_{pk}\right)\right] \quad (13)$$

3.4 SCCM Voltage Sensitivity Coefficients

This section contains concise formulations for node voltages and voltage sensitivity coefficients using an SCCM [17]. Further, SCCM permits the derivation of sensitivity coefficients $(S_{Ip})_{ij}$ and $(S_{Iq})_{ij}$ as functions of linear transverse characteristics (i.e., line resistance and reactance per kilometer) as represented by following equations:

$$(S_{Ip})_{ij} = -L_{ij}.r_{ij} \quad (14)$$

$$(S_{Iq})_{ij} = -L_{ij}.x_{ij} \quad (15)$$

3.5 Voltage Sensitivity Coefficients Calculations

The difference between voltage sensitivity coefficients using the CCM and SCCM is subsequently analyzed to assess their impact on voltage sensitivity analysis.

$$\Delta(S_{Ip})_{ij}\% = \frac{|(S_{Ip(SCCM)})_{ij}| - |(S_{Ip(CCM)})_{ij}|}{|(S_{Ip(SCCM)})_{ij}|} \cdot 100 \quad (16)$$

$$\Delta(S_{Iq})_{ij}\% = \frac{|(S_{Iq(SCCM)})_{ij}| - |(S_{Iq(CCM)})_{ij}|}{|(S_{Iq(SCCM)})_{ij}|} \cdot 100 \quad (17)$$

3.6 Three-Phase Modelling of DSTATCOM for Voltage Profile Improvement

The major mathematical equations indicating the reactive power compensation (Q_{FACT}) by DSTATCOM are given next [19]:

$$Q_{FACTa} = \left(\frac{V_{ia}^2}{X_L}\right) - \left(\frac{V_{ia} \cdot V_{fa}}{X_L}\right) cos\delta \tag{18}$$

$$Q_{FACTb} = \left(\frac{V_{ib}^2}{X_L}\right) - \left(\frac{V_{ib} \cdot V_{fb}}{X_L}\right) cos\delta \tag{19}$$

$$Q_{FACTc} = \left(\frac{V_{ic}^2}{X_L}\right) - \left(\frac{V_{ic} \cdot V_{fc}}{X_L}\right) cos\delta \tag{20}$$

where V_i is the bus voltage for each phase, i.e., phase-a, phase-b, phase-c, and V_f is the DSTATCOM voltage for each phase. X_L is the line reactance and δ is the phase angle displacement between V_i and V_f.

3.7 Flow Chart of Proposed Method

The flowchart of the proposed methodology is depicted in Fig. 2. This methodology, in general, considers reading the data from the DS (generators, loads, and lines), organizing the data, and power flow processing via the FBS technique [7]. The technique for computing the current and voltage at the node is repeated using the current node's layer number, the layer number of the node in the previous layer and connected to the referred node, and also the node numbers associated with the rest of the preceding layers [15]. The procedure that has been explained is the foundation for the formulation of the sensitivity factor [8].

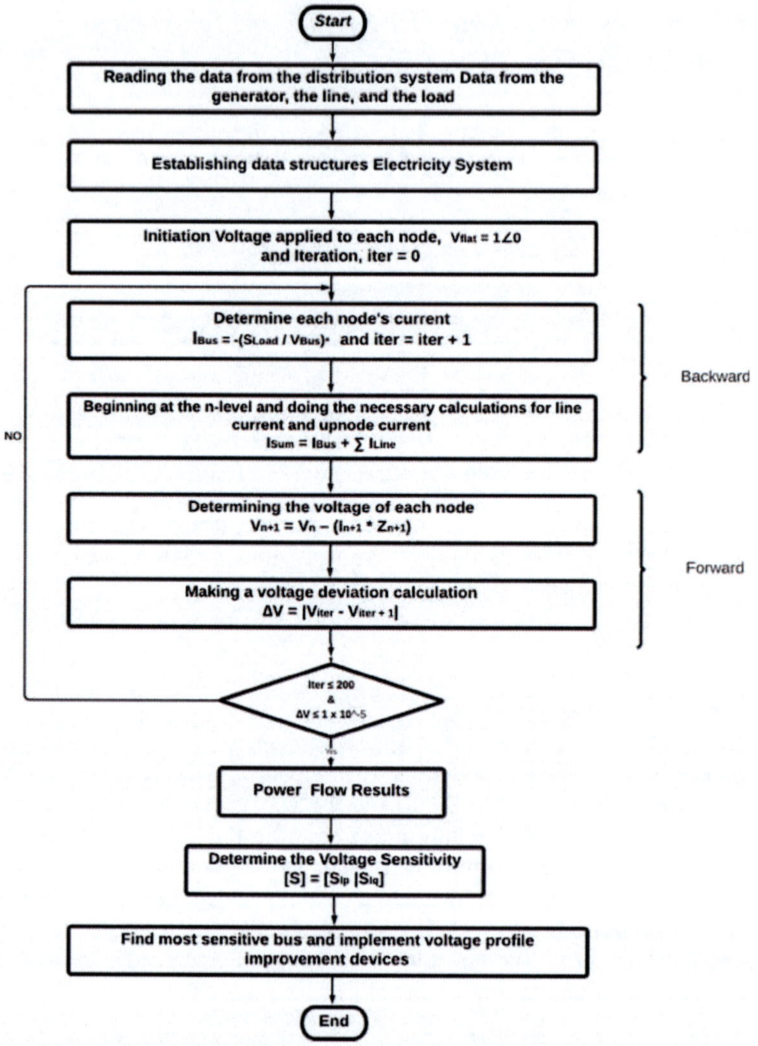

Fig. 2 Flow chart of proposed method

3.8 Data Formulation

The IEEE 33-node test DS is used in the simulation, and a maximum of 100 iterations and an error deviation of 1×10^{-5} are allowed. To carry out this study, a predetermined tolerance threshold of voltage magnitude is essential [6, 9]. This voltage magnitude tolerance has been set at $\pm 10\%$ [9], as seen in Fig. 3a. Here, the node voltages vary within the specified threshold. The voltage profile for the unbalanced

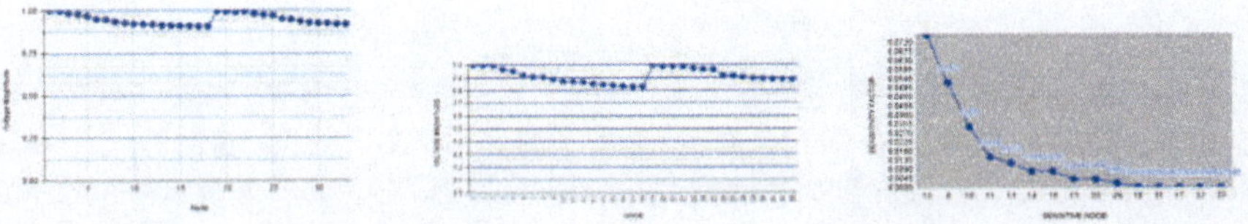

(a) (b) (c)

Fig. 3 **a** Standard case, **b** and **c**-Modified case

system is shown in Fig. 3b which highlights the violation of threshold limits by some of the buses. Subsequently, under such conditions, there is a need to optimally incorporate the voltage enhancement device, i.e., DSTATCOM [20–23] in DS to keep the system voltages within limits. The sensitivity analysis as described in previous sections is now carried out with data from modified (unbalanced) IEEE-33 bus DS. This analysis has been carried out for the buses which have exceeded the specified tolerance limit of $\pm 10\%$ for voltage magnitude and shown in Fig. 3b. These buses with bus numbers 9, 10, 11, 12, 13, 14, 15, 16, 17, 18, 29, 30, 31, 32, and 33 are identified as sensitive buses of modified RDS. Figure 3c represents the calculated sensitivity factors for the above buses, which identify buses 17, 18, 31, 32, and 33 are the most sensitive buses. Among the above five most sensitive buses, the most optimal bus is located in accordance with reference [16] for the installation of DSTATCOM for further analysis of the modified IEEE-33 bus DS.

4 Results and Discussion

In this study, the voltage profile of modified IEEE-33 bus DS is improved using DSTATCOM, which is generally placed near the load in DS. To accomplish this objective, firstly, load flow analysis for standard IEEE-33 bus RDS is carried out. This analysis provides the voltage profile of each bus, as shown in Fig. 4a, representing the graph between voltage magnitude and nodes of IEEE-33 bus RDS. Next, load flow analysis of IEEE-33 bus RDS under unbalanced conditions, as described in Sect. 4 has been done. The voltage profile in Fig. 3b shows that several buses, i.e., 9, 10, 11, 12, 13, 14, 15, 16, 17, 18, 29, 30, 31, 32, and 33, have exceeded the specified limit of $\pm 10\%$ for the nominal voltage. Therefore, these buses are now referred to as sensitive buses. Subsequently, the sensitivity factor analysis has been done on the above-mentioned weak buses, and the graphical representation is provided in Fig. 3c. Sensitivity factor analysis indicates that among these buses, the most sensitive buses are 17, 18, 31, 32, and 33 in unbalanced RDS. Thus, there is a need to incorporate FACTS devices like DSTATCOM in such a system to improve the overall voltage profile of the RDS. Now, to install DSTATCOM, the optimal location has been considered as per reference [16], which is found to be bus number 17. Therefore, DSTATCOM has been installed at bus 17 (Fig. 4a) to provide reactive power support

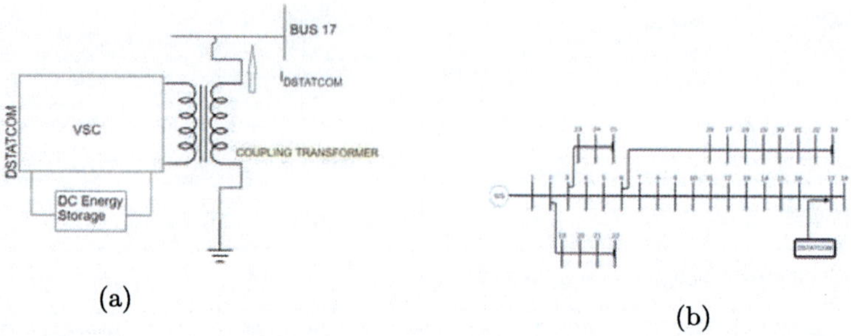

Fig. 4 DSTATCOM placement. **a** DSTATCOM connected to Bus-17 in IEEE-33 bus unbalanced RDS. **b** DSTATCOM placement at optimal location

in the modified RDS [11]. Further, the incorporation of DSTATCOM in the IEEE-33 bus system is shown in Fig. 4b.

To find the capacity of DSTATCOM, several scenarios that involve enhancing the reactive power capacity of bus 17 in the interval of 10% have been considered. To explain the process, two scenarios are explained next:

- $Q_{FACT} = 1.5Q_{17}$
- $Q_{FACT} = 1.1Q_{17}$

In the first scenario, the reactive power support at the weakest node 17 is being planted and is enhanced by installing a DSTATCOM with 50% of its original value in an unbalanced radial distribution system. Similarly, in the second scenario, reactive power support increased by 10% of its original value in the unbalanced RDS.

4.1 *Case-I:* $Q_{FACT} = 1.5Q_{17}$

A tolerance of ±10% in the nominal voltage is assumed here. Figure 5a depicts the magnitude of the voltage at each node; it can be concluded that there has been no violation of any limitations at any buses in this case.

Consequently, Fig. 5b depicts the sensitive nodes, and reactive power has been added to the node labeled as Q_{17}. This strategy will ensure that the reactive power requirements are satisfied; nevertheless, some of the sensitive nodes may experience an excess of reactive power, which serves no purpose and will begin to cause disturbances. To avoid these disturbances, in the next case, we have considered installing DSTATCOM, which provides 10% enhance reactive power support at bus 17. The comparison between the old sensitivity factor (when there was no incorporation of reactive power support) with the new sensitivity factor(after incorporation of reactive power support i.e., $Q_{FACT} = 1.5Q_{17}$) is shown in Fig. 5c.

4.2 Case-II: $Q_{FACT} = 1.1 Q_{17}$

In case II, the reactive power at the Q_{17} node is enhanced by 10% using DSTATCOM. The voltage profile shown in Fig. 6a clearly indicates that there is no violation of voltage limits on buses. Further, there will be the least possibility of excess reactive power at any of the buses that can cause disturbances in the RDS.

Therefore, the graph between the sensitivity factor and the corresponding sensitive node is shown in Fig. 6b. From Fig. 6b, superior outcomes have been achieved by installing a DSTATCOM which enhances reactive power by 10% of its nominal value at node Q_{17} of modified/unbalanced IEEE-33 bus RDS. This has subsequently improved the overall voltage profile of unbalanced RDS.

For finding the most optimum bus to install DSTATCOM, sensitivities at different conditions are considered in this study, as explained in the sections above. The graph shown in Fig. 6b shows the relationship between sensitivity factor 1 (after incorporating reactive power support, i.e., $Q_{FACT} = 1.5 Q_{17}$), sensitivity factor 2 (after incorporating reactive power support, i.e., $Q_{FACT} = 1.1 Q_{17}$), and sensitivity factor (old). The sensitivity factor (old) is calculated by collecting data from a three-phase unbalanced radial distribution system without DSTATCOM. Similarly, sensitivities for all the weak nodes are calculated for the case $Q_{FACT} = 1.1 Q_{17}$, and is shown in Fig. 6b. Figure 6c illustrates all the three sensitivity variables corresponding to all the weak nodes, i.e., node numbers 9, 10, 11,12, 13, 14, 15, 16, 17, 18, 29, 30, 31, 32, and

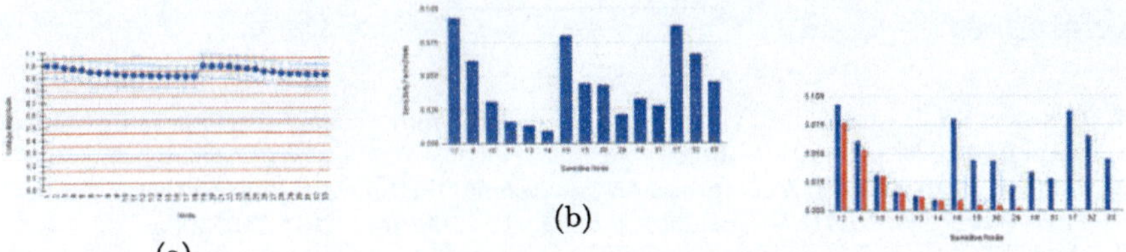

(a) (b) (c)

Fig. 5 **a**, **b**-Case: $Q_{FACT} = 1.5 Q_{17}$. **a** Nature of voltage magnitude and node after incorporating reactive power support. **b** Graph between sensitivity factor and sensitive node after incorporating reactive power support. **c** Graph between sensitivity factor(New) and sensitivity factor(Old)

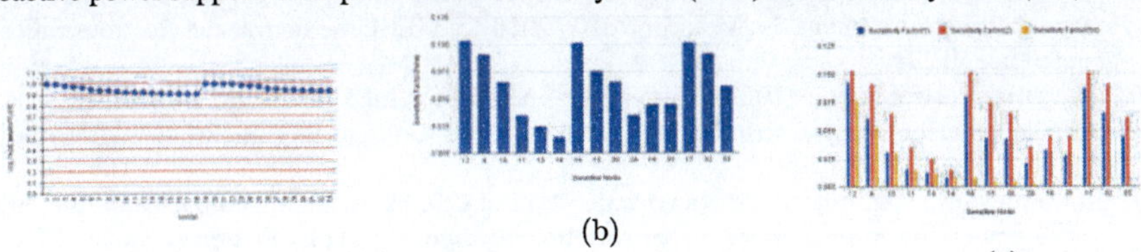

(a) (b) (c)

Fig. 6 **a**, **b**-Case: $Q_{FACT} = 1.1 Q_{17}$, **c**-Graph between sensitivity factors. **a** Nature of voltage magnitude and node after incorporating reactive power support. **b** Graph between sensitivity factor and sensitive node after incorporating reactive power support. **c** Graph between sensitivity factor(1), sensitivity factor(2) and sensitivity factor(Old)

33. Therefore, the analysis indicates that on installing DSTATCOM at node 17 so as to enhance reactive power by 10% results in the overall voltage profile improvement of modified RDS.

5 Conclusion

In this work, voltage profile enhancement for three-phase RDS is done under unbalanced conditions using sensitivity analysis and optimally incorporating DSTATCOM. Further, simple algebraic equations for determining and characterizing voltage sensitivity coefficients in relation to shifts in active and reactive power throughout every node of an RDS are proposed. The voltage sensitivity coefficients concerning the cumulative active and reactive components of the node currents are calculated by employing the CCM and the SCCM. Subsequently, the primary goal of providing reactive power support to underperforming buses as determined by sensitive analysis for an unbalanced load RDS is successfully achieved. The results indicate the overall voltage profile improvement of the modified IEEE-33 bus RDS.

The future scope of this work includes the extension of this study for higher-order systems. Further, the voltage profile analysis with distributed energy resources can be carried out for different IEEE systems.

References

1. Goswami SK, Basu SK (1991) Direct solution of distribution systems. IEE Proc 138:78. https://doi.org/10.1049/ip-c.1991.0010
2. Jabr RA, Dzafic I (2016) A compensation-based conic OPF for weakly meshed networks. IEEE Trans Power Syst 31:4167–4168. https://doi.org/10.1109/tpwrs.2015.2505508
3. Patel DK, Mansani S, Goswami AK (2022) Fast flexible direct power flow for unbalanced and balanced distribution systems. Int Trans Electr Energy Syst 2022:1–18. https://doi.org/10.1155/2022/2857358
4. Tiwari S, Vaddi N, Bhatlu S, Metta M (2020) Optimal power flow solution with nature-inspired antlion meta-heuristic algorithm. J Phys Conf Ser
5. Mahmoud K, Abdel-Akher M (2010) Efficient three-phase power-flow method for unbalanced radial distribution systems. In: Melecon 2010 - 2010 15th IEEE mediterranean electrotechnical conference. IEEE
6. Gianto R, Purwoharjono (2022) Trust-region method for load flow solution of three-phase unbalanced electric power distribution system. J Electr Comput Eng 2022:1–17. https://doi.org/10.1155/2022/5415300
7. Kawambwa S, Mwifunyi R, Mnyanghwalo D, et al (2021) An improved backward/forward sweep power flow method based on network tree depth for radial distribution systems. J Electr Syst Inf Technol 8. https://doi.org/10.1186/s43067-021-00031-0
8. Alzaareer K, Saad M, Mehrjerdi H, et al (2022) New voltage sensitivity analysis for smart distribution grids using analytical derivation: ABCD model. Int J Electr Power Energy Syst 137:107799. https://doi.org/10.1016/j.ijepes.2021.107799

9. Tyagi A, Kumar K, Ansari MA, Kumar B (2020) An efficient load flow solution for distribution system with addition of distributed generation using improved harmony search algorithms. J Electr Syst Inf Technol 7. https://doi.org/10.1186/s43067-020-00014-7
10. Carreno IL, Scaglione A, Saha SS, et al (2023) Log(v) 3LPF: a linear power flow formulation for unbalanced three-phase distribution systems. IEEE Trans Power Syst 38:100–113. https://doi.org/10.1109/tpwrs.2022.3166725
11. Tiwari S, Kumar A (2022) Optimal micro-PMUs placement with channel limits using dynamically controlled Taguchi binary particle swarm optimization. Electr Power Compon Syst 50:1072–1086. https://doi.org/10.1080/15325008.2022.2145392
12. Al-Sakkaf A, AlMuhaini M (2018) Power flow analysis of weakly meshed distribution network including DG. Eng Technol Appl Sci Res 8:3398–3404. https://doi.org/10.48084/etasr.2277
13. Kersting WH (1976) An application of ladder network theory to the solution of three-phase radial load-flow problems
14. Martinot E, Kristov L, Erickson JD (2015) Distribution system planning and innovation for distributed energy futures. Curr Sustain/Renew Energy Rep 2:47–54. https://doi.org/10.1007/s40518-015-0027-8
15. Alves G de O, Pereira JLR, Passos Filho JA (2020) A new unbalanced three-phase governor power flow formulation based on the current injections method. Int J Electr Power Energy Syst 123:106184. https://doi.org/10.1016/j.ijepes.2020.106184
16. Tahir MJ, Rasheed MB, Rahmat MK (2022) Optimal placement of capacitors in radial distribution grids via enhanced modified particle swarm optimization. Energies 15:2452. https://doi.org/10.3390/en15072452
17. Singh H, Bhuria V (2022) Optimal placement of DG and capacitor in radial distribution networks using differential evolution. Artificial Intelligence and Sustainable Computing. Springer Singapore, Singapore, pp 401–408
18. Tiwari S, Kumar A (2020) A robust Taguchi particle swarm optimization approach for network reconfiguration with distributed generation. In: 2020 IEEE 7th Uttar pradesh section international conference on electrical, electronics and computer engineering (UPCON). IEEE
19. Bagheri Tolabi H, Ali MH, Rizwan M (2015) Simultaneous reconfiguration, optimal placement of DSTATCOM, and photovoltaic array in a distribution system based on fuzzy-ACO approach. IEEE Trans Sustain Energy 6:210-218. https://doi.org/10.1109/tste.2014.2364230
20. Tiwari S, Kumar A, Sengupta S (2018) Voltage stability analysis with a novel hybrid controller using shunt and series combination. of FACTS device. In: 2018 3rd IEEE international conference on recent trends in electronics, information and communication technology (RTEICT). IEEE
21. Tamilselvan V, Jayabarathi T, Raghunathan T, Yang X-S (2018) Optimal capacitor placement in radial distribution systems using flower pollination algorithm. Alex Eng J 57:2775–2786. https://doi.org/10.1016/j.aej.2018.01.004
22. Tiwari S, Kumar A (2023) Advances and bibliographic analysis of particle swarm optimization applications in electrical power system: concepts and variants. Evol Intell 16:23–47. https://doi.org/10.1007/s12065-021-00661-3
23. JChen J, Hu W (2021) MATLAB simulation research on static var compensator. E3S Web Conf 256:01022. https://doi.org/10.1051/e3sconf/202125601022

A Case Study with Analysis for Photovoltaic Array Under Shaded Conditions

Sandeep Gupta, Tarun Varshney, T. Kranthi Kumar, and S. Anand

Abstract Solar power has been a major force in power generation in the modern world because it's one of the cleanest forms of energy and is easy to harvest and put to use today. The sun is abundantly everywhere on the earth, which is why the future of renewable energy is going to be more solar power. Shading conditions are generally a major problem in photovoltaic systems because they significantly affect the performance of PV systems, hence the need to study different shading conditions. The present work carries out a simulation of various PV modules connected with bypass diodes under different shading conditions by creating faults, these conditions are partially compared with the voltage, current, and power visualized from the scope over time in the form of a graph. The simulation software used in the present work is MATLAB Simulink.

Keywords PV cell · PV array · PV module · MATLAB simulink · Shading condition · Bypass diode

S. Gupta (✉)
Department of Electrical Engineering, Graphic Era (Deemed to be University), Dehradun, Uttarakhand, India
e-mail: jecsandeep@gmail.com

T. Varshney
Electrical Electronics and Communication Engineering, SET, Sharda University, Greater Noida, India
e-mail: tarun.varshney@sharda.ac.in

T. K. Kumar
Department of Science and Humanities, Sreenidhi Institute of Science and Technology, Hyderabad, India
e-mail: kranthi9614@gmail.com

S. Anand
Electrical and Electronics Engineering, Nitte Meenakshi Institute of Technology Yelahanka, Bangalore, India
e-mail: anand.s@nmit.ac.in

A. Kumar et al. (eds.), *Decarbonisation and Digitization of the Energy System*, Lecture Notes in Electrical Engineering 1099, https://doi.org/10.1007/978-981-99-7630-0_16

1 Introduction

Renewable sources provide energy that is clean, inexhaustible, and more affordable. They are differentiated from fossil fuels largely by their diversity, quantity, and capacity to be utilized wherever on Earth, but most crucially by their inability to create greenhouse gases or other damaging pollutants [1–3]. In contrast to fossil fuels, whose prices are usually down despite their present volatility, their expenses are also decreasing at a steady rate. Renewable energy sources continue to become the most cost-effective new technology for producing electricity in many regions of the world [4]. Renewable energy is the only way to enhance energy access for all developing countries. The IEA predicts that the global demand for electricity will climb by 70% by 2040, while its proportion of overall energy consumption will increase from 18 to 24%. Its expansion will be driven mostly by the growing economies of India, China, Africa, the Middle East, and South-East Asia [5].

The operating conditions and field factors are the main factors affecting the performance of a PV module, such as the geometric location, orientation towards the sun, temperature, and irradiation levels. Solar cells interconnected in parallel and series together form the PV array, these PV arrays are the most important hardware in a complete PV system because they harvest the sunlight which is converted into electricity. When there is shading on any PV module, it affects the entire performance of the PV system, but this shading effect can be reduced by connecting bypass diodes as several research has shown [6–8, 12]. In the present work, shading conditions in the PV module have been created and demonstrated by creating several faults and fairly comparing them with the output obtained from the scope during the simulations. In Simulink, the solar cell can be modelled with instruments that can implement any differential equation or algebraic relationship of a highly complex mathematical model [9, 10].

2 Mathematical Model

A photovoltaic cell can be represented by an analogous circuit. Several models are utilized in the literature, and they vary in complication and exactness, as well as how they operate under different conditions, Because of its simplicity and precision, the one-diode model is most often used to depict a PV cell [11, 13]. A photocurrent source (I_{ph}), a diode (D), an internal resistance (R_s), and a shunt resistor (R_h) are shown in Fig. 1. The shading effect on solar modules employing this diode is studied using the model with a bypass diode (D_b). Diode current (I_d), shunt current (I_h), and by-bass diode current (I_{db}) and (I_s) are basically the current going through the terminals. Current and Voltage are denoted as I and V respectively.

We know that under normal operating conditions, the open-circuit voltage (V_{oc}) of the photovoltaic module is 0.5–0.6 V at 25 °C, meanwhile the voltage in reverse bias operation can be up to −20 V [14–16].

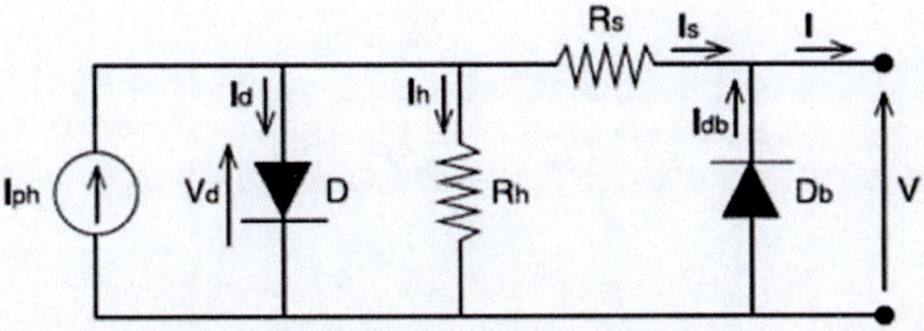

Fig. 1 Photovoltaic model with bypass diode

The current which is produced by the PV model is given by Kirchhoff's law:

$$I = I_{ph} - I_d - I_h + I_{db} \tag{1}$$

$$I_h = (V + I_s R_s)/R_h = (V + R_s(I - I_{db}))/R_h \tag{2}$$

$$I_{db} = I_{sat\ db}\left(e^{(-v/vt\ db)} - 1\right) \tag{3}$$

$$I = I_{ph} - I_{sat\ d}\left(e^{(vd/vtd)} - 1\right) - \left(V + R_s(I - I_{db})\ /R_h + I_{sat\ db}\left(e^{(-v/vt\ db)} - 1\right)\right. \tag{4}$$

I_{db} and V_t, which stand for current and thermal voltage, respectively, are the values that describe the bypass diode. Equations (1)–(4) provide a clear and compelling illustration of the fact that the current I flowing through the PV is a non-linear, implicit function of the voltage V, as well as the temperature and the amount of irradiation [6, 18]. Even yet, a non-linear system like this can be resolved in a computational environment like MATLAB, this way current (I) and voltage (V) non-linear relation can be obtained [17]. This relationship is shown by Eq. (5) [19, 20]. It employs the Lambert function W for the term, whose value is dependent on the voltage V and is given in Eq. (6). The current–voltage (I-V) properties of a typical silicon photovoltaic (PV) cell are shown in Fig. 2, which shows the cell working under normal conditions.

$$I = \frac{R_s(I_p + I_{sat\,d}) - V}{R_s + R_s} + I_{sat\,db}(e^{-v/vt\,db} - 1) - \frac{V_{t\_d}}{R_s}\text{Lambert W}(\theta) \tag{5}$$

with

$$\theta = \frac{(R_h R_s/(R_h + R_s))I_{sat\,d}e^{[R_h R_s(I_{ph\_}I_{sat\,d}) + R_h V/V_t\,d(R_h + R_s)]}}{V_{T\,d}} \tag{6}$$

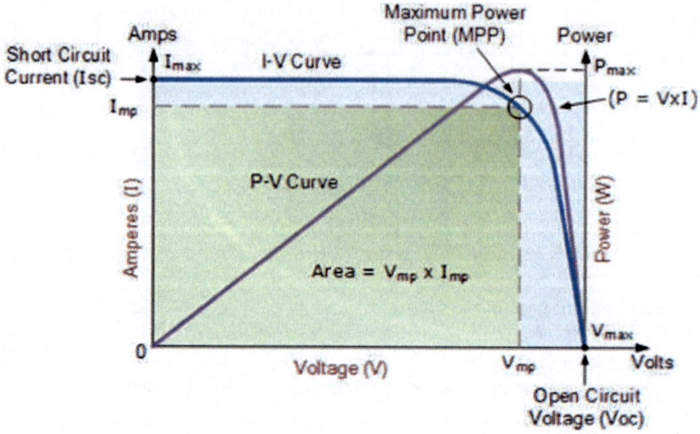

Fig. 2 Solar cell I-V typical curve [21]

3 Modelling of PV Array and PV Module

Equations (1)–(6) can be used to develop a model for the PV array and PV module. The present work is done on 6 PV modules in which 3 are connected in series in parallel to the other 3 which are also connected in series. The PV array chosen has a maximum power of 200 watts at 26.3 V and it's a KYOCERA solar KC 200GT PV module, the PV module specifications are given below in Table 1 [19].

The table shows the parameters that are used in the simulation of the PV modules whereby these parameters can be changed in the future depending on the simulation requirements.

Table 1 Kyocera solar KC 200GT module specifications

Maximum power, P_{mp}	200.143 W
Maximum power point voltage, V_{mp}	26.3 V
Maximum power point current, I_{mp}	7.61 A
Short-circuit current, I_{sc}	8.21 A
Open-circuit voltage, V_{oc}	32.9 V
Voltage/temp. coefficient, K_v	−0.35502%/°C
Current/temp. coefficient, K_I	0.06%/°C
The number of cells, N_s	54

3.1 Procedure for Modelling a PV Array Using MATLAB-Software

Step 1. Use the library browser to add the blocks to your Simulink file, PV array, constant block, diode, voltage measurement, current measurement, product block, X-Y graph, scope, RLC branch, and power GUI block.
Step 2. Once you have added all the blocks and components, set the parameters of the PV array block, choose KYOCERA SOLAR KC 200GT PV module
Step 3. Connect 3 PV modules in series and then connect another 3 PV modules in series but parallel to the other 3 PV modules, basically a 3 cross 2 PV array arrangement
Step 4. Connect the constant blocks to the irradiance and temperature of the PV module and set the parameter to 1000 watts and 25°C
Step 5. Connect the bypass diode across each module, and make sure in each string there is a blocking diode
Step 6. Current and voltage measurement blocks are connected to measure the current and voltage of the PV array, and the product block is connected to measure the power of the array.
Step 7. Connect the X-Y graphs
Step 8. Connect the scope to visualize our signals over time, and set the number of input ports to 3, these ports are used to connect voltage, current, and power signals.
Step 9. Create the shading conditions by creating a fault, line-to-line fault, and line-to-ground fault separately and compare
Step 10. Run the simulation for 15 ms

4 Simulation of PV Array Exposed to Different Shading Conditions

In this section, the different shading conditions will be explained and how the signals were gotten, the signal graphs will also be shown so that they can be easily compared. The signal graphs will also be shown so that they can be easily compared. The most common way to create shading on the PV array in Simulink is by changing the irradiance, that way one can see the difference and effects at several irradiances. There are other ways to create shading on the PV array when using the Simulink, this method is by creating faults, and they are listed below.

1. Shading by changing Irradiance:—In this condition, the irradiance value is changed in some PV panels. The 3rd and 6th PV irradiance is altered to 300 and 700 as shown in Fig. 3.
2. Line-to-ground fault:—In this condition, PV array is connected to the ground to create a fault as shown in Fig. 4.

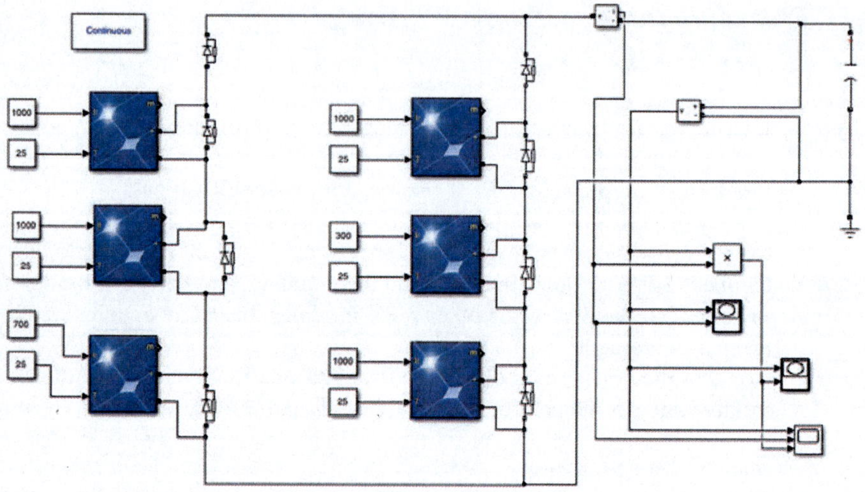

Fig. 3 PV array connection during the changing irradiance shading

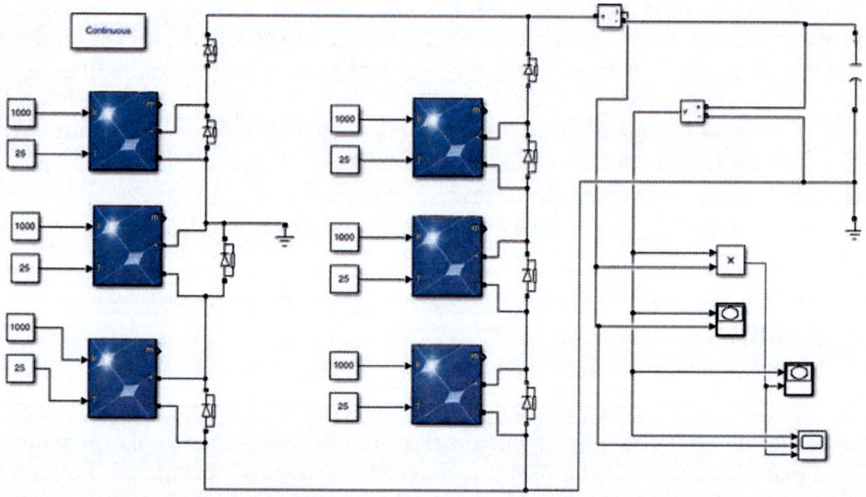

Fig. 4 PV array connections during line-to-ground fault condition

3. Line-to-line fault:—In this condition, a line is connected from one PV panel to another to create a fault as shown in Fig. 5.

Here, the irradiance in the third PV array of the first row and the second PV array in the second row has been changed to 700 and 300 respectively to create a shading in the Simulink software.

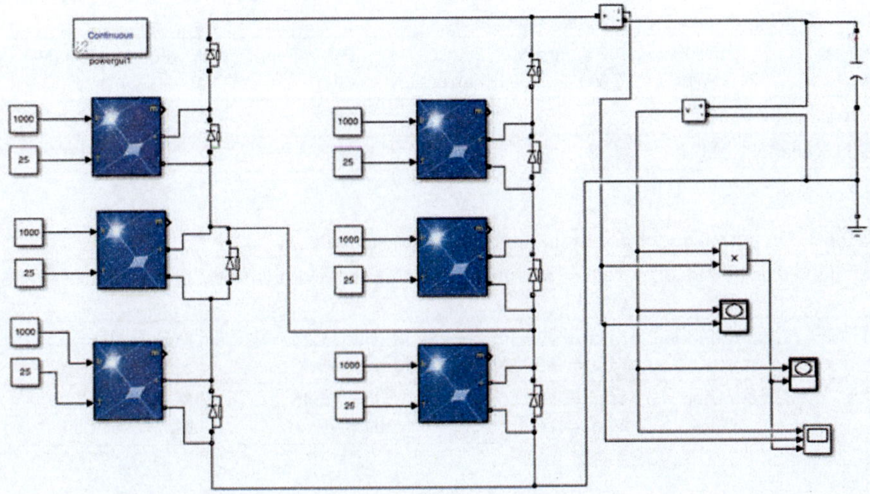

Fig. 5 PV array connections during line-to-line fault condition

The irradiation values have to be set back to 1000 W/M scale, to induce line-to-ground fault in string one with 67% mismatch, take the positive terminal of the second module to the ground which means the lower two modules are short-circuited and the top module or string one is the only module working properly throughout the entire voltage change. The simulation is run for 0.7 ms. Connect the positive terminal of the second module in string one to the positive terminal of the third module in string two. This will produce a line-to-line fault between the strings that has a 33% mismatch. Alternatively, you can connect the positive terminal of the second module in string two to the positive terminal of the third module in string one. Both of these faults are equivalent. In this case, a mismatch of 33 per cent indicates that just one module out of three is incorrect.

5 Result and Discussions

The results in Tables 2 and 3 shows the output graph which was obtained from the MATLAB SIMULINK. Table 2 shows the reading of the PV array under different irradiation, one can see that there is a significant change in the reading when the irradiation was reduced from 1000 to 500. Table 3 shows the output readings of the graph under different shading conditions namely line-to-line and line-to-ground faults, one can also observe the difference in output between the two faults.

Table 2 Output values for changing irradiance

Cases	Irradiance (W/m^2)	PV current (A)	Boost current (A)	PV voltage (V)	Bus voltage (V)	P_{pv} (W)
1st scenario	1000	38	9.8	30	90	1000
2nd scenario	500	18	7	30	70	500

Table 3 Output values for line-to-line and line-to-ground faults

Sr. no.	Condition/fault	Voltage measurement (V)	Current measurement (A)	Product output (W)
1	Line-to-ground	Max = 97.93 Min = 0.0000	Max = 16.45 Min = −3.116	Max = 593 Min = −3.046
2	Line-to-line	Max = 101.6 Min = 0.000	Max = 16.45 Min = −8.462	Max = 825.7 Min = −8.600

6 Conclusion

In this research work, it has been examined and shown how shading conditions can be created in the MATLAB SIMULINK software and how one can realize that shading conditions on a photovoltaic array make the output and power curve complicated by giving different and multiple maximum power points instead of a unique MPP. In addition to the simulation, bypass diodes were added and one can verify that the bypass diode helps in improving the output power in case of shading conditions in the PV array. The signal output has been compared in a tabular format for better understanding and comparisons. The results are shown with a case study for photovoltaic array under shaded conditions.

References

1. Gupta S, Sharma A (2018) Global scenario of solar photovoltaic (SPV) materials. international conference on advanced computational and communication paradigms (ICACCP). In: Lecture notes in electrical engineering (LNEE), vol 475, pp 126–133. Springer
2. Gupta S et al (2020) Wind energy potential, challenges with major technical issues. J Green Eng (JGE) (ISSN: 1904-4720) 10(12):12973–12987
3. Gupta S, Singh P, Swami RK (2022) Present wind energy market scenario in India. Smart energy and advancement in power technologies, Lecture notes in electrical engineering, vol 926, pp 637–647. Springer
4. Kannan N, Vakeesan D (2016) Solar energy for future world—a review. Renew Sustain Energy Rev 62:1092–1105
5. Nguyen DD, Lehman B (2006) Modeling and simulation of solar PV arrays under changing illumination conditions. In: 2006 IEEE COMPEL workshop, rensselaer polytechnic institute, Troy, NY, USA, July 16–19

6. Zebiri1 M, Mediouni1 M, Idadoub H (2018) Modelling and simulation of the shading effect on the performance of a photovoltaic module in the presence of the bypass diode. E3S Web Conf 37:06002
7. Hema S, Fathima K, Mary Saranya T, Vinodini D (2017) A IPSO tuned MPPT design for PV system under partially shaded condition. J Chem Pharmaceut Sci JCHPS Special Issue 8
8. Brown C (2021) Shading losses in PV system, and techniques to mitigate them. Solar PV Educ 101
9. Gupta S (2019) Maximum sunlight tracking using single axis solar panel prototype with simulation. Int J Innov Technol Explor Eng (ISSN 2278-3075) 8(7):221–225
10. Banu I-V, Istrate M, Asachi G (2012) Modelling and simulation of photovoltaic array. In: World energy system conference, WESC
11. Podder AK, Roy NK, Pota HR (2019) MPPT methods for solar PV systems: a critical review based on tracking nature. IET J Inst Eng Technol Rev ISSN 1752-1416
12. Xu L, Cheng R, Yang J (2020) A new MPPT technique for fast and efficient tracking under fast varying solar irradiation and load resistance. Int J Photo Energy 2020, Article ID 6535372, 18 p
13. Joshi S, Pandya V, Dhandhia A (2019) Simulation on MPPT based solar PV standalone system. PDPU J Energy Manage
14. Yadav D, Singh N (2014) Simulation of PV system with MPPT and boost converter. In: IEEE sponsored national conference on energy, power and intelligent control systems, 28th–29th March
15. Salman S, AI X, WU Z (2018) Design of P-&-O algorithm based MPPT charge controller for a stand alone 200 W PV system. Prot Control Mod Power Syst 3:25
16. Alik R, Josh A, Sutikno T (2016) A review on perturb and observe maximum power tracking in photovoltaic system. ISSN 3(3)
17. Gupta S, Dwivedi VK, Thakur P, Bansal RC (2023) Optimum relay coordination for reliability evaluation of distribution system with allocation of wind turbine generators. Int J Model Simul 1–16
18. Koutroulis E, Blaabjerg F (2012) A new technique for tracking the global maximum power point of PV arrays operating under partial-shading conditions. IEEE J Photovolt 2(2):184–190
19. Chin CS, Tan MK, Neelakantan P, Chua BL, Teo KTK (2011) Optimization of partially shaded PV array using fuzzy MPPT. In: 2011 IEEE colloquium on humanities, science and engineering research (CHUSER 2011), Dec 5–6
20. Sholapur S, Mohan KR, Narsimhegowda TR (2014) Boost converter topology for PV system with perturb and observe MPPT algorithm. IOSR J Electr Electron Eng (IOSR-JEEE) 9(4) II:50–56
21. Gupta S, Singh N, Singh SB (2019) Photovoltaic module designing with comparison of different MPPT techniques. In: International conference on manufacturing, advance computing, renewable energy and communication (MARC-2018), Lecture notes in electrical engineering (LNEE), vol 553, pp 121–131. Springer

Multi-objective Hybrid Optimal Algorithm for Distribution System Feeder Reconfiguration

M. Shobha and B. Datta

Abstract The power generated is transmitted through various networks, which include underground cables, power transformers, and other components. These losses are caused by the improper installation of generators and the transmission system overloading. The distribution system also has load-tapping capabilities. The uncertainty of the current voltage profile also leads to higher losses in the distribution system. Distribution system reconfiguration in a distribution system is a complex optimization problem. The location of generators is considered in the most advantageous manner possible. The placement of generators does not guarantee an improved performance index. The main factors that affect the placement of generators are size, type, and level of penetration. To overcome this in this paper, a Hybrid GSA-Tabu Multi-Objective Algorithm is proposed for feeder reconfiguration with optimal placement of DG in the distribution system in IEEE 33 bus system and its effectiveness is evaluated by comparing with existing LSA, ALO, CLAMS, and BBO-PSO algorithms.

Keywords GSA · Tabu · LSA · ALO · CLAMS BBO-PSO

1 Introduction

The distribution system plays a vital role in the electricity system and is responsible for controlling the distribution of electricity. Due to the dynamic nature of the distribution system, its importance has been acknowledged. The proper allocation and sizing of generators can help improve the system's technical performance. Distribution generation is often referred to as small-scale generation. It can be defined in different ways such as by the International Energy Agency. The Electric Power Research Institute defines distributed generation as those that are less than 50 kW. Different countries have different definitions of this type of generation.

M. Shobha (✉) · B. Datta
Department of Electrical Engineering, National Institute of Technology, Arunachal Pradesh 791112, India
e-mail: mshoba.phd20@nitap.ac.in

A. Kumar et al. (eds.), *Decarbonisation and Digitization of the Energy System*, Lecture Notes in Electrical Engineering 1099, https://doi.org/10.1007/978-981-99-7630-0_17

The power that's generated through distributed generation is transported through a network of distribution utilities, which includes power transformers and underground cables. Although the cost of distributing electricity is lower than the transmission system, the system's losses are still significant. The uncertainty of the voltage profile can affect the system's operations and cost. This causes the cost and power loss to increase. The main cause of technical losses is the power that's lost due to the conductor damage of the transmission and distribution lines. Network configuration refers to the process of changing the distribution system's configuration to reduce the losses and improve the system's overall performance. Network configuration can help reduce the losses by improving the operation of feeders. In this paper, a Hybrid GSA-Tabu Multi-Objective Algorithm is proposed for feeder reconfiguration with optimal placement of DG in the distribution system in IEEE 33 bus system, and its effectiveness is evaluated by comparing with existing LSA, ALO, CLAMS, and BBO-PSO algorithms.

1.1 Literature Review

The distribution system is typically composed of two or more generators and a transmission system. The configuration of this system can be changed radially to reduce its losses and improve its overall performance. Due to the nature of the distribution system's configuration, it needs to be optimized to maintain its efficiency. A distributed generator is a type of electric power source that's placed at the distribution level. It's widely used due to the environmental concerns when it comes to meeting the increasing power demands. A properly integrated distributed generator system can improve the efficiency of the distribution system and reduce power loss. A metaheuristic algorithm known as the Quasi-Oppositional Symbiotic Organisms Search was proposed by [1] to find the optimal placement of a distributed generator in radial distribution systems. The algorithm was tested and implemented on three different bus systems and showed good results. The proposed algorithm in [2] was tested and implemented on IEEE 13 bus system. It shows that the proposed algorithm can improve the reliability index, voltage stability index, and the cost of the distributed generator. A voltage stability analysis was performed in residential areas of the distribution system using rooftop photovoltaic generators by [3].

A probabilistic approach is proposed in [4] to consider the harmonic distortions of a distributed generator. In [5] and [6] used the optimal placement for distribution generator for voltage stability improvement. In [7] proposed an approach that takes into account the harmonic distortion and protection coordinates of distributed generators. A hybrid optimal method is proposed in [8] that takes into account the energy loss reduction of distributed generators and their optimal integration of them.

In [9] proposed a hybrid optimal method that takes into account the energy loss reduction and optimal integration of distributed generators. A metaheuristic and probabilistic approach are presented to find the optimal size and allocation of distributed generators in a distribution system [10]. Optimal placement of DG with uncertainties are presented in [11] and stochastic evaluation is presented in [12]. In [13] developed a new multi-objective function that can be used to represent the practical load models for optimal distribution generator allocation. The proposed multi-objective function is commonly used in standard radial distribution systems. A preference order ranking algorithm and a multi-objective PSO algorithm were also developed by [14] to find the optimal allocation of generators in a distribution system. Monte Carlo simulation was also used to formulate the uncertainties in a radial distribution system.

2 Proposed Hybrid GSA-Tabu Multi-objective Algorithm

In this paper, Hybrid GSA-Tabu Multi-Objective Algorithm is proposed for feeder reconfiguration with optimal placement of DG in the distribution system in IEEE 33 bus system and its effectiveness is evaluated by comparing with existing LSA, ALO, CLAMS, and BBO-PSO algorithms. The input parameters of the proposed algorithm are shown in Eq. (1).

$$X_i = \left[(V_1, P_{L1})^1, (V_2, P_{L2})^2, (V_3, P_{L3})^3, \ldots (V_n, P_{Ln})^n\right] \tag{1}$$

The force acted on the agent is given as

$$force^d(t) = \sum_{\substack{j \\ i \neq j}} force^d(t) rand_{ii} \tag{2}$$

The acceleration is given as

$$Acceleration\ \alpha_i^d = \frac{force_i^d(t)}{M_i(t)} \tag{3}$$

The velocity of the agent is given as

$$V_i^d(t+1) = rand_i \cdot \left[V_i^d\right] + \alpha_i^d(t) \tag{4}$$

The position of the agent is given as

$$X_i^d(t+1) = X_i^d(t) + V_i^d(t+1) \tag{5}$$

The flow chart of the proposed algorithm is shown below (Fig. 1).

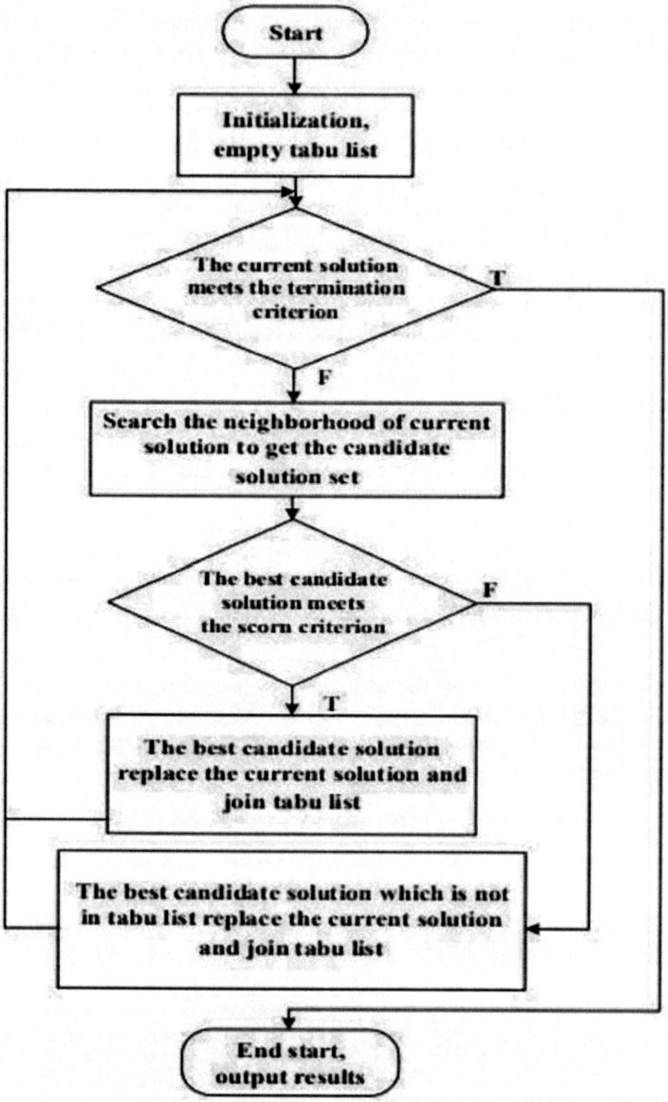

Fig. 1 Flow chart of the proposed algorithm

3 Results and Discussions

The proposed algorithm is implemented on IEEE 33 bus system under the following cases:

- Case 1: No Reconfiguration
- Case 2: Reconfiguration with tie switches and sectionalization
- Case 3: Only Optimal size of DG
- Case 4: Optimal Reconfiguration with Optimal Size of DG placement.

All these cases are implemented on Less, Moderate, Heavy-loaded cases, and Power losses and Voltage deviations are plotted. The power losses for different loaded cases with the proposed algorithm are shown in Fig. 2.

The Voltage Deviation for different loaded cases with the proposed algorithm are shown in Fig. 3.

The effectiveness of the proposed algorithm is evaluated by comparing it with LSA, ALO, CLAMS, and BBO-PSO algorithms and tabulated in Table 1.

Hence, in all the above cases, the proposed algorithm shows the mark improvement in power losses and voltage deviation.

4 Conclusion

In this paper, a Hybrid GSA-Tabu Multi-Objective Algorithm is proposed for feeder reconfiguration with optimal placement of DG in the distribution system. The proposed algorithm is implemented on IEEE 33 bus system. Detailed literature review is presented in the paper. Power losses and voltage deviations are analyses for less, moderate, and heavy loaded conditions are presented under four different cases. In Case 1: No Reconfiguration, In Case 2: Reconfiguration with tie switches and sectionalization, in Case 3: Only Optimal size of DG, and in Case 4: Optimal Reconfiguration with Optimal Size of DG placement is considered. Finally, the proposed algorithm's effectiveness is evaluated by comparing it with existing LSA, ALO, CLAMS, and BBO-PSO algorithms. In all the cases, proposed algorithm shows a mark improvement in power losses and voltage deviation.

Fig. 2 **a** Power loss under less load, **b** Power loss under moderate load, **c** Power loss under heavy load

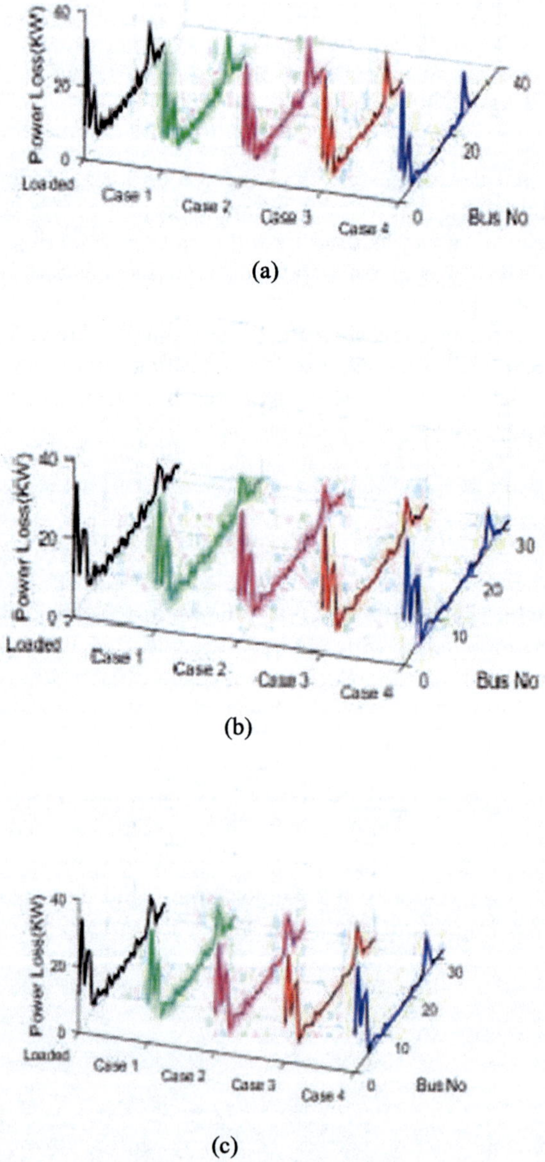

Fig. 3 **a** Voltage deviation under less load, **b** Voltage deviation under moderate load, **c** Voltage deviation under less load

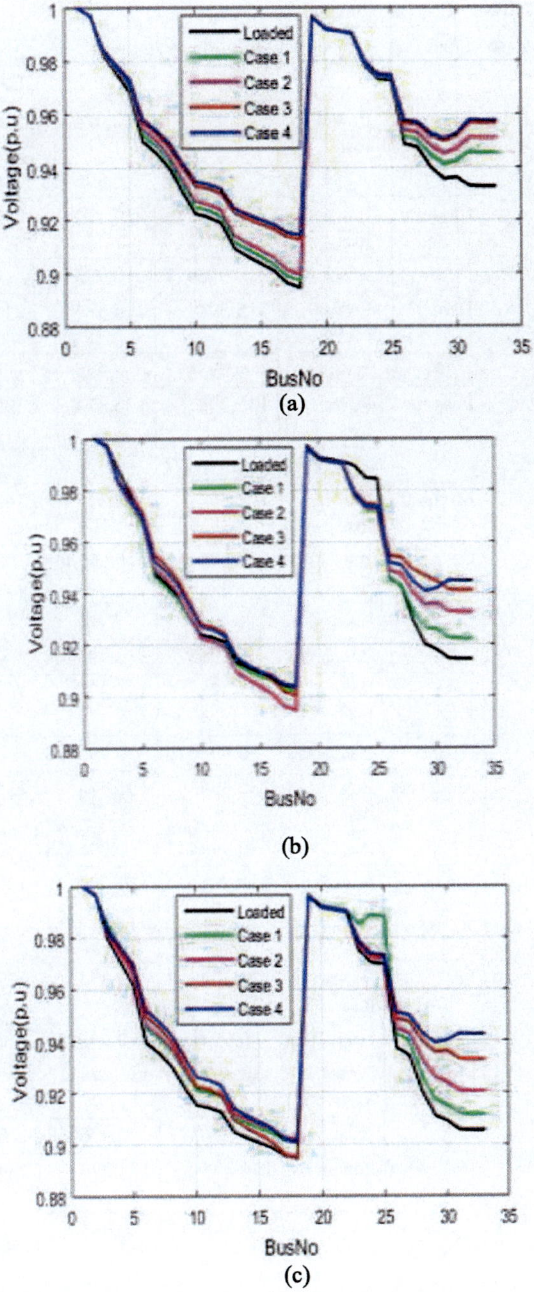

Table 1 Comparison analysis

Case	Load (%)	Objective	Solution techniques				
			LSA	ALO	CALMS	BBO-PSO	Proposed
1	Less	Power loss	210.094	210.074	210.054	210.034	210.0257
		Δv_{dev} (v)	0.996477	0.996468	0.996536	0.996678	0.997013
	Moderate	Power loss	210.394	210.374	210.354	210.294	210.2757
		Δv_{dev} (v)	0.995477	0.995468	0.995536	0.995678	0.996765
	Heavy	Power loss	215.394	215.374	215.354	215.294	215.1578
		Δv_{dev} (v)	0.993477	0.995468	0.995536	0.996778	0.997340
2	Less	Power loss	135.394	135.374	135.354	135.294	135.2176
		Δv_{dev} (v)	0.996477	0.996468	0.996536	0.996678	0.997096
	Moderate	Power loss	142.394	142.374	142.354	142.294	142.258
		Δv_{dev} (v)	0.995477	0.995468	0.995536	0.995678	0.996899
	Heavy	Power loss	145.394	145.374	145.354	145.294	145.287
		Δv_{dev} (v)	0.995377	0.995368	0.995436	0.995578	0.996756
3	Less	Power loss	131.394	131.374	131.354	131.294	131.1633
		Δv_{dev} (v)	0.996477	0.996468	0.996536	0.996678	0.997212
	Moderate	Power loss	135.394	135.374	135.354	135.294	135.2756
		Δv_{dev} (v)	0.994477	0.996468	0.996636	0.996878	0.997127
	Heavy	Power loss	139.394	139.374	139.354	139.294	139.2461
		Δv_{dev} (v)	0.995477	0.995468	0.995536	0.995678	0.996899
4	Less	Power loss	129.394	129.374	129.354	129.294	129.1418
		Δv_{dev} (v)	0.994497	0.996488	0.996656	0.996898	0.997146
	Moderate	Power loss	133.394	133.374	133.354	133.294	133.1782
		Δv_{dev} (v)	0.996377	0.996568	0.996536	0.996778	0.997029
	Heavy	Power loss	134.394	134.374	134.354	134.294	134.1452
		Δv_{dev} (v)	0.996277	0.996468	0.996536	0.996678	0.996995

References

1. Truong KH, Nallagownden P, Elamvazuthi I, Vo DN (2020) A quasi-oppositional-chaotic symbiotic organisms search algorithm for optimal allocation of DG in radial distribution networks. Appl Soft Comput J 88:106067. https://doi.org/10.1016/j.asoc.2020.106067
2. Zheng Y, Dong Z, Meng K, Yang H, Lai M, Wong KP (2017) Multi-objective distributed wind generation planning in an unbalanced distribution system. CSEE J Power Energy Syst 3(2):186–195. https://doi.org/10.17775/cseejpes.2017.0023
3. Shahnia F, Majumder R, Ghosh A, Ledwich G, Zare F (2011) Voltage imbalance analysis in residential low voltage distribution networks with rooftop PVs. Electr Power Syst Res 81(9):1805–1814. https://doi.org/10.1016/j.epsr.2011.05.001
4. Abdelsalam AA, El-Saadany EF (2013) Probabilistic approach for optimal planning of distributed generators with controlling harmonic distortions. IET Gener Transm Distrib 7(10):1105–1115. https://doi.org/10.1049/iet-gtd.2012.0769
5. Ettehadi M, Ghasemi H, Vaez-Zadeh S (2013) Voltage stability-based DG placement in distribution networks. IEEE Trans Power Deliv 28(1):171–178. https://doi.org/10.1109/TPWRD.2012.2214241
6. Georgilakis PS, Hatziargyriou ND (2013) Optimal distributed generation placement in power distribution networks: models, methods, and future research. IEEE Trans Power Syst 28(3):3420–3428. https://doi.org/10.1109/TPWRS.2012.2237043
7. Ravikumar Pandi V, Zeineldin HH, Xiao W (2013) Determining optimal location and size of distributed generation resources considering harmonic and protection coordination limits. IEEE Trans Power Syst 28(2):1245–1254. https://doi.org/10.1109/TPWRS.2012.2209687
8. Esmaeilian HR, Fadaeinedjad R (2015) Energy loss minimization in distribution systems utilizing an enhanced reconfiguration method integrating distributed generation. IEEE Syst J 9(4):1430–1439. https://doi.org/10.1109/JSYST.2014.2341579
9. Liu KY, Sheng W, Liu Y, Meng X, Liu Y (2015) Optimal sitting and sizing of DGs in distribution system considering time sequence characteristics of loads and DGs. Int J Electr Power Energy Syst 69:430–440. https://doi.org/10.1016/j.ijepes.2015.01.033
10. Gómez-González M, Ruiz-Rodriguez FJ, Jurado F (2015) Metaheuristic and probabilistic techniques for optimal allocation and size of biomass distributed generation in unbalanced radial systems. IET Renew Power Gener 9(6):653–659. https://doi.org/10.1049/iet-rpg.2014.0336
11. Nikmehr N, Najafi-Ravadanegh S (2015) Optimal operation of distributed generations in micro-grids under uncertainties in load and renewable power generation using heuristic algorithm. IET Renew Power Gener 9(8):982–990. https://doi.org/10.1049/iet-rpg.2014.0357
12. Silva ENM, Rodrigues AB, Da Guia Da Silva M (2016) Stochastic assessment of the impact of photovoltaic distributed generation on the power quality indices of distribution networks. Electr Power Syst Res 135:59–67. https://doi.org/10.1016/j.epsr.2016.03.006
13. Bohre AK, Agnihotri G, Dubey M (2016) Optimal sizing and sitting of DG with load models using soft computing techniques in practical distribution system. IET Gener Transm Distrib 10(11):2606–2621. https://doi.org/10.1049/iet-gtd.2015.1034
14. Gangwar P, Singh SN, Chakrabarti S (2019) Multi-objective planning model for multiphase distribution system under uncertainty considering reconfiguration. IET Renew Power Gener 13(12):2070–2083. https://doi.org/10.1049/iet-rpg.2019.0135

Monitoring and Control of Captive Generation Units in Presence of Grid Integrated Solar Power Plants

Sandeep Jangir and Srilatha Namilakonda

Abstract Industrial loads and large power consumers use various forms of Renewable Energy Sources (RES) apart from utility grid services, in order to reduce Running Maximum Demand (RMD). It also provides financial savings in utility bills with reduced consumption of energy from grid. Solar Power Plant (SPP)-based RES is the most abundant form of such energy and most widely used in world. In order to reduce down time of loads, industrial consumers install captive generation units with SPP-based RES to cater power requirements of these loads during grid failure. But the operations of these captive units are badly affected by SPP-based RES units, as soon as they start supporting the load in integrated system during grid failure. This paper analyzes the impact of SPP-based RES on captive generation unit and proposes a protection scheme using local reactive power units.

Keywords Renewable energy sources · Diesel generator set · Solar power plants · Captive generation

1 Introduction

Grid-interacted PV systems are designed to operate in parallel with Electric utility grid, captive generators, and Variable Loads [1]. In an integrated distribution system, which comprises of solar power plant, grid and captive generation, flow of active & reactive powers is an important concern [2]. A captive power plant, also called auto producer or embedded generation, is an electricity generation facility used and managed by an industrial or commercial energy user for their own energy consumption [3]. When the solar is integrated with the utility grid, reactive power drawn from grid is almost constant. However, the active power demand is reduced as the RES

S. Jangir (✉)
NRSC, Indian Space Research Organisation, Hyderabad, India
e-mail: sandeep_j@nrsc.gov.in

S. Namilakonda
Department of Electrical Engineering, Osmania University, Hyderabad, India

A. Kumar et al. (eds.), *Decarbonisation and Digitization of the Energy System*, Lecture Notes in Electrical Engineering 1099, https://doi.org/10.1007/978-981-99-7630-0_18

supplies maximum active power to the load. During grid failure when these RES are integrated with captive generation units, it impacts severely to power generated from captive source [4]. The amount of reactive power produced by generators must closely match with that being consumed, as most of the synchronous generators are defined by their power factor limits [5].

This paper addresses a typical problem of power system failure in an organization consisting of RES integrated with the grid and a captive generation unit i.e., Diesel Generator (DG) unit to support the load during grid failure with a defined power factor limit of 0.8. The DG has undergone a mechanical failure and leads to power failure when RES has been operated along with DG during grid failure. A strategy is proposed to protect this captive generation unit from outage by load variation and employing local reactive power units. MATLAB 2016 b software environment was used to simulate the problem at a reduced scale by integrating 100 kWp Solar Power Plant (SPP) with grid, captive generation unit (DG) and variable loads with capacitor banks to analyze the active and reactive power flow, source power factor.

2 Energy Consumption Trends

The SPPs are installed in the organization to cater to emergency energy demand in various phases. Figure 1 indicates the strategy implemented in commissioning of SPPs in the organization to mitigate the energy demand of the utility and saving of the energy charges in the utility bills. The increase in energy consumption has increased impact of RMD severely on grid and DG during this period as shown in Fig. 2. RMD observed during the year 2013 was 650 kVA and it has increased up to 1050 kVA in 2021.

3 Problem Statement

To analyze the source of the problem the survey of the major systems was done which support the distribution system in organization. It is understood that the entire loads and plants are distributed in two segments in parallel and are supported by running and standby systems equally. The initialization sequence of loads is detailed in Table 1.

The active power demand on grid during the peak hours was reduced to 10% and the reactive power demand of the inductive load was mitigated with capacitor banks. Figure 3 shows active and reactive power demand curve of a typical sunny day.

Initially the SPP is integrated with grid and feeds power to the loads. At the instant of grid failure, SPP gets disconnected from load. Captive generation units i.e., DG sets start and feed power to loads. Depending on the connection sequence of various components, active and reactive power demand on the system is minimal initially as the air conditioning loads (major active and reactive load) are not started by then.

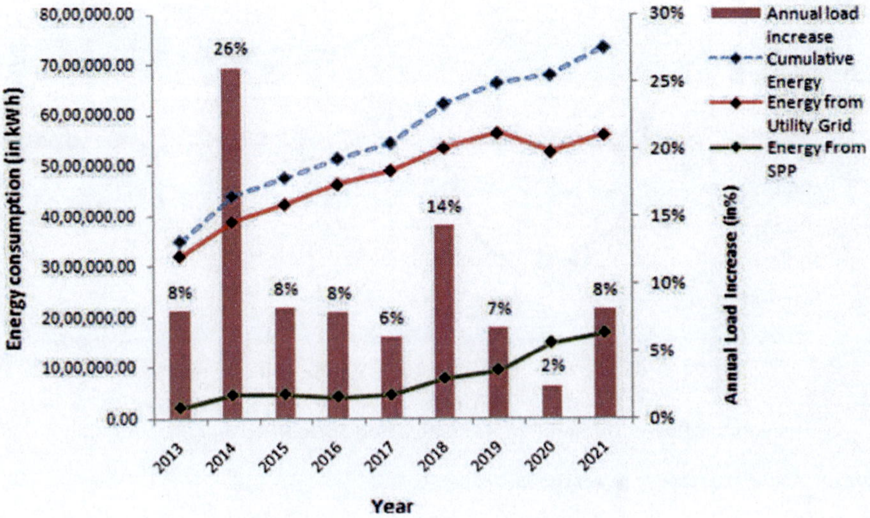

Fig. 1 Various energy and load trends, SPV installation from 2013 to 2021

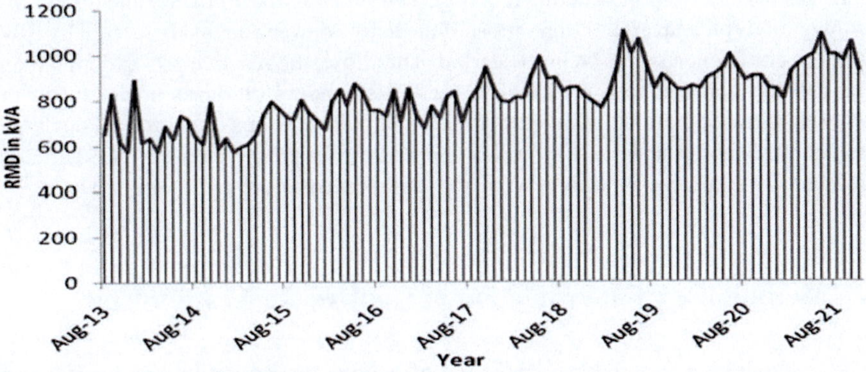

Fig. 2 Running maximum demand (RMD) from 2013 to 2021

Table 1 Load initialization on distribution system

Sl. no.	Description of the system	Initialization sequence
1	Lighting load (minor load)	Immediately
2	UPS and server loads (moderate load)	Immediately
3	Air Conditioning load (major load)	After 3 min
4	DG sets support to load during grid failure	Within 30 s
5	Start of solar power plants	45 s

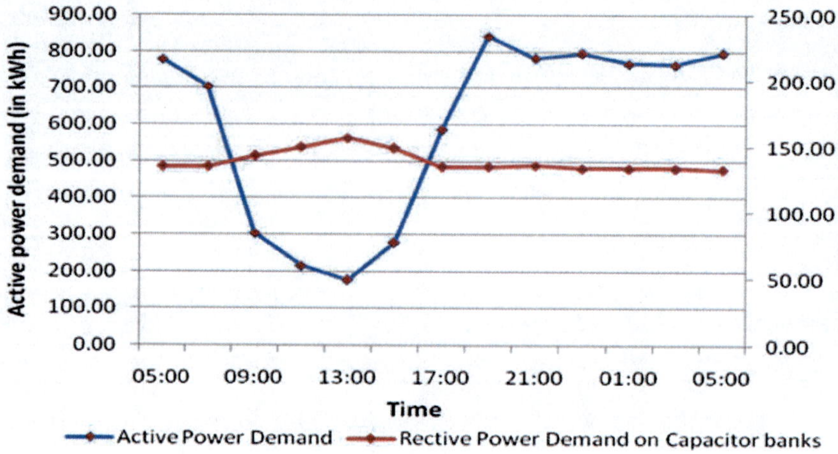

Fig. 3 Active and reactive power demand curve

The active power from the DG feeds most of the lighting and UPS loads initially and thus less active power demand is observed on DG sets. But since the air conditioning loads pick up after 3 min., most of the active power demand is catered by SPP that has commenced after DGs are started. Therefore, the reactive power demand on the DG set increases to a very high value. This condition reduces power factor of DG set below operational range i.e., to 0.6 and voltage drops drastically. This large unbalance in active and reactive power demand causes pressure on the DG shaft and causes failure of gear assembly, AVR and isolation transformer.

4 Methodology for Protection of Captive Generation Unit

The methodology adopted for protection of captive generation unit in presence of RES is to monitor the power factor of the DG and maintain the value within permissible range under all operating conditions [6]. The task of maintaining the power factor has been implemented through employing capacitor banks. The details of the monitoring and protection of DG are presented in block diagrams shown in Fig. 4. A scaled-down model of the organization is simulated in MATLAB using a 100kWp SPP in presence of variable load environment [7] as indicated in Fig. 5. The distribution system consists of RES, captive power source, grid and capacitor banks. The variable load represents the industrial load.

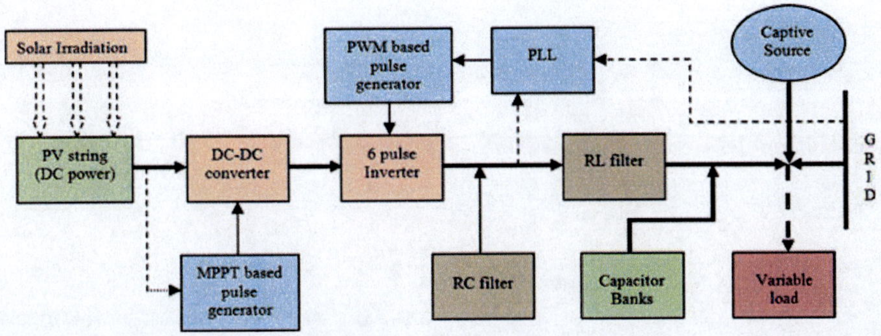

Fig. 4 Block diagram of the simulation setup

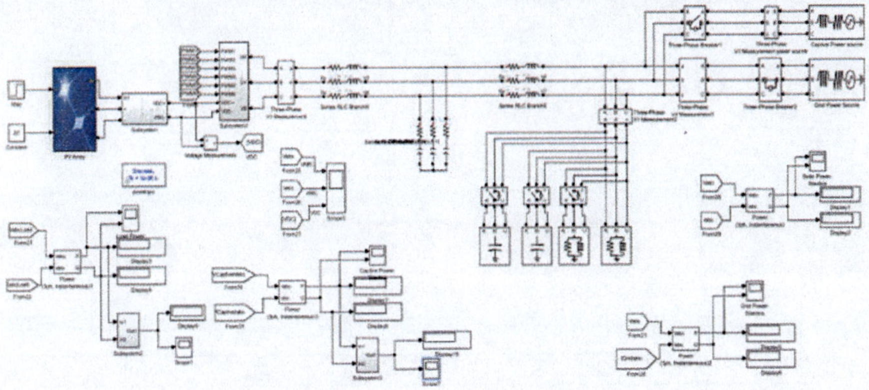

Fig. 5 Problem formulation in MATLAB for protection and control of captive power source

5 Results and Analysis

The results are analyzed with respect to the performance of SPP. The DC power characteristics are observed in variable temperature and irradiation environment. When temperature is increased from 25 °C to 50 °C @0.15 s., V_{DC} reduces significantly, also when irradiation is reduced from 1000 W/m^2 to 500 W/m^2 at 0.2 s., I_{DC} reduces significantly (Fig. 6).

The AC output current I_{AC} of SPP has minimal impact with change in temperature but there is significant impact with change in irradiation. Also, there is significant drop of active power due to variations in temperature and irradiation (Figs. 7 and 8).

When 100kWp SPP is connected with 100 kW, 10kVAr fixed load, 100 kW, 75kVAr captive power source, grid, 20 kW, 10kVAr variable load and 2 nos 10 and 20 kVAr capacitor banks, it is observed that SPP supplies maximum active power to the load, and minimal active power is drawn from grid. It supplies very minimal reactive power to the load and draws maximum reactive power from grid (Fig. 9).

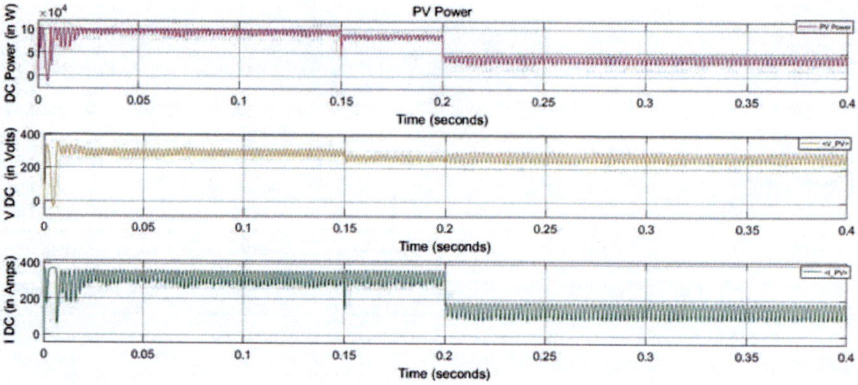

Fig. 6 Performance of SPP with variations in temperature and irradiation

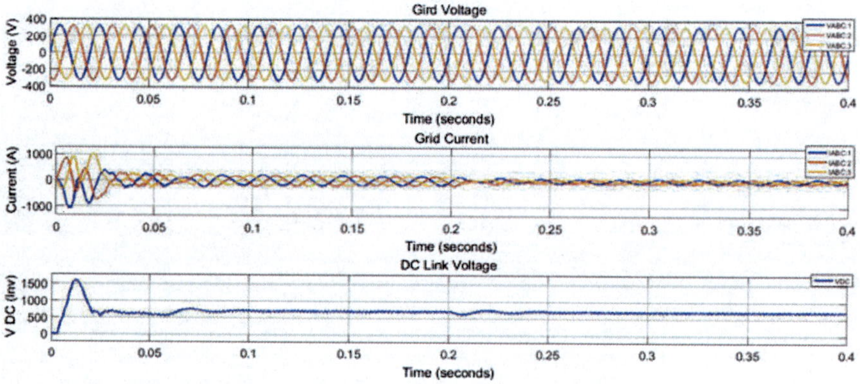

Fig. 7 V_{AC}, I_{AC} of SPP for with variations in temperature and irradiation

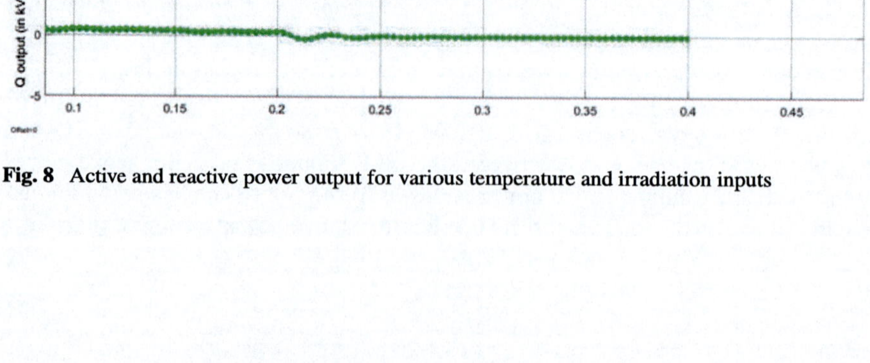

Fig. 8 Active and reactive power output for various temperature and irradiation inputs

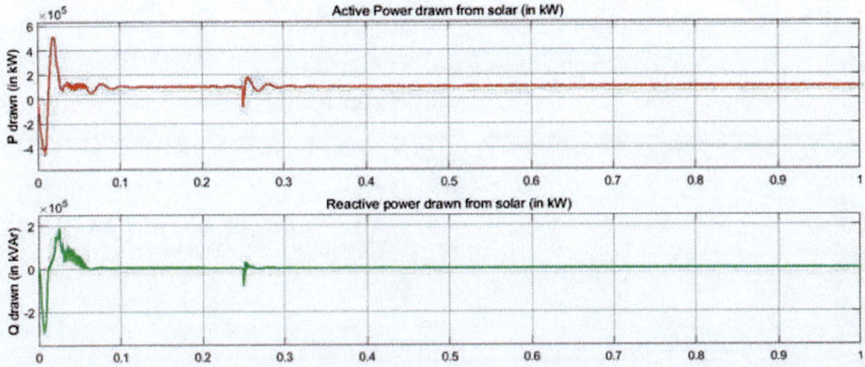

Fig. 9 P and Q output of SPP during integration with grid, captive generation and variable loads

At 0.2 s grid gets disconnected and captive generation unit is switched on at 0.25 s. At this moment maximum reactive power is drawn from captive power source and maximum active power is drawn from solar power plant (Figs. 10, 11 and 12).

Now, 0.4 s. 20 kW, 10 kVAr load added in the system. As the SPP is of 100 kWp the entire 20 kW, 10 kVAr active and reactive power demand of load is transferred to the captive power source. Subsequently, at 0.5 s a capacitor bank of 10 kVAr is added in the system and the reactive power demand on the captive power source is dropped from 10 kVAr to 3.50 kVAr. Another capacitor bank of 5 kVAr is added in the system at 0.6 s and the reactive power demand is further reduced. The load power factor is approx. 0.98 at initial and reduced to near 0.96 when load is added in the system at 0.4 s. The power factor is improved when capacitor band is added at 0.5 s and 0.6 s to 0.98 and 0.99 respectively.

The power factor on captive generation characteristics with active and reactive power demand as PF is near zero initially as there is zero active and reactive power demand initially. When the active and reactive power demand increases at 0.4 s, power

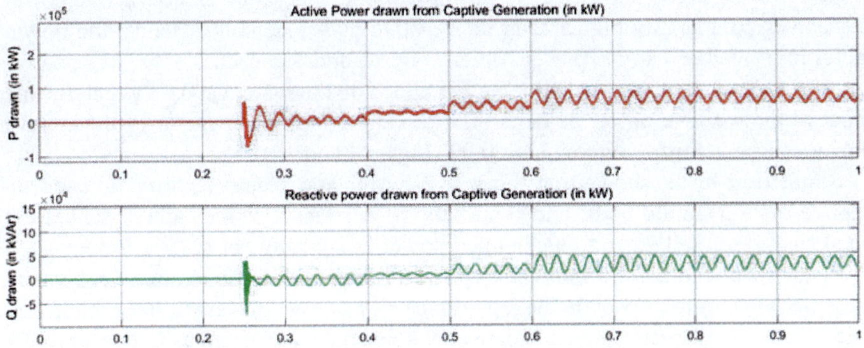

Fig. 10 P and Q generation from Captive unit from 0.25 s after grid failure at 0.2 s

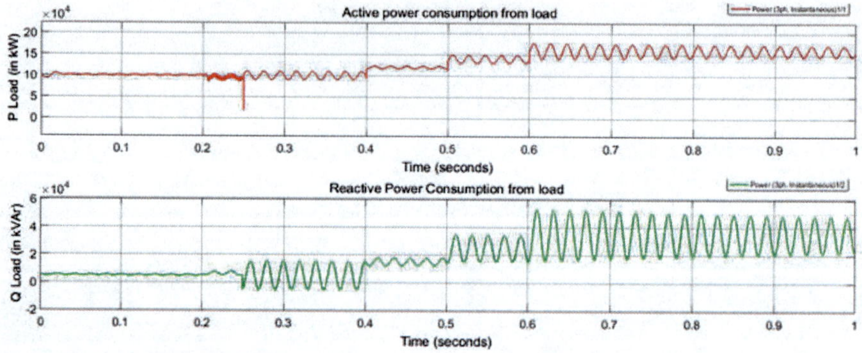

Fig. 11 Active and reactive power demand from load after grid failure

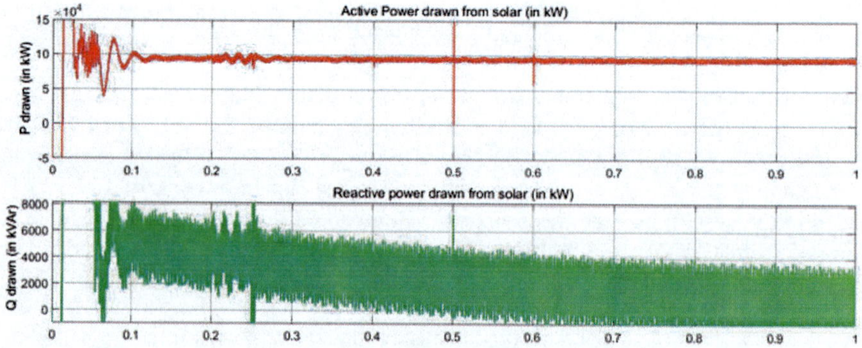

Fig. 12 P and Q generation from Solar power during grid failure

factor improves to 0.95 (in view of less active power demand). When capacitor bank was added in the system at 0.5 s, the reactive power demand was reduced on the captive power source but the active power demand was same. Hence, active power demand is comparably higher than the reactive power demand. Hence, the power factor improved on captive power source (Fig. 13 and Table 2).

Subsequently, another capacitor bank is added on the system at 0.6 s which further reduced the reactive power demand on the system and power factor of the captive power source is further improved to 0.99 (Figs. 14 and 15).

Simulation result shows that the active power and reactive power demand on captive system should be balanced, i.e., the power factor of the captive generation plant should not fall down to the defined power factor range of the generator.

Addition of more and more active power generation in the captive tied system may provide severe impact on the power factor of captive power source, which may intern damage the rotating mechanism of the mechanical system as indicated in the case study of the project.

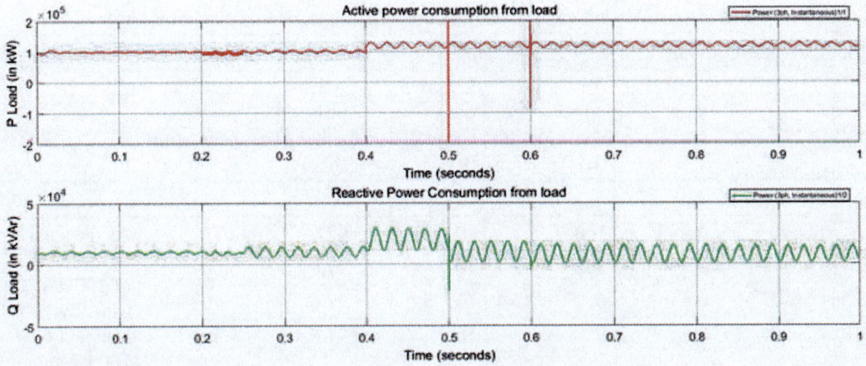

Fig. 13 Load consumption during grid failure with reactive power support

Table 2 PV characteristics with captive generation, variable loads and capacitor banks

Sl. no	Description	Time switching time (s)	Power consumed Active power consumption (kW)	Reactive power consumption (kVAr)	Power generated Active power generation (kW)	Reactive power generation (kVAr)	Remarks
A	Solar power						
1	Load 1	Initial	100.0	10	97.6	4.72	As the solar power plant is of 100 kW_p capacity. Hence it fed 100 kWand 4.72 kAVr to load
2	Load 2	0.4	20	10	97.6	3.00	
3	Cap.-1	0.5	0	10	97.6	2.50	
4	Cap.-2	0.6	0	10	97.6	1.80	
	Max. value		120.00	20	97.6	4.72	
B	Grid power (off @ 0.2 s)						
1	Load 1	Initial	100	10	7.20	2.70	The grid shares minimal P&Q power initially due to filters circuit. Grid is off at 0.2 s
2	Load 2	0.4	20	10	0	0	
3	Cap.-1	0.5	0	10	0	0	
4	Cap.-2	0.6	0	20	0	0	
			120	20	7.20	2.70	
B	Captive power (starts @ 0.25 s)						
1	Load 1	0.25	100	10	0	2.40	The grid shares minimal P&Q power initially due to filter circuit. Cap. Gen. starts at 0.25s
2	Load 2	0.4	20	10	23.00	10.00	
3	Cap.-1	0.5	0	10	23.00	3.50	
4	Cap.-2	0.6	0	20	23.00	2.00	
			120	20	23.00	10.00	

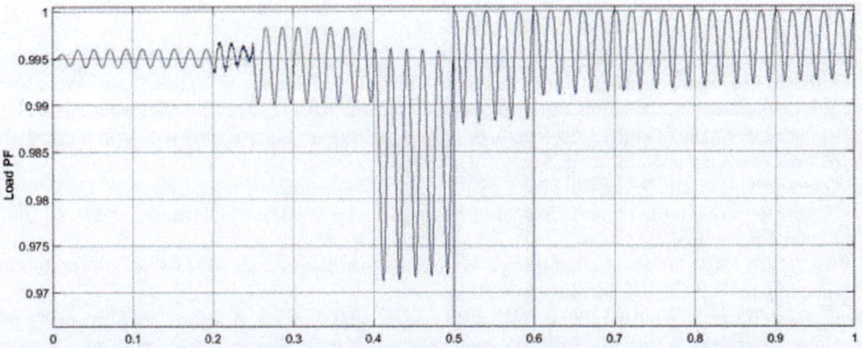

Fig. 14 Load power factor characteristics with reactive power support during grid failure

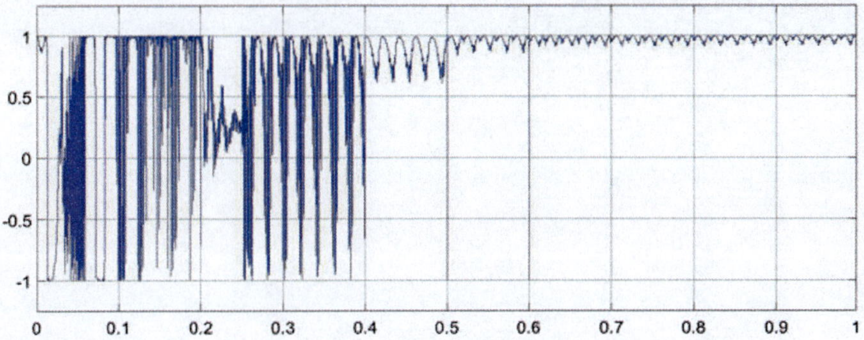

Fig. 15 Captive power source power factor characteristics with reactive power support

6 Conclusion

In an integrated distribution system where RES and captive power source are connected active power and reactive power demand on captive system should be balanced, i.e., the power factor of the captive generation plant should not be fall to the defined power factor range of the generator. Addition of more and more active power generation in the captive tied system may provide severe impact on the power factor of captive power source, which may intern damage the rotating mechanism of the mechanical system as indicated in the case study of this paper.

References

1. Simeon AM, Wanjekeche T, Hamatwi E (2019) Impacts of increased integration of Wind and Solar generators on the Namibian grid power losses. In: 2019 IEEE PES/IAS Power Africa
2. Mahela OP, Shaik AG (2013) Detection of power quality events associated with grid integration of 100kW solar PV plant. IEEE Trans 4(2):302–313
3. Ganesan S, Ramesh V, Umashankar S (2017) School of electrical engineering, VIT University, Vellore, 632014, India Hybrid Control of Microgrid with PV, Diesel Generator and Bess. Int J Renew Energy Res 7(3)
4. Jaivant NA (2015) Integration of grid connected photovoltaic system with active power filtering functionality. GSES Technical Paper (2015)
5. Cabrera-Tobar A (2019) Eduard Bullich-Massagué and Mònica Aragüés-Peñalba and Oriol Gomis-Bellmunt. Active and Reactive Power Control of a PV Generator for Grid Code Compliance Energies 12:3872. https://doi.org/10.3390/en12203872[9]AndyLeonIEEEPESChicagoChapterDecember12th,2018ReactivePowerCompensationforSolarPowerPlants,IEEEbooks
6. Ambrosio DD, Spataro W, Rongo R (2013) 2.7 Genetic algorithms, optimization and evolutionary modeling. Elsevier Ltd
7. Author Lms Sol., Environmental impact on PV based solar power plant, SPP Environmental impact, race lab 2018

Forecasting of Daily Average Power Demand for the Chhattisgarh State of India

Tamal Chatterjee and Baidyanath Bag

Abstract In this article, forecasting of daily average power demand for Chhattisgarh state of India is done by employing the statistical model of time-series analysis. In time-series analysis, the more the availability of historical data, the more accurate will be the prediction. Forecast of demand distribution is computed by model simulation for best fit as concluded by data analytics. The trend of the average power demand data is then mapped over the time horizon by employing visualization tools. Thereafter, forecasting of power demand can be done for any date in future. The traditional statistical method of time-series analysis has been chosen over state-of-the-art automated machine learning (ML) forecasting methods or the manual method of expert forecasts primarily due to its simplicity, speed, excellent visualization and mono-attribute data input. The time-series analysis model has produced a well-visualized forecast with components analysis done in a lucid way that is easy to understand even for a layman.

Keywords Mean absolute percentage error · Time-series forecasting · Power demand forecast for Chhattisgarh India

1 Introduction

Forecasting is one of the important tasks of data science that assists organizations with respect to capacity planning, strategic decision-making, goal setting, anomaly detection and course correction. Power demand forecast is a critical document for the planners in the Indian electricity domain both at the business end as well as for the strategists. The power generators use the power demand forecasts for their short-term and medium-term planning purpose channelized by the coordinating agency (SLDC,

T. Chatterjee (✉) · B. Bag
Department of Electrical Engineering, National Institute of Technology, Raipur, Chhattisgarh, India
e-mail: chatru_1977@yahoo.co.in

B. Bag
e-mail: bbag.ee@nitrr.ac.in

A. Kumar et al. (eds.), *Decarbonisation and Digitization of the Energy System*, Lecture Notes in Electrical Engineering 1099, https://doi.org/10.1007/978-981-99-7630-0_19

RLDC, NLDC/POSOCO/Grid-India). The central planner CEA supplemented by CERC advises the power ministry on the long-term infrastructure augmentation of the National Power portfolio. Hence power demand forecast plays very important role in national power domain. The more the accuracy of the forecast better is the planning and the resultant economic development down the line in every sector. Needless to mention that power sector is the lifeline of the economic development of the country and a key driver in the GDP growth. Electricity demand is function of a variety of factors including weather conditions, calendar effect, economic development and power prices [1]. Several models have been developed to forecast the electrical power's demand over a time period and, of late, it has been accelerated on account of the artificial intelligence (AI) powered data analytics-based research that has been predominant. Modeling the long run when the variables are non-stationary is less cumbersome when done by AI [2]. Testing multiple hypotheses is akin to fitting multiple predictors in regression model. Overfitting is frequent, hence data reduction and validation of model hold the key. Statistical estimation is usually model-based [3]. Refer the flow chart in Fig. 1 which showcases the lifecycle of time-series forecasting. The features of a time series are—seasonality, trend changes, outliers and holiday effects. Trend function checks out non-periodic changes in dataset of time series. Seasonality relates to periodic changes (e.g., weekly and yearly). Holiday effects relate to potential non-regular pattern over a day or more [4]. Electricity demand forecasts require modeling of dynamic non-linear relation between the various attributes affecting electrical power usage, and then taking into account long-term patterns in the attributes brought about by economic, demographic and technological development, including climate change. Peak demand forecasts are important for infrastructure planning, whereas short-term forecasts are crucial for system scheduling. Time-series forecasting lifecycle has three stages—modeling, simulating and forecasting; evaluation [5]. One of the popular concepts in the forecast modeling is time-series analysis. Data analytics tools like Python have made these models quick thinking and reasonably accurate. Multiple seasonal patterns are best handled by univariate statistical forecasting techniques. However, any cross-series information cannot be deciphered by univariate type of time-series forecasting. With the advent of technology, upgraded sensors and data storage capabilities are offering superior sampling rates (sub-hourly, hourly and daily) [6]. Short-term load forecasting is useful in real-time generation scheduling, load frequency control, load dispatch, load flow planning, workaround for potential system overload; and involves forecasts of demand in terms of days. Medium-term load forecasting is in terms of months, and helps in scheduling preventive maintenance of units, allowing organizations to plan for optimum raw material and low-cost fuel procurement. Long-term load forecast is in terms of years; and usually helps the organizations to schedule unit expansions or budget for major equipment installation [7]. Section 2 of this article elaborates on the concept of time-series analysis starting with decision dilemma for the user (business) followed by forecasting concepts involving regression analysis and statistical model of time-series analysis. Section 3 delves on fitting the model, whereas Sect. 4 discusses model validation and Sect. 5 elaborates the model evaluation. Section 6 focuses on forecasting by the use of model and Sect. 7 contains the

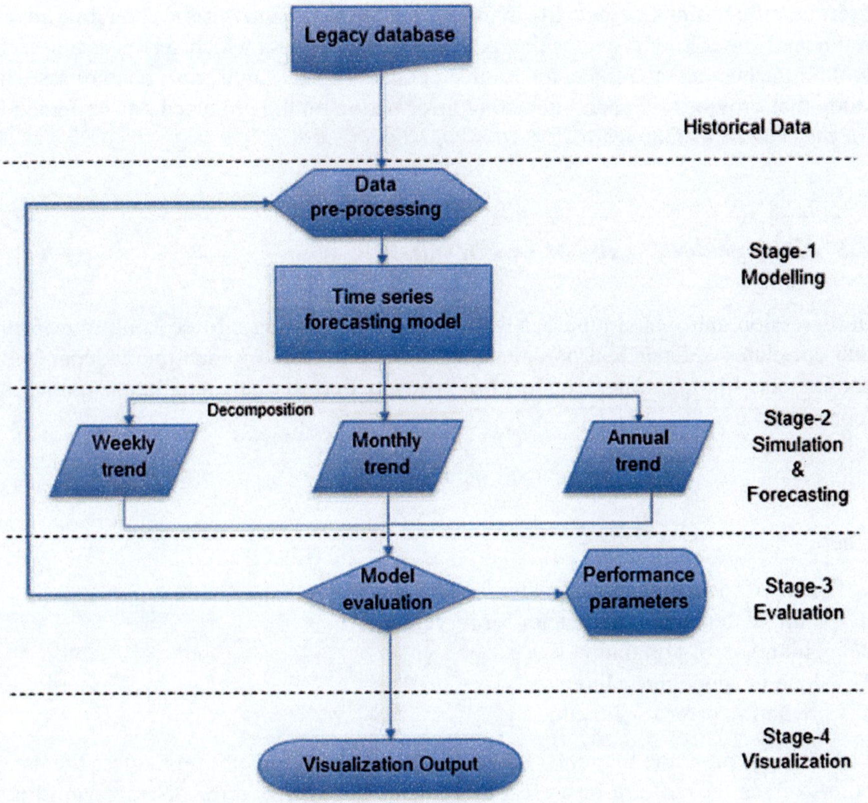

Fig. 1 Flow chart—time-series forecasting

results obtained by using the real-time data of user in the forecasting model. Section 8 elucidates the conclusion and is followed by the references used in literature survey.

2 Concept of Time-Series Analysis

2.1 Decision Dilemma

Decision-making is so critical to business strategists that on daily basis, various decisions that are made by the businesses worldwide, determine the future of companies. Decision-making involves information gathered about economic environment, financial status and the competitive landscape of market. This information is derived from data gathered over certain time horizon and processed by way of statistical analysis. When the statistical data is structured in a way to enable decision making, it becomes useful for the business, especially in foreseeing the future growth and economic environment. Forecasting provides the requisite statistical data which serves as a useful tool for the business to project the future to come. This research paper encompasses a study that provides rich and interesting information on the projected power demand for the state of Chhattisgarh.

2.2 *Regression Analysis in Forecasting*

In regression analysis, a model is developed by generating a forecasting trend line that correlates a dependent parameter vis-à-vis one or more than one independent parameters. In slope-intercept form, regression line through population points is represented by,

$$y_i = \beta_0 + \beta_1 x_i + \epsilon \tag{1}$$

where

x_i value of independent parameter for i^{th} value
y_i value of dependent parameter for i^{th} value
β_0 y intercept of population line
β_1 slope for population line
ϵ prediction error for i^{th} value

The regression line may miss some data points of the scatter plot unless the data points of scatter plot are in perfect alignment. In (1), ϵ is error of regression line during fitting such points. To check if the regression line is a good fit, the preferable method is to test the model using past data points (x and y). In this approach, the independent parameter (x values) is placed in regression model and evaluated output value ($\hat{y}$) is computed for every x value. The predictions of data points ($\hat{y}$) are then respectively checked with y values to calculate the difference. The difference $y - \hat{y}$ is the residual (pointwise error of the regression line). Residuals help to locate outliers and have a critical role to play in regression analysis which ultimately refines the forecasting output greatly. Graphical plots of the residuals can help in finding out,

1. absence of linearity,
2. non-homogeneous variation of error, and
3. whether error terms are independent.

The mandatory pre-conditions for regression analysis are homogeneity in error variation, independence of error terms and normal distribution of error terms. Error term represents changes beyond rudimentary analysis that can not be fitted in model; and is assumed to be normally distributed. Outliers lie away from the rest of the points; and often cause large residuals. Outliers can be result of mis-recorded data; and exert a pull on the regression line. Investigation of outliers helps determine if they can be retained or regression equation needs to be re-evaluated excluding them. Estimate's standard error represents standard deviation of regression model's error and can be adopted for deducing the error of model. Sum of Squares of Error (E_{ss}) is a prerequisite for estimate's standard error (s_e).

$$E_{ss} = \sum (y - \hat{y})^2 \quad (2)$$

$$s_e = \sqrt{\frac{E_{ss}}{n - 2}} \quad (3)$$

where,

n is the population count
s_e is standard error of the estimate
y is dependent parameter
$\hat{y}$ is predicted data point value of y
E_{ss} is the Sum of Squares of Error

To validate the assumption of normal data being used in regression analysis, approximately 68% of error points (residuals) shall be in $0 \pm 1s_e$ range and 95% of error points (residuals) shall be within $0 \pm 2s_e$ [2]. In time-series forecasting, the regression results are extrapolated to formulate data points beyond the domain of independent variable. In time-series analysis, data is gathered on specific attributes for a defined time period viz. 15 min block, hourly, daily, weekly, fortnightly, monthly, quarterly, semi-annually, annually, etc. which may display trend, cyclicality, seasonality and irregularity in the output. Trend displays the general direction of data over a long time duration when plotted in the form of scatter diagram or any similar form. A forecast fits a trend line through the scatter plot of time-series data and then predicts forthcoming data points using trend line's equation. To forecast the future data points using a trend line, solve for $\hat{y}$ after inserting the specified time period into the trend line equation. Equation of the trend line is obtained as follows:

$$y = b_0 x + b_1 \quad (4)$$

where,

x is the value of independent parameter
y is the value of dependent parameter

$$b_0 = \frac{\sum y}{n} - b_1 \frac{\sum x}{n} \quad (5)$$

$$b_1 = \frac{\sum xy - \frac{\sum x \sum y}{n}}{\sum x^2 - \frac{(\sum x)^2}{n}} \quad (6)$$

Prediction in regression analysis is always an estimate and not a guaranteed outcome. Hence confidence intervals and prediction intervals are required to place the predicted data point inside the context of inferential estimation. Since the predicted value might deviate slightly from the actual value, a provision of confidence interval is proposed as a range for the mean y value. A prediction interval pertains to allowance for a single datapoint; and is wider than confidence interval because it has to accommodate a wider gamut of individual values.

2.3 *Time-Series Analysis in Forecasting*

Forecasting is widely used in the strategic decision-making by businesses world over. Time-series data is the information gathered on a specific parameter (also called as a variable) over a time period. Time-series forecasting attempts to evaluate and assimilate the patterns, cycles and trends over time, or using historical data information to predict the future. Smoothing, regression trend analysis technique and decomposition are some of the common time-series analysis methods available to the modern-day data scientists. Time-series data generally comprises four elements: trend, cyclical variation, seasonal variation and irregularity; but it is not necessary for time-series data to have these elements every time. Trend is the representation of the general direction of time-series data over long time duration. Cycles are patterns through which time-series data moves between peak and troughs over a time period; usually, of more than a year. Seasonal effects are comparatively shorter in cycle, often measured by the month. Irregular fluctuations are fast variations in time-series data, which occur at time frames smaller than seasonal effects. Stationary time-series data don't exhibit any of the trend, cyclical and seasonal effects. Forecasting stationary data involves analysis of only such non-uniform, uneven and rapidly varying fluctuation effects. Overall error measurement can be achieved by any of the mean absolute error (MAE), mean absolute deviation (MAD), root mean square error (RMSE), mean percentage error (MPE) and mean absolute percentage error (MAPE) techniques. Techniques applicable on stationary data forecasting are called smoothing techniques and they smoothen out the uneven and rapidly varying fluctuations in the data. Common smoothing techniques are:

1. naïve forecasting models,
2. averaging models,
3. exponential smoothing.

Naïve forecasting models presume that incumbent data can be the best representative of predictions for future [2].

Averaging models are constructed by computing mean of data over certain time period; and using the mean so obtained, to forecast the successive time period. An elementary averaging model, viz. the simple average model uses average of the values of historical data to predict outcome of time period t. A moving average model considers an average of historical set of data points that are re-evaluated for every set of updated time period under consideration. Moving averages do not alter or adapt for trends, cycles or seasonality. A moving average model assigns certain weights for historical time periods in its calculations.

Exponential smoothing technique uses an exponentially decreasing weight values on the historical time periods in the forecast. Exponential smoothing technique follows the equation,

$$F_{t+1} = \alpha X_t + (1 - \alpha) F_t \tag{7}$$

where,

F_{t+1} is forecast for successive time instant $(t + 1)$
F_t is forecast for current time instant t
X_t is actual value for current time instant
α is exponential smoothing constant

2.4 Trend Analysis in Forecasting

Trend analysis can be done by way of the following models—Linear Regression, Quadratic Regression, Holt's Two-Parameter Exponential Smoothing Method.

Linear regression uses simple regression concept in deducing the trend line represented as,

$$Y_i = \beta_0 + \beta_1 X_{ti} + \epsilon_i \tag{8}$$

where,

Y_i is time-series data point for period i
X_{ti} is i^{th} time period

The quadratic regression considers a trend analysis to arrive at a quadratic expression represented as,

$$Y_i = \beta_0 + \beta_1 X_{ti} + \beta_2 X_{ti}^2 + \epsilon_i \tag{9}$$

where,

Y_i is time-series data point for period i
X_{ti} is ith time period
X_{ti}^2 is square of ith time period
β_1 and β_2 are the smoothing weights

In Holt's method, weights (β) are used to smoothen the trend; akin to technique used by single exponential smoothing method (α) [2].

2.5 Seasonal Effects in Forecasting and Decomposition

Seasonal effects are similar repetitive projections in data patterns that occur in time periods lower than 12 months. Decomposition is a well-known technique for filtering out effects of seasonality. The decomposition methodology utilizes *T.C.S.I* multiplicative model as a base.

Where,

T is trend
C is cyclical
S is seasonal
I is irregularity

Decomposition process entails the following steps.

Step-1: Estimate the trend

The seasonal effects are usually adjusted so that they average to 0 for an additive decomposition or they average to 1 for a multiplicative decomposition. For trend estimation, either a smoothing procedure such as moving averages is employed; or the trend is modeled with a regression equation.

Step-2: De-trending the series

For an additive decomposition, de-trending is done by subtracting the trend estimates from the series. For a multiplicative decomposition, it is done by dividing the series by the trend values.

Step-3: Seasonal factors estimation

Averaging the de-trended values for a specific season is accomplished.

Step-4: Determining random (irregular) component

For the additive model, random = series – trend – seasonal. For the multiplicative model, random = series/(trend*seasonal)

Winters' three-parameter exponential smoothing technique introduces γ as the weight of seasonality to imprint seasonal projections in forecasting. Using the aforesaid three weights α, β and γ, that represent factors of smoothing value, trend value and seasonal value for observations, forecast is accomplished [2].

2.6 Autoregression in Forecasting

When correlation occurs among error terms in regression model, the problem of autocorrelation crops up in time-series data. Ways to overcome this problem are either to use additional independent variables or to transform the independent variable.

Transforming variables can be achieved by first differences approach, percentage changes approach and autoregression models.

2.7 *Seasonal Autoregressive Integrated Moving Average Model (SARIMA)*

SARIMA is an enhancement of ARIMA (Autoregressive integrated moving average) model in which seasonality is also considered as a factor affecting the model fitment. ARIMA model is a common statistical tool for univariate data, that uses time-series forecasting encompassing autoregression, trend analysis, moving average technique, and presumes the time-series data not to be stationary (statistical parameters are not time varying) while considering that hardly any seasonality exists. It captures the standard temporal dependencies that are unique to a time-series data and its predictions can be represented algebraically as follows:

Predicted y_t = Constant + Linear combination lags of y (upto p lags) + Linear combination of lagged forecast errors (upto q lags)

In equation form above statement is represented as,

$$\hat{y}_t = \alpha + \beta 1yt - 1 + \beta 2yt - 2 + \ldots \beta pyt - p + \epsilon t + \Phi 1\epsilon t - 1 + \Phi 2\epsilon t - 2 + \cdots + \Phi q\epsilon t - q$$

where,

$\hat{y}_t$	is predicted dependent variable at instant t
α	is the constant parameter
β_1, β_2, etc.	are coefficients of lagged terms y_{t-1}, y_{t-2}, etc.
ϵ_t	is the error term corresponding to time t
Φ_1, Φ_2, etc.	are coefficients of error terms ϵ_{t-1}, ϵ_{t-2}, etc.

3 Fitting the SARIMA Model

SARIMA model, which is based on identifying the relationship between past observations and future predictions, is characterized by key parameters - p, d, q and S which represent the autoregressive, integrated (difference), moving average and seasonal orders respectively. The time-series dataset is split 80:20 into two sets, one for training and other for testing. The training set comprising of date stamps up to 31-Mar-2022 is employed in model fitting. The steps to fit SARIMA model using p, d, q *and* S parameters are as follows:

Step-1: Check for stationarity/Convert to stationary time series

Initial step in fitting SARIMA model is to deduce if time-series is stationary. A stationary time-series has fairly constant statistical parameters (viz. mean and variance) over time. If a time-series is found to be non-stationary, it is to be made stationary by taking the difference between consecutive observations until it becomes stationary. The number of times the difference operation is performed is the value of parameter '*d*'. Value of *d* is deduced from the outcome of ADF test (Augmented Dickey-Fuller test) which determines the factor that makes data stationary. Null hypothesis of 'a unit root is present in the time series' is tested by ADF and success indicates non-stationarity. A significant indicator of stationarity of test series is the *p*-value. The probability of getting a test statistic as extreme or more extreme than the one computed from available data is *p*-value. However, here assumption is that the null hypothesis of non-stationarity is true. It may be noted that *p*-value is different from the *p* parameter of SARIMA model. The *p*-value is representative of the strength of evidence against the null hypothesis of non-stationarity. A smaller *p*-value indicates stronger evidence against non-stationarity and supports the hypothesis of stationarity. If *p*-value of ADF test is lower than 0.05 significance level, null hypothesis is rejected and the time series is considered stationary. If time series is found to be stationary then assign $d = 0$. If time series is non-stationary, further analysis such as differencing or seasonal differencing may be necessary to make the time series stationary before fitting SARIMA model. To achieve stationarity, if differencing is required once, then $d = 1$, if twice then $d = 2$ and so on. The ADF stationarity test outcome is as follows:

ADF Statistic: −3.636009

p-value: 0.005109

Conclusion: The time-series is stationary.

Step-2: Identify order of differencing (*d*)

Parameter *d* represents the degree of differencing essential to make time series stationary. In differencing the differences between consecutive/interval-based readings in the time series is computed. Thus, *d* determines how many times the differencing is applied to the data to make it stationary. From Fig. 2, it is evident that the statistical parameters have levelized after first differencing. Hence value of *d* comes out to be 1.

Step-3: Identify the autoregressive (AR) Term (*p*)

The autoregressive term captures the relationship between the previous observations and the current observation. The parameter *p* is the number of lagged values that will be used as predictors in the model. A higher value of *p* means that the model will use more lagged values as predictors. Order of AR term is determined visually from the PACF plot of the differenced time series. If the PACF plot shows significant spikes up to *p* lags, then order of AR term is *p*. This is achieved by analyzing PACF plots. A PACF plot is a type of plot used in time-series analysis that shows partial correlation between a time series and its own lagged values, after accounting for

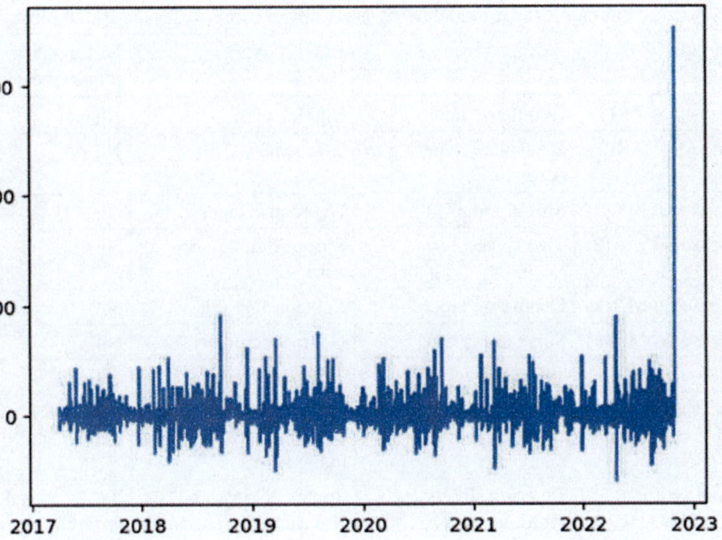

Fig. 2 Plot of first differences

the correlation at shorter lags. Thus a PACF plot shows how much a time series at time t is correlated with its own values at times t–k, after removing the effects of the intermediate values between t and t–k. The PACF bar chart has lag (or time difference) on x-axis and partial correlation coefficient is assigned to y-axis. The partial correlation coefficient has a range of –1 to 1, where 1 means perfect positive correlation, 0 means no correlation, and –1 points to a perfect negative correlation. In a PACF plot, the correlation coefficient at lag 0 is always 1, since a time series is always perfectly correlated with itself at the same time. The PACF plot helps identify appropriate order of autoregressive (AR) model for a time series. The number of lagged values of the time series that are included in the model decides the order of AR. Term p is deduced from Fig. 4 by correlating with the chart of Table 1.

Step-4: Identify Moving Average (MA) Term (q)

Moving average term captures the relationship between the previous forecast errors and the current observation. The parameter q denotes order of moving average (MA) component, which captures the influence of past error terms on the current value. In other words, it is the number of lagged error terms that will be used as predictors in the model. A higher value of q means that the model will use more lagged error terms as predictors. Order of MA term is determined by looking at the ACF plot of the differenced time series. If the ACF plot shows significant spikes up to q lags, then order of MA term is q. An ACF plot is a type of plot used in time-series analysis that shows the correlation between a time series and its own lagged values. In other words, an ACF plot shows how much a time series at time t is correlated with its own values at instants t–1, t–2, t–3 and so on. The ACF plot is a bar chart with the lag (or

Table 1 Identification chart for model fit vis-à-vis differencing, ACF and PACF plots

Model type (fit)	ACF pattern	PACF pattern	Differencing requirement
Non-seasonal AR(1)	Geometric taper	Single spike	Nil
Non-seasonal AR(2)	Sinusoidal, tapering to zero	Two spikes	Nil
Non-seasonal MA(1)	Single spike	Geometric taper	Nil
Non-seasonal MA(2)	Two spikes	Sinusoidal, tapering to zero	Nil
Non-seasonal MA(k)	Geometric taper	Spike at lag of k	Nil
Non-seasonal MA(k)	Spike at lag of k	Geometric taper	Nil
ARMA	Tail off to zero	Tail off to zero	Nil
Non-seasonal ARIMA	Do not tail off, but stay close to zero over many lags	Do not tail off, but stay close to zero over many lags	Yes (d = 1 for linear trend)
Seasonal (without trend) ARIMA (k, 0, 0)X(k, 0, 0)$_S$	Tapering slowly at multiples of S	Spikes at lag k and S (for *S*th differences)	Yes (seasonal differencing of lag S)
Seasonal (with trend) ARIMA (0, 0, k)X(0, 0, k)$_S$	Spike at lag k, S, S $\pm$ k	Non-seasonal tapering at lag = k and seasonal taperings at intervals S	Yes (seasonal and if required, non-seasonal differencing)
Seasonal (with trend) ARIMA (1, d, 0)X(0, 1, 1)$_S$	Non-seasonal behavior: tapering pattern in early lags Seasonal behavior: cluster of spikes at lag S	Non-seasonal behavior: spike at lag 1 and S–1 Seasonal behavior: taperings at intervals S	Yes (seasonal and if required, non-seasonal differencing)

time difference) on *x*-axis and correlation coefficient on *y*-axis. The autocorrelation coefficient has a range of –1 to 1, where 1 means perfect positive correlation, 0 means no correlation, and –1 points to a perfect negative correlation. In an ACF plot, the correlation coefficient at lag 0 is always 1, since a time series is always perfectly correlated with itself at the same time. The ACF plot can help identify the presence of autocorrelation in a time series. Term *q* is deduced from Fig. 3 by correlating with the chart of Table-1.

Step-5: Identify seasonality factor (*S*)

For weekly model, let us look at corresponding ACF and PACF (Figs. 3a and 4a) to check out lags 7, 14 and so on. In the ACF, there's a clear negative spike at lag 7 and then not much else. The PACF tapers in multiples of S; that is the PACF has significant lags at 7, 14, 21 and so on. This matches the attributes of a seasonal weekly MA(1) component. Since we are looking at 7th differences, the model we adopt for the weekly series is SARIMA $(0, 1, 1)_7$.

Fig. 3 ACF Plot

Seasonality	ACF Plot
S=7	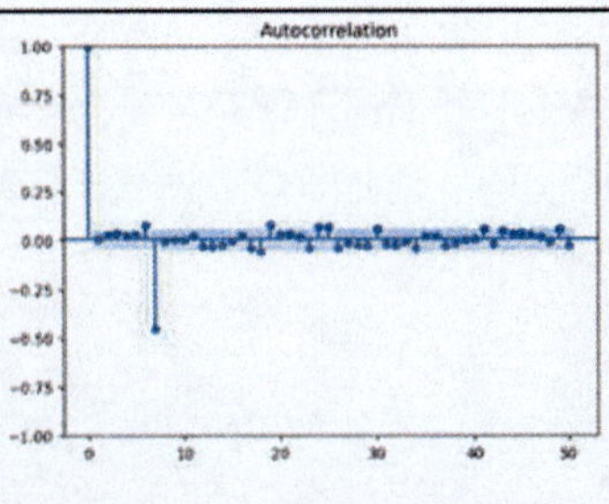 a
S=30	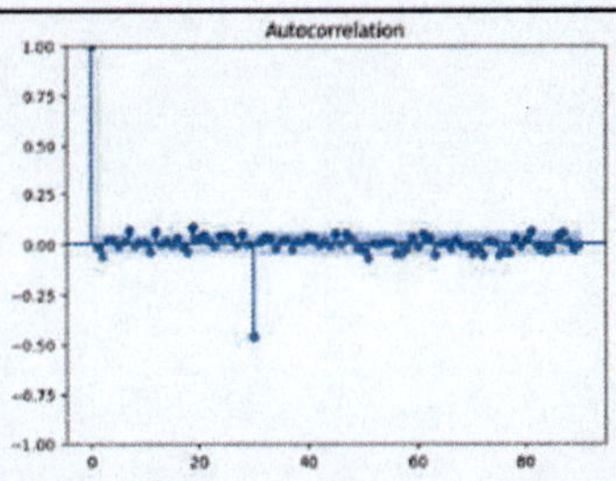 b
S=12 (yrly)	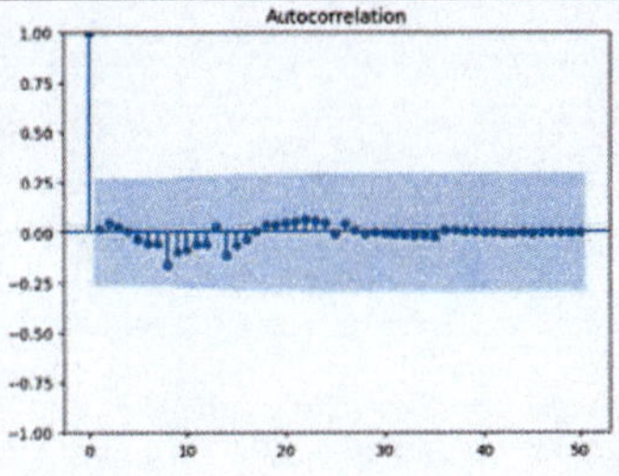 c

For month-wise yearly model, let us look at corresponding ACF and PACF (Figs. 3b and 4b) to check out lags 30, 60 and so on. In the ACF, there is a clear negative spike at lag 30 and then not much else. The PACF tapers in multiples of S; that is the PACF has significant lags at 30, 60, 90 and so on. This matches the attributes of a seasonal weekly MA(1) component. Since we are looking at 30th differences, the model we adopt for the weekly series is SARIMA $(0, 1, 1)_{30}$.

For annual trend model, let us look at corresponding ACF and PACF (Fig. 3c and Fig. 4c) to check out lags 12 24, and so on. The data has been converted to 12 months lag representing a year with differencing done twice ($d = 2$). In the ACF, there's

Fig. 4 PACF Plot

Sea-son-ality	**PACF Plot**
S=7	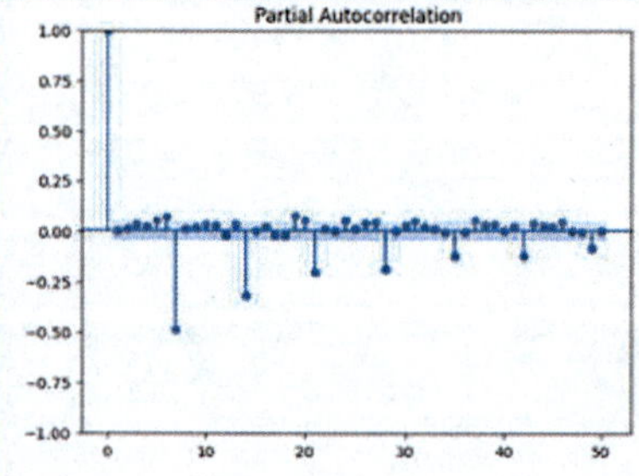 a
S=30	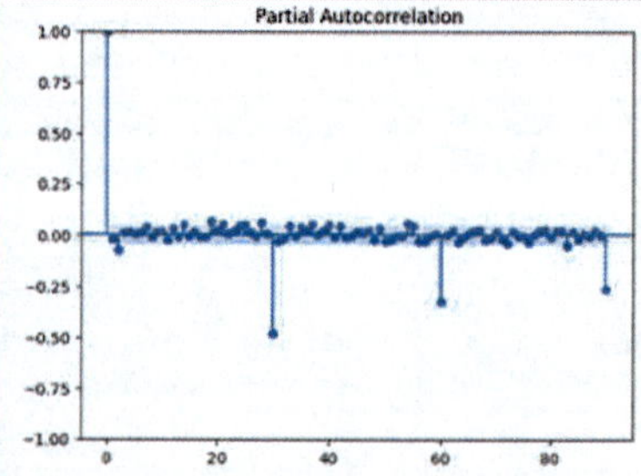 b
S=12 (yrly)	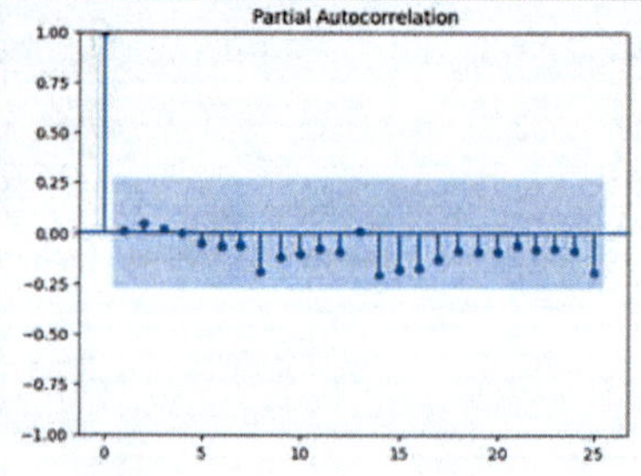 c

a clear seasonal tapering with $S = 12$. The PACF does not have any clear spike or seasonal trend. Hence, as listed in Table 1, the model most suitable here is ARIMA(1, 2, 0)X(0, 1, 1)$_S$, which is a combination of non-seasonal and seasonal model.

ACF and PACF Plots for the weekly, monthly and annual trend models.

4 Validating the SARIMA Model

After fitting the SARIMA model using training data, it is validated using balance of the data post 31-March-2022 which forms the test dataset. This translates into 20% of time-series dataset being utilized for testing. In testing phase, model generates predictions for test data series (as a forecast) which is then compared for accuracy with the actual test dataset. Figure 5a–c are the plots of the legacy datasets corresponding to the three models—trend, weekly seasonality and yearly (month-wise) seasonality. Figure 6 depicts the validation for test dataset apart from generating a forecast for next 12 months horizon too. From Fig. 6, it is apparent that model fits data well and can be used for forecasting future values.

5 Evaluating the SARIMA Model

The model's accuracy is evaluated with the help of following metrics:

Mean Absolute Percentage Error (MAPE) metric: MAPE as a metric, is easy to understand and interpret, and it provides a measure of relative error, which can be compared across different time series with different scales or units of measurement. It is a common measure of forecasting accuracy that calculates average absolute percentage difference between forecasted values and actual values in a time series. It is expressed as a percentage, and a lower MAPE indicates better forecasting accuracy. The formula for calculating MAPE is,

$$\text{MAPE} = \frac{1}{n} \times \sum \left| \frac{(y_t - \hat{y}_t)}{y_t} \right| \times 100\%$$

where,

y_t is actual value at instant t
$\hat{y}_t$ is forecast at instant t
n is number of observations

Root Mean Square Error (RMSE) metric: It is the square root of MSE (Mean Square Error) metric. It is useful for evaluating the performance of the model when the error is large. Its calculation is done as follows:

$$\text{RMSE} = \sqrt{\frac{1}{n} \times \sum (y_t - \hat{y}_t)^2}$$

Mean Absolute Error (MAE) metric: It measures the absolute difference between actual values and the predictions. It gives an idea of how far are forecasts from the predictions. It is calculated as follows:

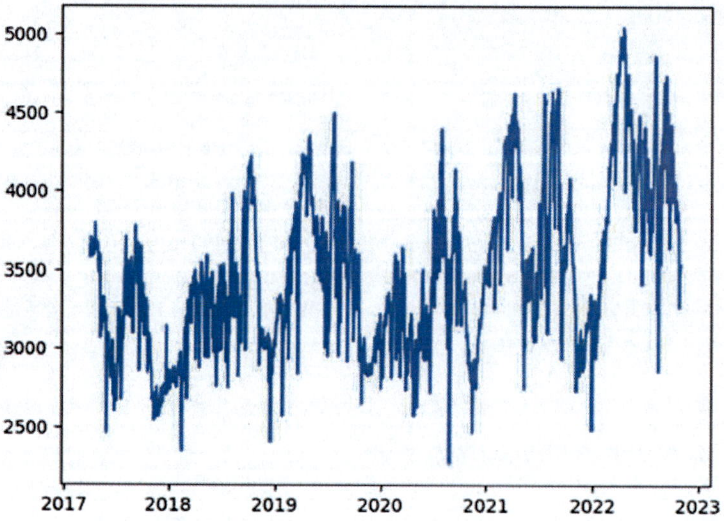

a. Plot of trend (annual) time series dataset without forecasting (*x*-axis label being years and *y*-axis label being Demand in MW)

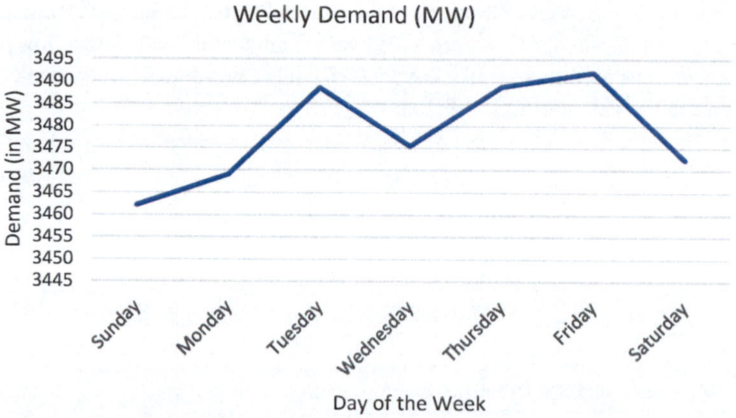

b. Plot of weekly (day wise) time series dataset without forecasting

Fig. 5 **a** Plot of trend (annual) time-series dataset without forecasting (*x*-axis label being years and *y*-axis label being demand in MW). **b** Plot of weekly (day-wise) time-series dataset without forecasting. **c** Plot of yearly (month-wise) time-series dataset without forecasting

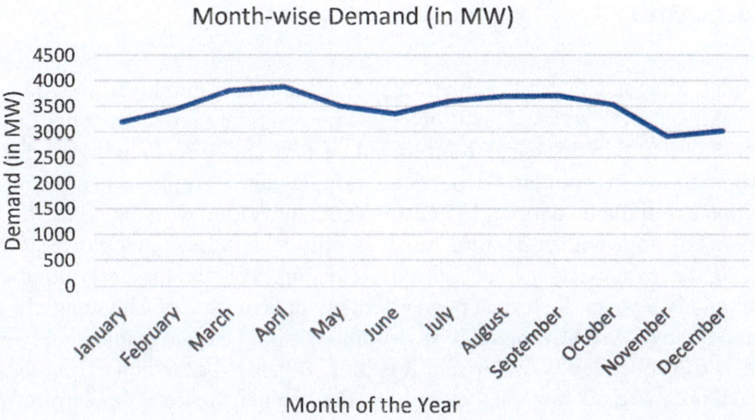

Fig. 5 (continued)

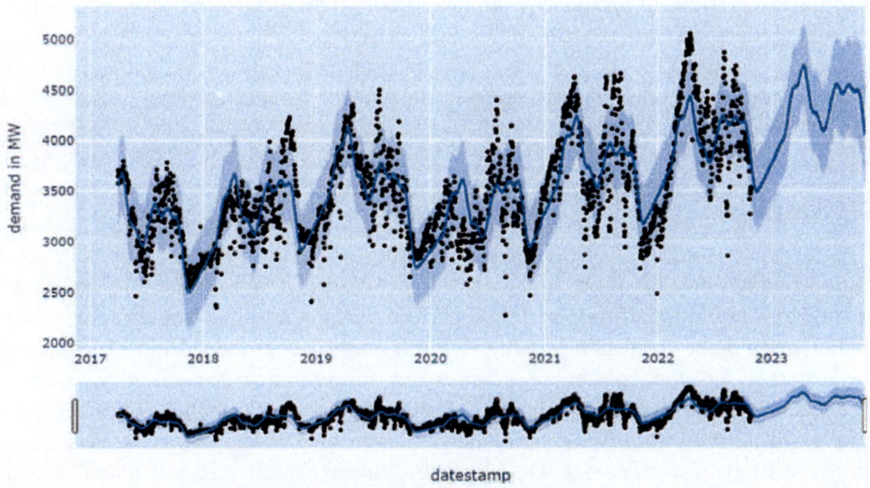

Fig. 6 Trend analysis and projection of average power demand

$$\text{MAE} = \frac{1}{n} \times \sum \left| \left(y_t - \hat{y}_t \right) \right|$$

The evaluation results are tabulated in the form of a comparison table as follows:

6 Forecasting

The model, after satisfactory evaluation and validation, is employed for forecasting future values of time series. This is achieved by extending the independent variable (date or date stamp in this case) to future values (extrapolating to one-year duration) and keeping the corresponding dependent variable values empty. When the model is run on this extrapolated dataset, forecast is generated along with the predicted variations as an envelope with upper limit and lower limit; sans the scatter plot. Scatter plot represents the legacy data's visualization. Refer Fig. 6 for the forecast output of past approximately 5 years' historical power demand data in state of Chhattisgarh, which is carried out by SARIMA model with decomposition. Daily average power demand is used as the dependent variable and it is plotted date-wise. When extrapolation of the model is carried out beyond the available data points, the trend has a uniform rate. Uncertainty in trend is gauged by taking generative model forward; and it is assumed that future will have the same mean frequency and value of rate changes as in the past. The model has superiority in fitting history and so the error value drops. However, if projected forward, it produces wide uncertainty intervals. Further decomposition outputs of this model are illustrated as trend, weekly seasonality variation and yearly seasonality components in Fig. 7a–c. The benefit of this model is that it provides view of individual component (decomposition) of the forecast separately [8].

7 Results

Figure 5 shows the plot of original dataset before forecasting in which Fig. 5a corresponds to trend (annual) time series; Fig. 5b corresponds to the weekly (day-wise) model; and Fig. 5c corresponds to the yearly (month-wise) model. Figure 6 captures the scatter plot of the daily average power demand with date pegged on the x-axis. The forecast of the optimum value of demand for the next year (blue line) has also been projected with the worst case and best case of predictions shown in shaded envelope (light blue in color). The upper and lower ceiling values denote the best case and worst case values which take care of the deviations from the optimum values of forecast. While Fig. 6 is the regression line of the forecast, Fig. 7 decomposes it into the forecast of trend, weekly seasonality and annual (month-wise) seasonality plots. Figure 7a is the annual trend line of power demand for the historical and forecast period; Fig. 7b is the weekly seasonality of the power demand and Fig. 7c is the month-wise seasonality of the power demand. To arrive at the actual/predicted value of a specific date, we need to recompose (add up) the three corresponding values from the decomposition plots of Fig. 7. Figures 3 and 4 respectively depict the ACF and PACF plots for the three decomposition models. Table 2 has the comparison of the MAPE, RMSE and MAE metrics for the decomposition models. The RMSE and MAE metrics of weekly trend model come out to be similar on account of low number of datasets in the series.

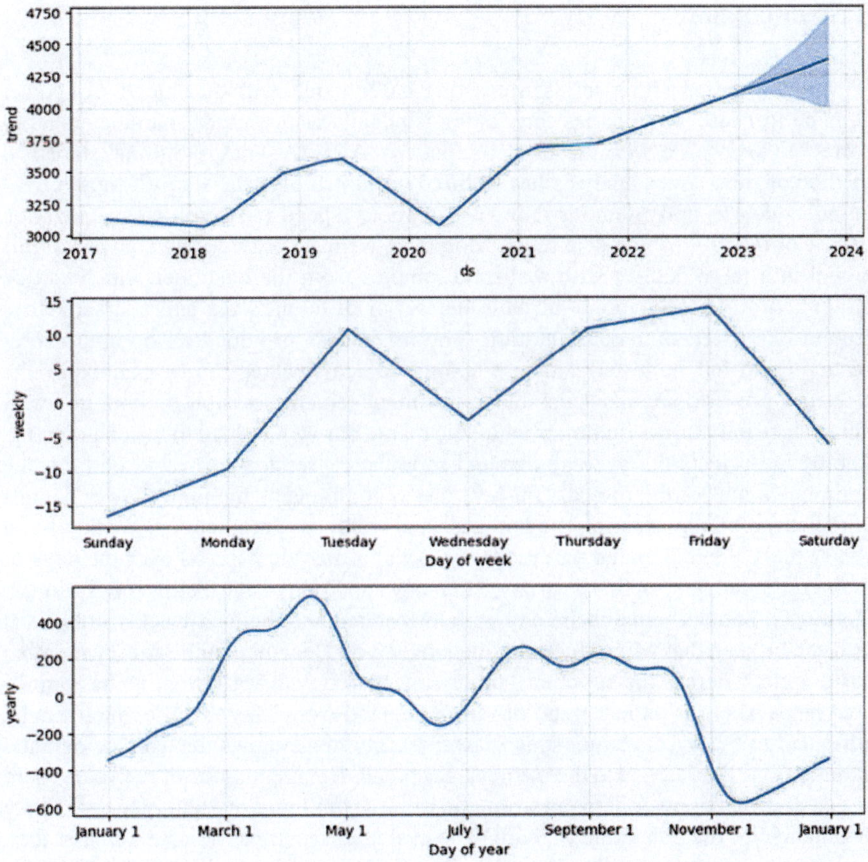

Fig. 7 Trend, weekly and yearly seasonality of average power demand. **a** (figure on top): Annual seasonality trend (with *x*-axis bearing date stamp and *y*-axis as demand in MW). **b** (figure in middle): Weekly seasonality (with % load variation from baseline value plotted on *y*-axis). **c** (at bottom): Month-wise seasonality (with load variation in MW from baseline value plotted on *y*-axis)

Table 2 Comparison of model evaluation metrics

Evaluation metric	Trend (annual) model	Weekly trend model	Monthly trend model
MAPE (%)	6.74	2.9	3.57
RMSE (MW)	298.7	100.9	128.4
MAE (MW)	232.8	100.9	123.2

8 Conclusion

We have presented a time-series forecasting-based model with the objective of power demand forecast. Time-series forecasting is adaptive since model parameters auto-correct to keep pace with the changing pattern. ARIMA—the traditional statistical method of time-series analysis has justified its choice over the state-of-the-art ML-based models by aptly demonstrating its simplicity, speed and excellent visualization in case of mono-attribute data input. Moreover, we have provided an analysis of this model in a large dataset with real-time samples from the customer which can be re-applied on another mono-attribute dataset of different sizes and characteristics without any need of reconfiguration. The model has provided an optimum forecasting trend for the tested database and the forecast values can be utilized by the electrical power domain's stakeholders for management decision-making in developing electricity infrastructure in a holistic manner as well as load dispatch planning. The trend projects the average demand in linear regression form and outlines the optimum values of the forecast for next one year alongside the upper/lower bounds of the envelope representing the best case and worst case predictions. Weekly trend shows the average demand variations vis-à-vis mean value spread over the days of the week. It can be seen that Tuesday, Thursday and Friday have distinct peaks during the weeks. Sunday remains the day with lowest demand, being a weekly holiday. It also can be seen that winter (comprising November, December and January months) has troughs whereas the summer (comprising March and April months) has prominent peak. There is a clear trend of fall in demand from May to July which can be attributed to the onset of monsoon season. Yearly trend shows the average demand variations with reference to the mean value spread over the months of the year. There is a clear drop in power demand from mid of 2019 until mid of 2020, which is likely attributable to the pandemic (Covid). The evaluation metrics indicate a higher than expected error value possibly because of the heterogeneity resulting in higher noise in the time series on account of the Covid time period.

References

1. Black K (6th edn) Business statistics for contemporary decision making. Wiley Plus
2. Harris R, Sollis R (2003) Applied time series modelling and forecasting. John Wiley & Sons
3. Harrell FE Jr (2001) Regression modelling strategies: with applications to linear models, logistic regression, and survival analysis. Springer
4. Taylor SJ, Letham B (2017) Forecasting at scale. Peer J Preprints 5:e3190v2
5. Hyndman RJ, Fan S (2015) Monash electricity forecasting model. Monash University, Business and Economic Forecasting Unit
6. Bandara K, Bergmeir C, Hewamalage H (2020) LSTM-msnet: leveraging forecasts on sets of related time series with multiple seasonal patterns. IEEE Trans Neural Netw Learn Syst
7. Dewangan F, Abdelaziz AY, Biswal M (2023) Load forecasting models in smart grid using smart meter information: a review. Energies 16(3):1404

8. Kumar SR, Binod S, Ranjan NJ, Shashikant (2021) Short/medium term solar power forecasting of Chhattisgarh state of India using modified TLBO optimized ELM. Eng Sci Technol Int J. Elsevier

Power Management Strategy for Battery-Supercapacitor-Based HESS in a Residential Grid-Connected PV System

Nibha Kumari, Sourabh Ghosh, Asheesh K. Singh, Navneet K. Singh, and S. N. Singh

Abstract The rapid growth in electric vehicles and other residential loads requires renewable energy sources-assisted infrastructure to downsize the load on the utility grid. It becomes necessary to introduce grid-scale energy storage systems (ESSs) with renewable energy sources to increase the reliability of the power supply for critical loads. These ESSs may be electrochemical batteries, supercapacitors, flywheels, fuel cells, etc. The life cycle of these ESSs, especially for electrochemical batteries, starts to degrade during large transients during charging or discharging. This calls for hybrid energy storage systems (HESSs) which can reduce battery degradation of costly ESSs by incorporating supercapacitors. In this paper, a battery pack is combined with a supercapacitor to constitute a HESS for the residential grid-connected PV system. This configuration is preferable due to the high density of energy of the battery pack, and the high density of power of the supercapacitor. Moreover, a simplified power management scheme is proposed for effective coordination between the sources of the HESS. The proposed approach is proved to be effective by validation on the MATLAB® Simulink platform by analyzing the characteristics of the HESS under two different demand/generation scenarios.

Keywords Hybrid energy storage systems · Power management algorithm · Renewable energy sources · Battery degradation

---

N. Kumari · S. Ghosh (✉) · A. K. Singh · N. K. Singh
Motilal Nehru National Institute of Technology Allahabad, Prayagraj, India
e-mail: ghoshsourabh27@yahoo.com

S. N. Singh
Atal Bihari Vajpayee-Indian Institute of Information Technology and Management, Gwalior, India

A. Kumar et al. (eds.), *Decarbonisation and Digitization of the Energy System*, Lecture Notes in Electrical Engineering 1099, https://doi.org/10.1007/978-981-99-7630-0_20

1 Introduction

The intermittent generation of power through renewable energy can be overcome by employing appropriate Energy Storage Systems (ESSs) either at the common point or at the DC link. It is the case when renewable power generation is unable to satisfy the demand and the deficient power gets supplied by the installed ESSs. Generally, these ESSs are electrochemical and possess high-energy densities. However, during a sudden change in generation or load demand, it is probable to get a high-frequency transient deficit or excess of power. This condition should be avoided as a high frequency transient current will lead to battery degradation. The life-cycle degradation of electrochemical batteries may be slow or fast depending on the frequency of such transient occurrences. It is known that electrochemical batteries have low power densities and high-energy densities. In the literature, such a configuration is termed a Hybrid Energy Storage System (HESS). The most widely used HESS uses a battery pack and a supercapacitor connected via power converters at the DC link of any system [1]. In this hybrid configuration, the battery pack charges/discharges with low-frequency mean power and the supercapacitor charges/discharges with high-frequency transient power. In other words, the supercapacitor acts as a buffer to protect the battery pack from frequent and transient power demand, and, thus, improves battery life.

A standalone PV system with HESS and loads suffers from battery degradation due to the negligence of the states of the battery and the supercapacitor [2]. A power management algorithm for a DC microgrid and HESS is proposed in [3], with stability analysis of power converters using small-signal transfer functions. A centralized energy management scheme for HESS is proposed in [4], without determining the issues of power quality. A model predictive controller is employed for regulation of voltage of the DC link and synchronization of grid in the presence of HESS [5]. A real-time power-sharing in HESS predictive optimization algorithm is proposed, based on the prediction and probability of state trajectories along with system losses [6]. A simple and effective real-time supervisory energy management system is implemented using the fuzzy logic controller for HESS [7]. This technique suffers from poor adaptive correction for its control systems. A joint control strategy is developed for standalone PV systems, HESS, and residential loads, which focuses mainly on DC link voltage regulation and neglects the states of the battery and supercapacitor [8]. A rule-based power management scheme is proposed for HESS, which is described to select an appropriate parameter for various modes [9]. A case study of residential zero energy building is presented by considering the self-utilization of RESs and HESSs followed by an economic assessment [10].

A simplified power management algorithm (PMA) is proposed for battery-supercapacitor and is validated using simulation case studies. This prioritizes the utilization of the renewable power generated, battery, and supercapacitor before the utility grid. The primary contributions of this paper are as follows.

(1) A simplified power management algorithm for battery-supercapacitor HESS is proposed for a PV system connected with grid.

(2) The Energy storage is connected at the DC link via fully active converter configuration using separate controllers.
(3) Stable DC link voltage, bidirectional power flow capability, and self-utilization of renewable generation and energy storage systems.
(4) Simulation studies on both excess and deficit renewable power validate the proposed power management algorithm.

The remaining paper is organized into different sections can be given as. A short literature review is discussed in Sect. 1. The architecture of system and design parameters are described in Sect. 2. The simplified power management algorithm is proposed in Sect. 3. The analysis of the simulation results for two different situations is discussed in Sect. 4. Finally, the paper is concluded in Sect. 5.

2 Grid-Connected PV System with HESS

The configuration of the PV-connected grid system with HESS is illustrated in Fig. 1. It contains a PV, battery, supercapacitor, and variable resistive loads at the DC link. The PV system is integrated with the battery and supercapacitor via a boost converter. The most widely utilized technique is perturbation and observation technique for the PV system to achieve the maximum power point. It is assumed that the dc link voltage should be regulated at 480 V, irrespective of the variations in generation and load demand. The DC bus is connected to the grid via a three-phase inverter. This inverter is accountable for the control and regulation of the DC link voltage. A voltage controller is used to maintain the dc bus voltage nearly reference value. It is assumed that the system is free of parasitic losses, and the conventional proportional-integral control method is employed for each power converter in the system. An active parallel bidirectional converter configuration of the battery and the supercapacitor is used for better utilization and efficiency. This topology is used to control the power flow and voltage, separately at different levels. Both bidirectional converters can work in buck and boost mode, where the mode is decided through PMA.

In this paper, a PV array of 250 Kw is considered with a short-circuit current of 6.09 A and an open-circuit voltage of 85.3 V. The rated voltage of the battery is taken as 300 V, having capacity of 500Ah and the supercapacitor is rated at 35 V having capacitance of 58F. The residential loads are modeled as purely resistive loads of 20 Kw each. The supply frequency is taken as 50 Hz, and for grid-synchronization purposes, a PLL (phase-locked loop) is used. A detailed specification of the system is given in Table 1.

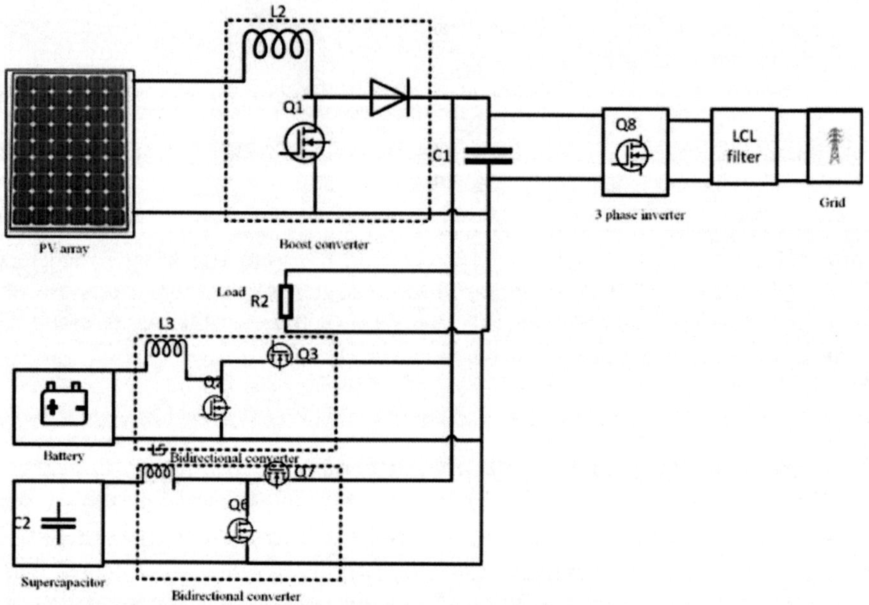

Fig. 1 Configuration of the grid-connected PV system with HESS

Table 1 System parameters for the grid-connected PV system with HESS

Parameter	Value
Rated power of PV array	250 kW
Nominal voltage of battery	300 V
Nominal voltage of supercapacitor	35 V
Cut-off frequency of the low pass filter	5 Hz
Inductor in the battery-side converter	20 mH
Inductor in the supercapacitor-side converter	5 mH
Capacitor in the battery-side converter	500 μF
Capacitor in the supercapacitor-side converter	220 μF

3 Power Management Algorithm

The main purpose of the proposed PMA is to maintain the DC link voltage while considering the SOC of the supercapacitor and battery. The proposed PMA also considers renewable power generation and load requirements. The self-sufficiency of the system gets encouraged as the PMA prioritizes PV generation followed by HESS, for load demand fulfillment.

The deficit power is supplied by the utility, if PV is not generating sufficient power and HESS sources have reached below their minimum state of charge (SOC) limits.

Similarly, the excess power is injected into the utility, if PV is generating excess power and HESS sources have reached above their maximum SOC limits. The prime goal of the HESS is used to share the stress on the battery with the companion supercapacitor, and reduce the battery degradation. The proposed flow chart of PMA for HESS, consid ering the state of charge of the supercapacitor and battery, is illustrated in Fig. 3, that ensures no deep charge or discharge of the battery. The difference in generated PV power with load power gives rise to two cases as mentioned here. (1) Excess power mode (PPV > pLoad). (2) Deficit power mode (PPV < pLoad). An average current is extracted through LPF using cut-off frequency w_c (2*pi*5rad/sec). The transient current is achieved by excluding the average current from the total current as shown in Fig. 2. The power equation should be maintained to get a stable system as given through Eqs. (1)–(5).

$$P_g(t) + P_{pv}(t) + P_B(t) + P_{sc}(t) - P_l(t) = P_t(t) \tag{1}$$

$$P_l(t) = P_{avg}(t) + P_{tran}(t) = V_{dc} i_t(t) \tag{2}$$

$$i(t) = \frac{P_{avg}(t)}{V_{dc}} + \frac{P_{tran}(t)}{V_{dc}} = i_{avg}(t) + i_{tran}(t) \tag{3}$$

$$i_{avg}(s) = \frac{\omega_c}{s + \omega_c} i_t(s) \tag{4}$$

$$i_{tran}(s) = \left(1 - \frac{\omega_c}{s + \omega_c}\right) i_t(s) \tag{5}$$

4 Results and Discussion

The efficacy of the described PMA is validated by simulation on the MATLAB® Simulink platform by analyzing the characteristics of the PV array and HESS under two different demand/generation scenarios. The proposed model is assessed under both variable generation and load demand. The simulation results show that the DC bus voltage is well regulated at the time of sudden change in generation and demand. The variation in generated power through PV and load gives rise to two modes, i.e., excess power mode (battery and super capacitor charges), and deficit power mode (battery, super capacitor discharge).

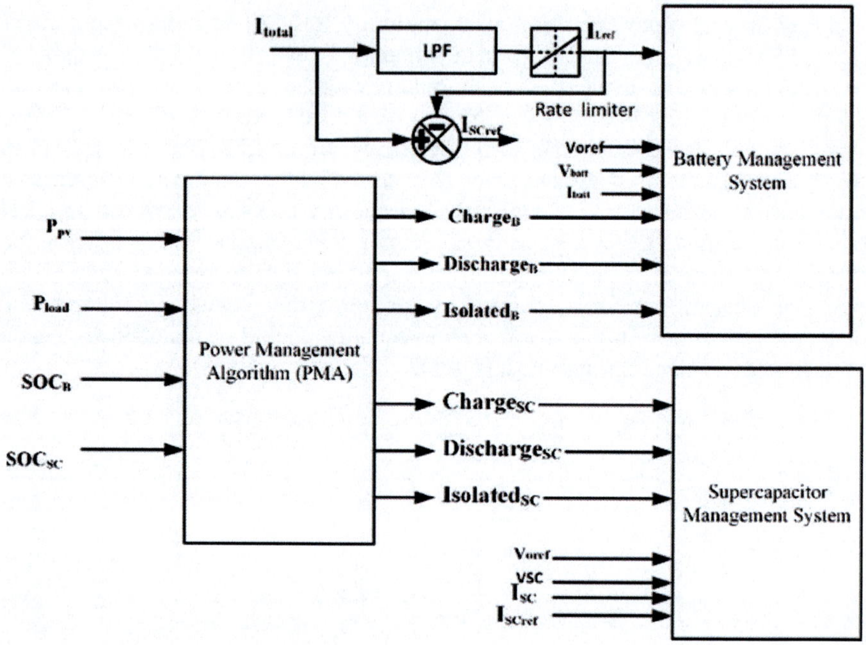

Fig. 2 Proposed PMA for the HESS

4.1 *Case I: Fixed Load with Step Variation in Generation*

A sudden change in solar irradiance is created every 0.5 s which leads to change in solar PV generation. It is evident from the results shown in Fig. 4., that excess PV generation serves the load, and remaining power is used for charging HESS followed by injection to the grid. Similarly, deficit PV generation leads to discharging of the HESS to serve the required load power, and remaining power is taken from the grid. Also, the supercapacitor takes the high-frequency dynamics in power, and the battery is left with high-frequency variations in the power.

4.2 *Case II: Fixed Generation with Step Variation in Load*

A sudden change in load demand is created every 0.5 s which leads to change in load current. It is evident from the results depicted in Fig. 5, that deficit PV generation leads to discharging of the HESS to serve the required load power, and remaining power is taken through the grid. Similarly, excess PV generation serves the load, and remaining power is used for charging HESS followed by injection into the grid.

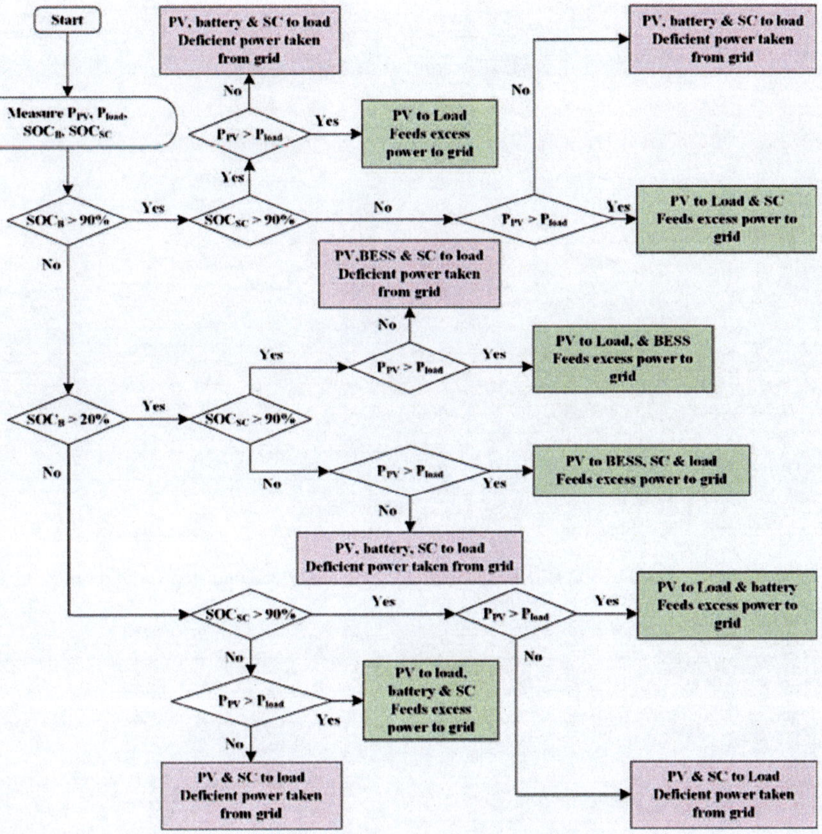

Fig. 3 Rule-based PMA for the HESS considering excess and deficit power

Also, the supercapacitor takes the high-frequency dynamics in the power, and the battery is left with high-frequency variations in the power.

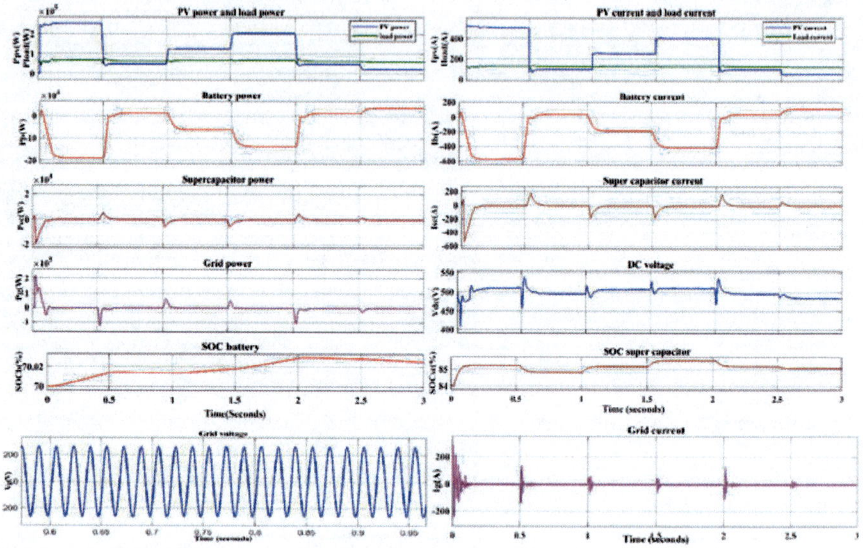

Fig. 4 Characteristics of PV, HESS, and grid during a fixed load and step variation in generation

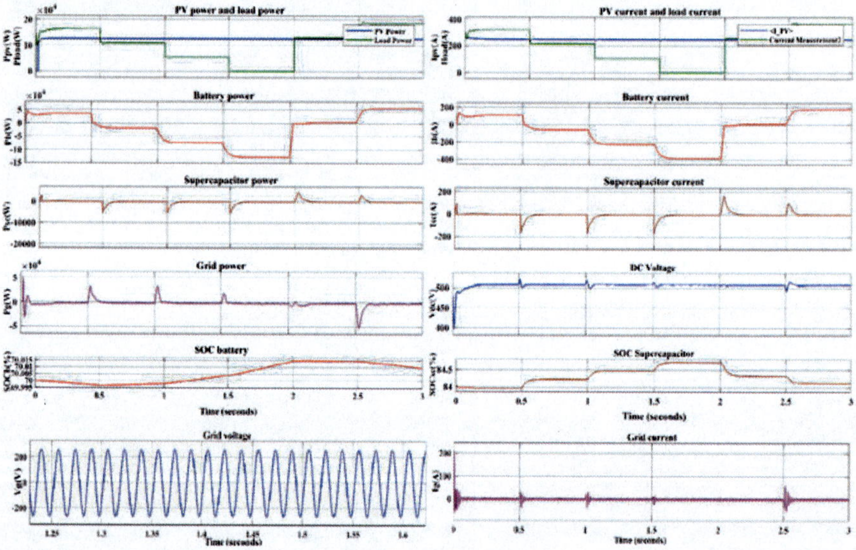

Fig. 5 Characteristics of PV, HESS, and grid during a fixed generation and step variation in load

5 Conclusion

This paper demonstrates a simplified power management scheme for PV connected to grid system with a battery-supercapacitor HESS. The proposed power management scheme ensures better DC link voltage regulation, smoother operation of the battery pack, and reduced burden on the utility grid. The battery pack and the supercapacitor allow bidirectional power flow due to a fully active parallel configuration and separate control structures. The PV connected to grid system employing HESS is simulated on MATLAB® Simulink platform under two scenarios, i.e., variable generation with fixed load, and fixed generation with variable load. The proposed scheme is easy to implement, requires lower computation effort, maintains system stability, and improves the lifetime of the battery pack by handling quick dynamics occurring within 0.1 s.

Acknowledgements This work is supported by the Ministry of Education, Government of India under the Scheme for Promotion of Academic and Research Collaboration (SPARC) with the project titled "E-Mobility: An Electricity Grid Perspective (P1542)".

References

1. Tummuru NR et al (2019) Control strategy for AC-DC microgrid with hybrid energy storage under different operating modes. Electr Power Energy Syst 104(2):807–816
2. Kollimalla SK et al (2014) Design and analysis of novel control strategy for battery and supercapacitor storage system. IEEE Trans Sustain Energy 5(4):37–44
3. Kotra S, Mishra MK (2015) Control algorithm for a PV based hybrid microgrid. In: 2015 Annual IEEE India conference. IEEE, pp 1–6
4. Pannala S et al (2020) Effective control and management scheme for isolated and grid connected DC microgrid. EEE Trans Ind Appl 56(6):1–14
5. Chen H et al (2021) Model predictive control based real-time energy management for HESS. CSEE J Power Energy Syst 7(4):862–874
6. Zhai C et al (2020) A novel predictive energy management strategy for EV prediction. IEEE Trans Vehicul Technol 69(11):12559–12569
7. Yin H et al (2016) An adaptive fuzzy logic-based EMS on battery/ultra capacitor hybrid electric vehicles. IEEE Trans Transp Electrificat 2(3):300–311
8. Manandhar U et a (2018) Validation of faster joint control strategy for battery and super capacitor-based energy storage system. IEEE Trans Power Electron 65(4):3286–3295
9. Bharatee A et al (2022) A power management scheme for grid-connected PV integrated with HESS. J Mod Power Syst Clean Energy 10(4)
10. Argyrou MC et al (2021) A novel power management algorithm for a residential grid-connected PV system with battery-supercapacitor storage for increased self-consumption and self-sufficiency. Energy Convers Manage 246(11)

Techno-Economic and Sensitivity Analysis of Standalone Hybrid Energy System Using HOMER Software: A Case Study of Kanur Village in India

Subhash Yadav, Pradeep Kumar, and Ashwani Kumar

Abstract This paper presents a techno-economic evaluation of an optimally designed isolated microgrid for a remote village 'Kanur,' Maharashtra, India. The microgrid is designed using HOMER consisting of a Solar Photovoltaic (PV), Wind turbine (WT), Diesel generator (DG), and Battery energy storage (BES). The optimal design is evaluated based on the net present cost (NPC) and cost of energy (COE). Additionally, a sensitivity analysis is also presented for the variation of NPC and COE with the nominal discount rate (NDR) and inflation rates. The results show that the system can be operated with 0.0% capacity storage, i.e., the design offers sufficient generation to meet the load demand. The optimal configuration PV/WT/DG/BES/converter provides the lowest NPC 569,275 $ and COE 0.157 $/kWh among six other feasible configurations to meet the load demand. This configuration with DG provides the most economical COE and lowest excess energy generation at 28.6%. The sensitivity analysis shows that with change in NDR and expected inflation rate, NPC and COE significantly vary. It shows that it is difficult to evaluate the system performance of the isolated microgrid at the design state. However, a conservative choice of the NDR and inflation may seem optimal at the initial stage, but may lead to difficulties in the operational stages later. Thus, the selection of these parameters should be done after the sensitivity analysis.

Keywords Hybrid energy system · Sensitivity analysis · HOMER software

S. Yadav (✉)
JSS Academy of Technical Education Noida, Noida, India
e-mail: subhash05754@gmail.com

P. Kumar · A. Kumar
National Institute of Technology Kurukshetra, Kurukshetra, India
e-mail: pradeepkumar@ieee.org

A. Kumar
e-mail: ashwani.k.sharma@nitkkr.ac.in

A. Kumar et al. (eds.), *Decarbonisation and Digitization of the Energy System*, Lecture Notes in Electrical Engineering 1099, https://doi.org/10.1007/978-981-99-7630-0_21

1 Introduction

Electricity is vital in daily activities, improving human development and the quality of life. In India, a large population in remote or rural areas has no access to electricity supply [1]. The availability of an electric grid in these areas is either practically difficult or uneconomical [2]. As a result, a diesel generator (DG) is a prevalent means to meet electrical load demand in these areas [3]. However, it is uneconomical and polluting in nature. Renewable energy sources (RESs) in isolated modes have recently become popular for electrification in rural and remote areas [4]. The stochastic nature of RESs makes providing a secure and reliable power supply difficult, whereas the system needs stable operation between generation and load demand [5].

Based on the body of the literature reviewed, Table 1 summarizes some isolated and grid-connected microgrids designed to electrify the area mentioned [3, 4, 6–18]. Several objectives are considered, such as minimization of net present cost (NPC), cost of energy (COE), or techno-economic and environmental analysis to design the microgrids. The studies do not include evaluation of excess energy generation at optimal system design. Moreover, the design and analysis of the microgrids involve considering uncertain factors such as load variations, price variations of fuel or energy, and other design parameters considered. Therefore, the sensitivity analysis becomes vital. The sensitivity analysis of the microgrid performance with load and price is already available in the literature [9, 10], however, the sensitivity analysis of the variation of nominal discount rate (NDR) and inflation is not considered. Since the microgrid is designed to operate for a long time, the economic planning is essential. Any variation in the nominal discount and inflation rate helps to evaluate the risk factor related to investment, operation, and profitability [14]. The nominal discount and inflation rate are also assumed constant or vary within the range at the outset of the planning study. But, these factors are uncertain in the future. Thus, in microgrids, this directly impacts the NPC, COE, and optimal system components design. Therefore, the sensitivity analysis of the variation of nominal discount and inflation rate becomes essential.

This research paper presents a technical and economical analysis of a microgrid design using hybrid optimization of multiple energy resources (HOMER). The microgrid study is based on the design of an isolated microgrid for a remote village 'Kanur,' Maharashtra, India, located at 16° 1′ 13.3″ N latitude and 74° 5′ 48.98″ E longitude. The microgrid consists of a solar photovoltaic (PV) and wind turbine (WT), as the site has sufficient wind and solar energy potential, discussed later. A battery energy storage system (BES) and DG or their combination is also used to address the variable nature of the RESs and electrical load demand. The RESs and energy storage are collectively termed a hybrid energy system [5]. The technical analysis aims to identify the optimal microgrid configuration which can satisfy the energy demand in a reliable, sustainable, and economical combination of BES and DG. On the other hand, the economical analysis provides a sensitivity analysis based on the nominal discount and inflate rate variation. The result shows that the optimal design for an isolated microgrid meets the load demand with 0.0% capacity shortage,

Table 1 Literature review based on related research work

References	Configuration	On-grid/off-grid	Location	Year	Objective(s)#								
					O1	O2	O3	O4	O5	O6	O7	O8	O9
[3]	PV/DG/BG	On-grid	Nigeria	2021	✓	✓	✓						
[4]	Hydro/PV/DG /BES	Off-grid	Iraq	2019	✓	✓	✓	✓			✓		
[6]	PV/WT/DG/BES	Off-grid	Bangladesh	2022	✓	✓		✓	✓				
[7]	PV/DG	Off-grid	Nepal	2022				✓	✓	✓			
[8]	PV/DG/PHS	Off-grid	Bangladesh	2021	✓	✓		✓			✓		
[9]	PV/WT/BG/DG/BES	Off-grid	Bihar, India	2021		✓						✓	
[10]	PV/DG/BES	Off-grid	Zambia	2023		✓		✓			✓		✓
[11]	PV/DG/BES	Off-grid	Andaman & Nicobar, Lakshadweep, India	2021	✓	✓	✓						
[12]	PV/DG/BES	Off-grid	Iraq	2022	✓			✓			✓		
[13]	PV/DG	Off-grid	Sharjah	2020		✓		✓			✓		
[14]	PV/DG/BES	Off-grid	Benin	2020	✓	✓	✓	✓					
[15]	PV/WT/ Hydro /DG/BES	Off-grid	Malaysia	2015			✓	✓		✓	✓		
[16]	PV/WT/DG	Off-grid	Colombia	2016	✓	✓	✓				✓		
[17]	WT/PV/DG/BES	Off-grid	Egypt	2020	✓	✓		✓			✓		
[18]	Hydro/PV/WT/ DG/BES	Off-grid	Southern Cameroons	2019		✓	✓						

#O1: Minimize NPC, O4: CO_2/GHG Emission, O7: Optimal Sizing
O2: Minimize COE, O5: Voltage and frequency stability, O8: Reliability
O3: Techno-economic analysis, O6: Fuel Consumption, O9: Life Cycle Cost

i.e., 100% of reliability of power supply. The excess energy (waste energy) generation is also low at 28.6% of total generation.

The rest of the paper is organized as follows: Sect. 2 presents the methodology and materials. Results and discussions are presented in Sect. 3. Finally, the conclusions are drawn in Sect. 4.

2 Methodology and Material

The HOMER software is a powerful tool for optimal component sizing, techno-economic analysis, and sensitivity analysis in microgrids. HOMER performs optimization and provides optimal microgrid configuration, design, and ranking based on NPC and COE. HOMER performs simulation, optimization, and sensitivity analysis [19].

2.1 Generation Potential at Site

The study site's wind speed (V_w), solar irradiation (G_i), and ambient temperature (T_a) data are downloaded from the NASA surface metrology. This data is input to the HOMER. The monthly average of V_w, G_i, clearness index, and temperature are depicted in Fig. 1. The average V_w is 5.55 m/s, the annual average G_i is 5.17 kWh/m^2/day, the average clearness index is 0.55, and the annual average temperature is 25.02 °C. The electrical load demand is estimated hourly for Kanur village, which has 257 households with domestic, community, commercial, and agricultural loads. The hourly load demand is presented in Fig. 1d. The peak load demand observed in August is 40.43 kW. Moreover, HOMER software applies more realistic random variability factors with a day-to-day variation of 5% and time steps variability of 10% for load demand. After applying the random variability factor, peak load demand rises, and load factor falls are 50.31 kW, and 0.43, respectively.

2.2 System Configuration and Economics

Figure 2 shows the layout of the proposed isolated microgrids with PV, WT, DG, BES, and converter. In this configuration, BES, WT, and PV are connected to the DC bus, while DG and electrical load are connected to the AC bus. The converter converts DC power into AC. The lifetime of project is considered 20 years with an annual discount rate of 6.5% and an inflation rate of 6% in India as on November 2022 [20].

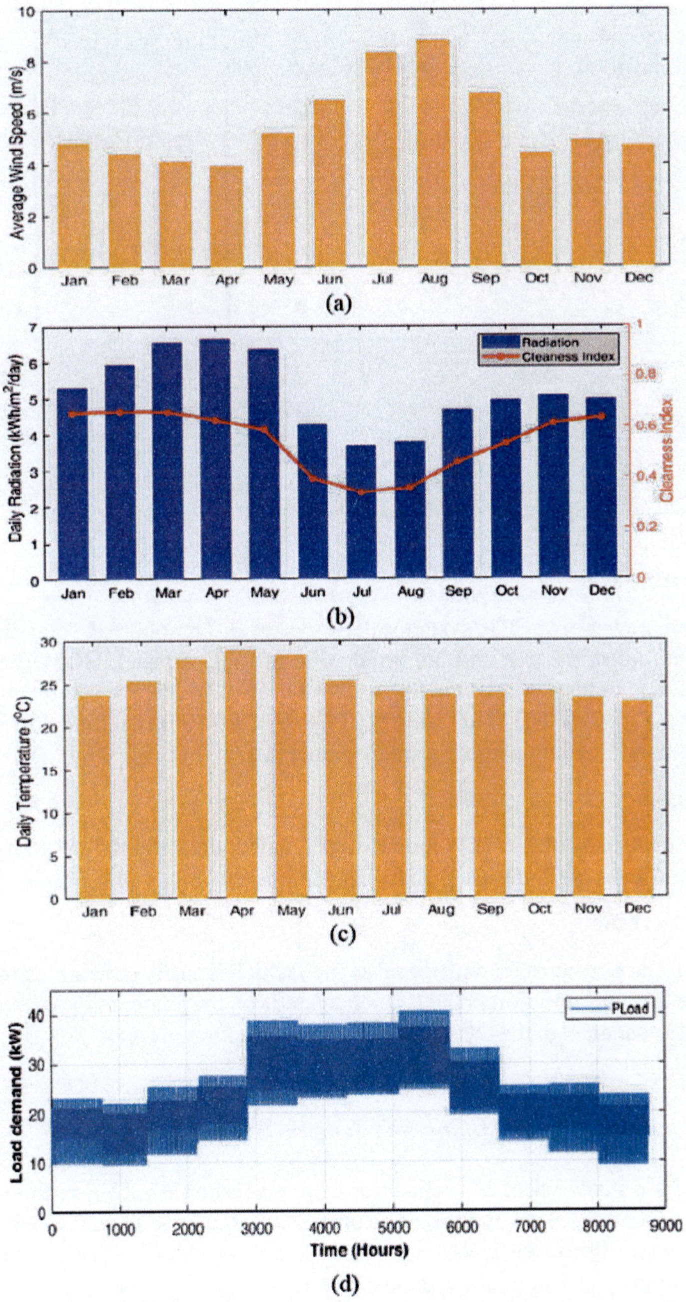

Fig. 1 **a** Monthly average V_w, **b**, monthly average G_i and clearness index, **c** monthly average ambient temperature, and **d** load demand for Kanur village

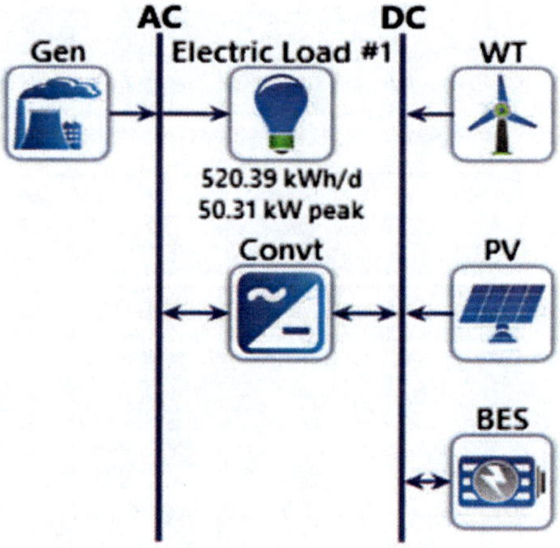

Fig. 2 Proposed isolated microgrid configuration

2.3 *Modeling of Microgrid Components*

Solar Photovoltaic

The output power from PV systems depends on the amount of solar irradiation available in the study area and the orientation of the PV panel. HOMER software calculates the output power of the PV system [4, 21, 22]. The PV cell temperature is calculated as [4, 22]. The study uses the PEIMAR SG300MBF model of solar PV. Its techno-economical parameters are available in [21–24].

Wind Turbine

The study uses an Aeolos-V1kW model of WT, calculated as in [25–27]. The techno-economical parameters are taken from [21, 23].

Diesel Generator

The DG produces electricity with diesel as fuel, which is easily transported to isolated areas. The fuel consumption in the DG set depends on the amount of power produced and the rated capacity of the DG set. The fuel consumption in DG (FC_{DG}) is evaluated as [11, 28].

$$FC_{DG} = f_1 C_{DG} + f_2 P_{DG} \tag{1}$$

where f_1 shows coefficient of fuel curve intercept (L/hr/kW), C_{DG} represents rated capacity of DG (kW), f_2 shows the slope of fuel curve (L/hr/kW), and P_{DG} represents produced power from DG (kW).

The efficiency of DG is evaluated as

$$\eta_{DG} = \frac{3600 \times P_{DG}}{\rho_{fuel} \times FC_{DG} \times LHV_{fuel}} \tag{2}$$

where ρ_{fuel} represents diesel fuel (DF) density and LHV_{fuel} is the lower heating value of DF. The study uses autosized genset model of DG. Its techno-economical parameters are available in [11, 28].

2.4 Battery Energy Storage

The kinetic battery model named "EnerSys PowerSafe SBS 150F" is used in this study. The battery charging and discharging model is described below.

In battery charging operations, the BES stores surplus energy. However, when electrical load demand exceeds power generation, BES discharges to satisfy the load requirement, and the battery state of charge (SOC_{bt}) is evaluated as [29–31].

$$SOC_{bt}(t) = SOC_{bt}(t-1) \times (1-\psi) + ((P_{PV}(t) + P_{WT}(t)) \times \eta_{con} + P_{DG} - P_{load}(t)) \times \eta_{chg} \quad (3)$$

$$SOC_{bt}(t) = SOC_{bt}(t-1) \times (1-\psi) - (P_{load}(t) - P_{DG}(t) - (P_{PV}(t) + P_{WT}(t)) \times \eta_{con})/\eta_{dhg} \quad (4)$$

where ψ shows battery self-discharge rate, η_{chg} is battery charging efficiency, η_{con} represents converter efficiency, and P_{load} is electrical load demand.

The minimum SOC_{bt} level ($SOC_{bt}^{\min}$) of a battery depends on its depth of discharge (DOD_{bt}) [23, 30]. The BES specification and economic parameters are given in [23].

$$SOC_{bt}^{\min} = (1 - DOD_{bt}) \times C_{bt} \quad (5)$$

where C_{bt} is the nominal capacity of battery.

Converter

The total converter capacity P_{cont} is evaluated based on peak load demand (P_{load}^{Pk})[21, 31] as described in (6). Its techno-economical parameters are depicted in [21].

$$P_{cont}(t) = P_{load}^{Pk}(t)/\eta_{cont} \quad (6)$$

2.5 Economic Modeling

The HOMER's primary concern is minimizing operating costs and determining the optimal system combinations. It performs economic analysis to minimize the total NPC and COE. The capital recovery factor (CRF) converts total annualized system cost into life cycle cost, also known as NPC. The CRF is the function of annual real interest rate and project lifetime. The NPC and CRF are evaluated as [4, 22].

$$NPC = {C_{Anul,Totl}}/{CRF(i_{Rr}, L_p)} \tag{7}$$

$$CRF(i_{Rr}, L_p) = \left(i_{Rr}(1 + i_{Rr})^{L_p}/((1 + i_{Rr})^{L_p} - 1)\right) \tag{8}$$

where $C_{Anul,Tot}$ shows total annualized cost ($/year), L_p is lifetime of project (year), and i_{Rr} represents annual real interest rate (%).

The COE is defined as the ratio of total annual cost and total energy (E_{Totl}) served to the electrical load in a year, as [22, 32].

$$COE = {C_{Anul,Totl}}/{E_{Totl}} \tag{9}$$

3 Result and Discussion

The proposed isolated microgrid comprises WT, solar PV, and DG as primary energy sources with BES. The optimization determines the optimal configuration to minimize NPC and COE using HOMER software. The sensitivity analysis is as well performed to examine the impact of uncertain parameters on the system performance and COE. The results are presented in the dollar to make the comparison of design costs of the proposed system easier with the designs available in the literature.

3.1 *Optimization Results*

The result obtained from HOMER is shown in Table 2. The HOMER cycle charging dispatch strategy used in this optimization result. The capacity shortage of 0%. The HOMER provides seven possible configurations from above mentioned resources. The most optimal configuration with the lowest NPC 569,275 $ and COE 0.157 $/kWh is PV/WT/DG/BES/Converter.

Table 2 Summary of optimization result for optimal configuration

Architecture					Cost		System				
PV (kW)	WT	DG (kW)	BES	Converter (kW)	NPC ($)	COE ($)	Ren Frac (%)	Total Fuel (L/yr)	Cap Short (%)	Excess Elec (%)	Unmet load (%)
110	40	56.0	288	52.5	$569,275	$0.157	94.5	3,060	0.0435	28.6	0.0194
158	37		588	54.6	$748,155	$0.207	100	0	0.0969	41.1	0.0631
312		56.0	432	69.1	$793,816	$0.220	94.1	3,230	0.0996	60.7	0.0542
	151	56.0	400	59.4	$971,486	$0.269	88.6	6,218	0.0989	46.1	0.0771
566			866	68.4	$1.34M	$0.369	100	0	0.0975	77.9	0.0722
	271		956	69.1	$1.62M	$0.447	100	0	0.0997	68.4	0.0668
		56.0	1,540	54.6	$2.62M	$0.726	0	61,162	0.0515	0	0.0465

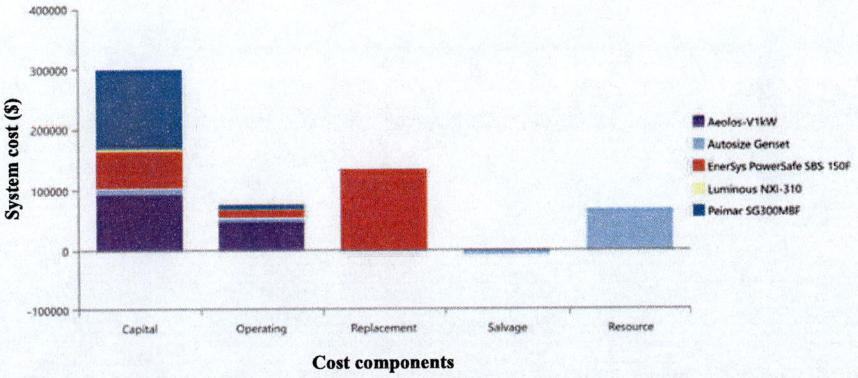

Fig. 3 Cost components summary of optimal hybrid configuration PV/WT/DG/BES/Converter

The optimal design has 140 kW of PV, 40 units of WT with each unit of 1kW, i.e., 40 kW, DG capacity of 56 kW, BES with 288 units, and converter capacity of 52.5 kW. The excess energy generation from optimal configuration is also 28.6%, the lowest among all configurations. The DG/BES/converter configuration has 0.0% excess energy generation. However, it is due to controlled generation from DG, but COE is 0.726 $/kWh which is very high, and the renewable energy fraction is 0%. The next second most optimal configuration system is PV/WT/BES/converter with a 100% renewable energy fraction, but NPC is 748155 $, COE is 0.207 $/kWh, and excess energy generation is 41.1%, which is comparatively high. The cost components summary of most optimal hybrid configuration is depicted in Fig. 3. It shows that capital cost has the highest share in cost components, followed by replacement cost and operating cost.

Moreover, the highest capital cost investment is required for solar PV, followed by WT, Battery, DG, and inverter in the same order. The replacement cost occurs only for BES because the battery life is 5 years and needs three times replacements during the project's full completion. The cash flow outline for the most optimal hybrid system year-wise is shown in Fig. 4. It also shows that maximum cost occurs in capital cost followed by replacement cost, fuel cost, and operating cost, respectively. The replacement cost occurs due to battery replacement at the end of 5, 10, and 15 years.

The output power from the solar PV system is shown in Fig. 5. It shows that the maximum and minimum output power from solar PV is 118 kW and 0 kW, respectively. The PV penetration of 99.4% is observed with 4425 h/year operation. The total annual electricity production and mean output power from the PV system are 188,893 kWh/year and 518 kWh/day, respectively, with a capacity factor of 19.6%.

The output power from the WT system is depicted in Fig. 6. The maximum electrical output power from the WT system is 60.3 kW, and wind penetration is 52.2% with 8011 h/year hours of operation. The total annual electricity production

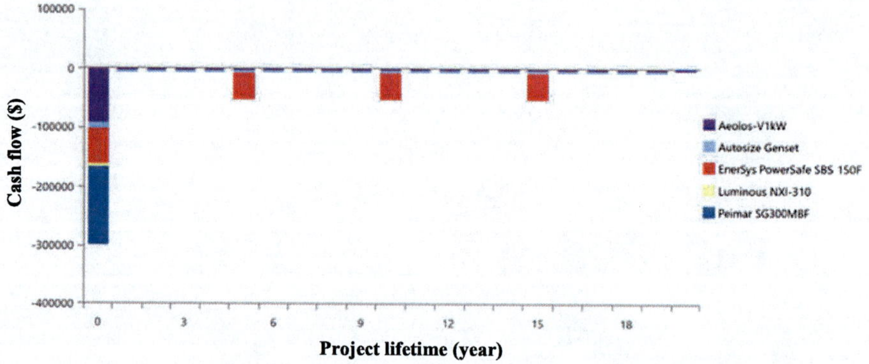

Fig. 4 Cash flow summary for optimal hybrid PV/WT/DG/BES/Converter system components

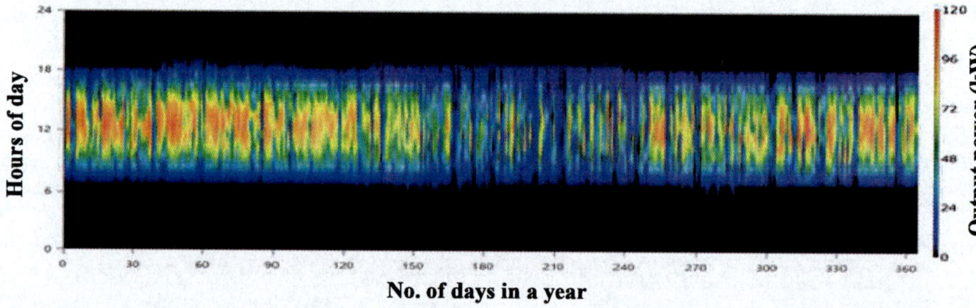

Fig. 5 Output power from solar PV system

and mean output power from the WT system are 99,128 kWh/year and 11.3 kW, respectively, with a capacity factor of 28.3%.

The SOC_{bt} of BES is presented in Fig. 7, showing that BES charging and discharging occur within the permissible level. The annual throughput power from BES is 83157 kWh/year. The converter output power in the inverter and rectifier

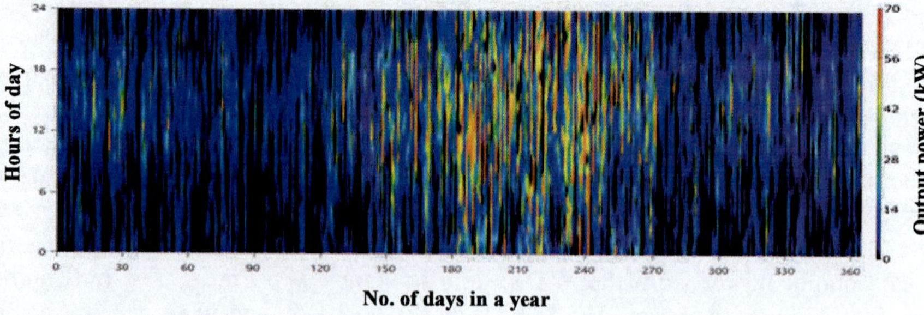

Fig. 6 Output power from the WT system

mode is shown in Fig. 8. The annual operation hours of the converter are 8579 h/year. The energy input and output from the converter are 204,956 kWh/year and 184,461 kWh/year, respectively, with a capacity factor of 40.1%.

The power generation from DG on particular days is shown in Fig. 9. The total annual electricity production from DG is 10538 kWh/year, with the consumption of 3060 L of diesel fuel. The maximum and minimum electrical outputs are 56 and 14 kW, respectively.

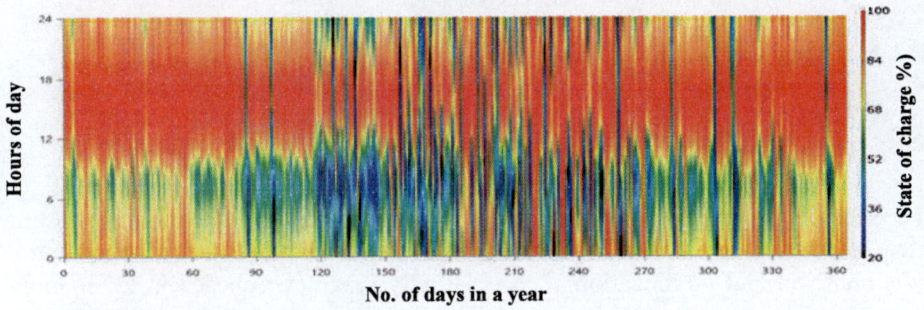

Fig. 7 State of charge (%) of BES

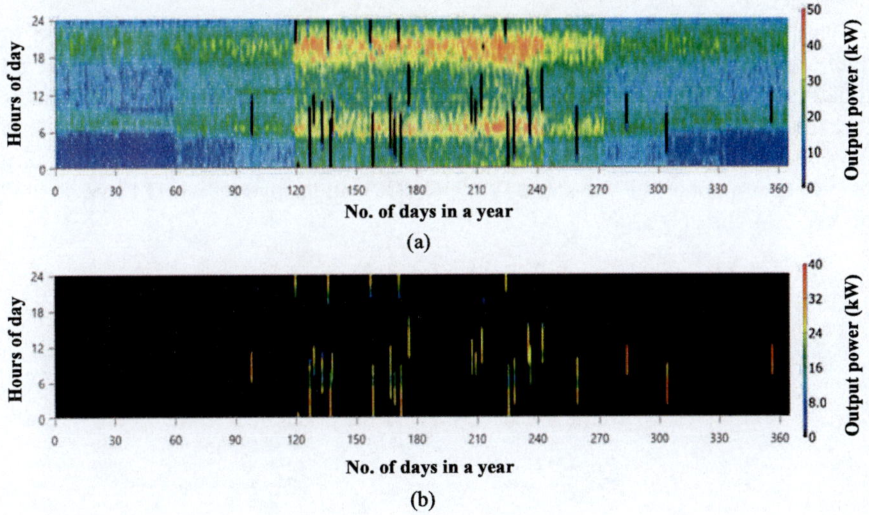

Fig. 8 Converter output power **a** inverter, and **b** rectifier

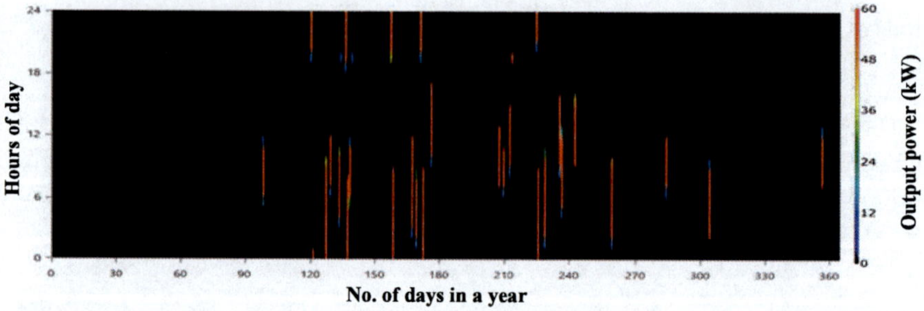

Fig. 9 Power generation from DG

3.2 *Sensitivity Analysis*

The sensitivity analysis shows the impact of variation in techno-economical parameters on system performance. The analysis is performed after the implementation of simulation and optimization. The sensitivity parameter chosen is the nominal discount and expected inflation rate, which highly impacts the system's economics. It is based on the market scenario. The sensitivity of the different parameters is evaluated considering the rest of the system parameters constant while varying only one parameter.

The list of economic configurations with parameter variations is displayed in Table 3, which shows that with the change in the nominal discount and expected inflation rate, NPC, cost components, and COE vary significantly. The table shows that the most economical COE can be obtained at 0.146 \$/kWh when the NDR is 6.5% and the expected inflation rate is 7.5%. At this rate of COE, the optimal capacity of system components PV system, WT, DG, BES, and converter are 111 kW, 41 kW, 56 kW, 284 units 60.8 kW.

Table 3 Sensitivity result for hybrid PV/WT/DG/BES configurations

Sensitivity		Architecture					Cost				
NominalDiscountRate (%)	ExpectedInflationRate (%)	PV (kW)	WT	DG (kW)	BES	Converter (kW)	NPC ($)	COE ($)	Operating cost ($/yr)	Initial capital ($)	O&M ($/yr)
6.50	6.50	114	39	56.0	286	55.7	$581,842	$0.153	$13,972	$302,407	$3,870
7.50	6.50	109	40	56.0	288	57.9	$558,510	$0.162	$14,318	$298,542	$3,937
7.00	6.50	110	40	56.0	288	52.5	$569,333	$0.157	$14,154	$299,749	$3,924
8.00	6.50	109	40	56.0	288	57.9	$547,057	$0.166	$14,345	$298,542	$3,941
6.50	6.00	110	40	56.0	288	52.5	$569,275	$0.157	$14,154	$299,749	$3,924
7.50	6.00	109	40	56.0	288	57.9	$546,900	$0.166	$14,345	$298,542	$3,941
7.00	6.00	109	40	56.0	288	57.9	$558,399	$0.162	$14,318	$298,542	$3,937
8.00	6.00	109	40	56.0	288	57.9	$536,082	$0.171	$14,369	$298,542	$3,945
6.50	7.50	111	41	56.0	284	60.8	$611,385	$0.146	$13,943	$303,324	$3,971
7.50	7.50	114	39	56.0	286	55.7	$581,842	$0.153	$13,972	$302,407	$3,870
7.00	7.50	113	40	56.0	284	61.0	$595,111	$0.149	$13,871	$303,668	$3,919
8.00	7.50	110	40	56.0	288	52.5	$569,449	$0.157	$14,153	$299,749	$3,924
6.50	7.00	113	40	56.0	284	61.0	$595,175	$0.149	$13,871	$303,668	$3,919
7.50	7.00	110	40	56.0	288	52.5	$569,391	$0.157	$14,153	$299,749	$3,924
7.00	7.00	114	39	56.0	286	55.7	$581,842	$0.153	$13,972	$302,407	$3,870
8.00	7.00	109	40	56.0	288	57.9	$558,621	$0.162	$14,318	$298,542	$3,937

The variation of the individual economic parameter with nominal discount and inflation rate is depicted in Fig. 10 and Fig. 11, respectively. In both the figures, it is observed that the drawing inference for the variation of the operating cost, initial capital cost, and operation and maintenance (O & M) cost is complex, as the combination of RESs and energy storage components also varies with the discount and inflation rate. Figure 10 shows that the design at an inflation rate of 7%, also shows a peculiar behavior in terms of variation of cost with the linear variation of the inflation rate. In all other cases, the NPC and COE increase and decrease with the inflation rate and discount rate, respectively. At 7% inflation, the NPC of the system increases, whereas the COE decreases. A similar variation is observed for different discount rates in Fig. 11 for the sensitivity of NPC and COE with the inflation rates. Here, at a 6% NDR, the NPC is relatively high, and COE is relatively low. At all other inflation and discount rate variations, the variation is easily predicted. Thus, from the analysis, it can be observed that if the designs consider a NDR of 6% and an inflation rate of 7%, the cost of the system remains relatively low. If the practical values deviate from these values, the operation of the microgrid has to be modified, as it will increase the system operating cost. This also impacts system planning. As a result, the choice of a suitable NDR and inflation rate becomes critical for the designers.

4 Conclusion

This paper proposes an optimal design and sensitivity analysis of a hybrid energy system-based isolated microgrid for reliable operation and minimization of COE. The optimal capacity of microgrid components WT, PV, DG, BES, and converter is obtained with the minimization of COE. The minimized COE is obtained at 0.157 \$/kWh, subject to a capacity shortage of 0.0% reliability constraints. The significant contributions drawn from the work are:

- The optimal design of a hybrid energy system for an isolated microgrid is performed, which meets the load demand with 0.0% capacity shortage, i.e., 100% of reliability of power supply.
- The most optimal hybrid energy system is PV/WT/DG/BES/converter, which has the lowest excess energy (waste energy) generation, only 28.6%.
- The second most optimal system configuration is PV/WT/BES/converter with a COE of 0.207 \$/kWh, and excess energy is 41.1%; moreover, it provides 100% renewable energy fraction but COE and excess energy generation rise by 24.15%, and 12.5% compared to the most optimal system. The remaining other feasible optimal configuration suggested by HOMER, as per Table 2, is that the COE and excess energy generation is much higher. Thus it concludes that using DG with a RES, i.e., a hybrid energy system, provides the most economical COE and lowest excess energy generation.

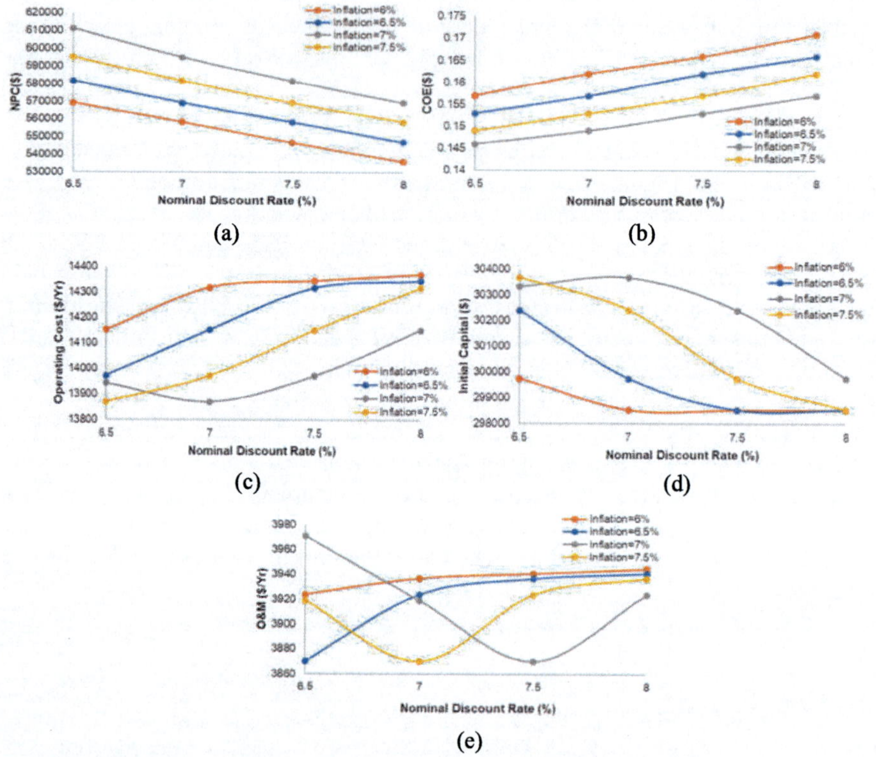

Fig. 10 Sensitivity of economic parameters **a** NPC ($), **b** COE ($), **c** operating cost ($/year), **d** Initial capital cost ($), **e** O&M ($/year) with nominal discount effect

- Sensitivity analysis shows that considering suitable values of the discount rate and the inflation rate is important. They impact the NPC and COE considerably. If the selection of the values is conservative at the initial stage, a significant variation may be observed at the operational stage. In present study, a NDR of 6% and an inflation rate of 7% keep the operation at a minimum COE, while minor variation in these values leads to a large change in operational costs.

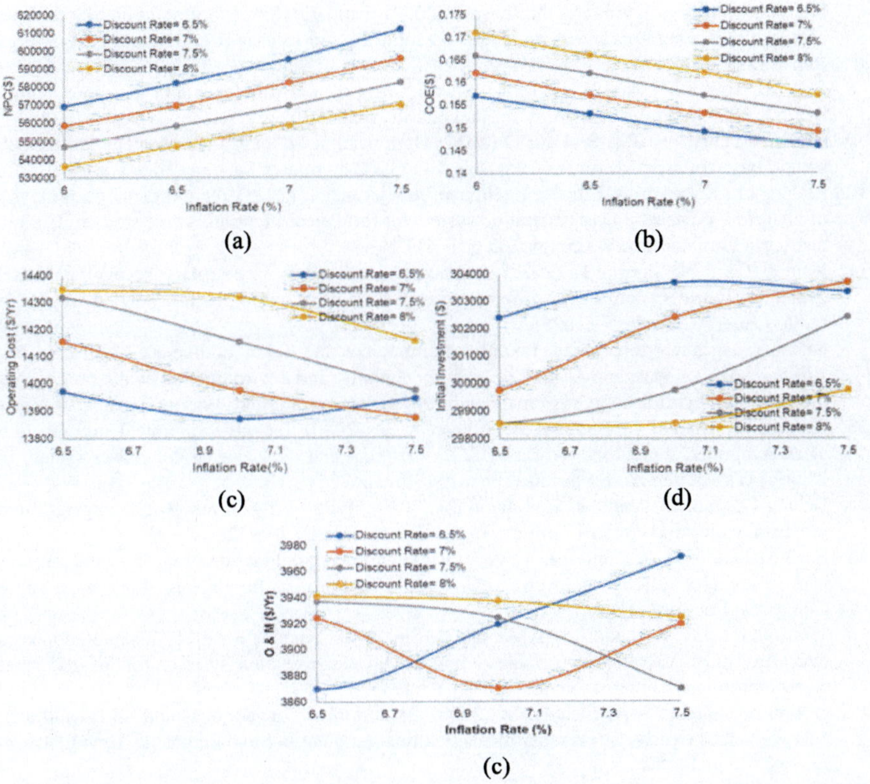

Fig. 11 Sensitivity of economic parameters **a** NPC ($), **b** COE ($), **c** operating cost ($/year), **d** initial capital cost ($), **e** O&M ($/year) with inflation rate

References

1. Chambon CL, Karia T, Sandwell P, Hallett JP (2020) Techno-economic assessment of biomass gasification-based mini-grids for productive energy applications: the case of rural India. Renew Energy 154:432–444
2. Sen R, Bhattacharyya SC (2014) Off-grid electricity generation with renewable energy technologies in India: an application of HOMER. Renew Energy 62:388–398
3. Oladapo S, Yakubu J, Oluwafemi T (2021) Analysis of backup power supply for unreliable grid using hybrid solar PV/diesel/ biogas system. Energy 227:120506
4. Saleh A, Faridun Mohammad Tajuddin Naim R, Mohd A, Azralmukmin R, Makbul AM (2019) Optimization and sensitivity analysis of standalone hybrid energy systems for rural electrification : a case study of Iraq. Renew Energy 138:775–792
5. Ribo-Perez D, Herraiz-Canete A, Alfonso-Solar D, Carlos Vargas-Salgado TG-N (2021) Modelling biomass gasifiers in hybrid renewable energy microgrids ; a complete procedure for enabling gasifiers simulation in HOMER. Renew Energy 174:501–512
6. Shezan Sk A, Ishraque F Md, Muyeen SM, Abu-Siada A, Saidur R, Ali MM, Rashid MM (2022) Selection of the best dispatch strategy considering techno-economic and system stability analysis with optimal sizing. Energy Strateg Rev 43:100923

7. Raj S, Bajracharya I, Ananda R, Bhave P (2022) Estimation of air pollutant emissions from captive diesel generators and its mitigation potential through microgrid and solar energy. Energy Rep 8:3251–3262
8. Das BK, Hasan M, Rashid F (2021) Optimal sizing of a grid-independent PV/diesel/pump-hydro hybrid system: a case study in Bangladesh. Sustain Energy Technol Assess 44:100997
9. Kumar G, Guerrero JM, Prakash O (2021) Optimisation of solar/wind/bio-generator/diesel/ battery based microgrids for rural areas: a PSO-GWO approach. Sustain Cities Soc 67:102723
10. Mulenga E, Kabanshi A, Mupeta H, Ndiaye M, Nyirenda E (2023) Techno-economic analysis of off-grid PV-Diesel power generation system for rural electrification: a case study of Chilubi district in Zambia. Renew Energy 203:601–611
11. Kumar P, Pal N, Sharma H (2022) Techno-economic analysis of solar photovoltaic/diesel generator hybrid system using different energy storage technologies for isolated islands of India. J Energy Storage 41:102965
12. Saleh A, Faridun Mohammad Tajuddin Naim R, Zidane Tekai Eddine K, Chun-Lien Su, Alrubaie AJK, Alwazzan MJ (2022) Techno-economic and environmental evaluation of PV/ diesel/battery hybrid energy system usingimproved dispatch strategy. Energy Rep 8:6794–6814
13. Salameh T, Ghenai C, Merabet A, Alkasrawi M (2020) Techno-economical optimization of an integrated standalone hybrid solar PV tracking and diesel generator power system in Khorfakkan, United Arab Emirates. Energy 190:116475
14. Delano O, Odou T, Bhandari R, Adamou R (2020) Hybrid off-grid renewable power system for sustainable rural electrifi cation in Benin. Renew Energy 145:1266–1279
15. Basir MR, Jidin R, Pasupuleti J, Azwa S (2015) Optimal combination of solar , wind , micro-hydro and diesel systems based on actual seasonal load profiles for a resort island in the South China Sea. Energy 1–18
16. Mamaghani AH, Escandon SAA, Najafi B, Shirazi A, Rinaldi F (2016) Techno-economic feasibility of photovoltaic, wind, diesel and hybrid electrification systems for off-grid rural electrification in Colombia. Renew Energy 97:293–305
17. Osman M, Farahat MA, Elsayed M (2020) Operation of conventional and unconventional energy sources to drive a reverse osmosis desalination plant in Sinai Peninsula, Egypt. Renew Energy 145:141–152
18. Muh E, Tabet F (2019) Comparative analysis of hybrid renewable energy systems for off-grid applications in Southern Cameroons. Renew Energy 135:41–54
19. Shahzad MK, Zahid A, Rashid Tu, Rehan MA, Ali M, Ahmad M (2017) Techno-economic feasibility analysis of a solar-biomass off grid system for the electri fi cation of remote rural areas in Pakistan using HOMER software. Renew Energy 106:264–273
20. Reserve bank of India-RBI bulletin. https://m.rbi.org.in//Scripts/BS_ViewBulletin.aspx?id=21468. Accessed 30 Dec 2022
21. Singh S, Singh M, Kaushik SC (2016) Feasibility study of an islanded microgrid in rural area consisting of PV, wind, biomass and battery energy storage system. Energy Convers Manage 128:178–190
22. Kumar P, Pal N, Sharma H (2022) Optimization and techno-economic analysis of a solar photovoltaic/biomass/diesel/battery hybrid off-grid power generation system for rural remote electrification in eastern India. Energy 247:123560
23. Askarzadeh A, dos Santos Coelho L (2015) A novel framework for optimization of a grid independent hybrid renewable energy system: a case study of Iran. Sol Energy 112:383–396
24. Yong Y, Li R (2020) Techno-economic optimization of an off-grid solar/wind/battery hybrid system with a novel multi-objective differential evolution algorithm. Energies 13:1–16
25. Guezgouz M, Jurasz J, Bekkouche B, Ma T, Javed MS, Kies A (2019) Optimal hybrid pumped hydro-battery storage scheme for off-grid renewable energy systems. Energy Convers Manage 199:112046
26. Xu X, Hu W, Cao D, Huang Q, Chen C, Chen Z (2020) Optimized sizing of a standalone PV-wind-hydropower station with pumped-storage installation hybrid energy system. Renew Energy 147:1418–1431

27. Jumare IA, Bhandari R, Zerga A (2020) Assessment of a decentralized grid- connected photovoltaic (PV)/wind/biogas hybrid power system in northern. Energy Sustain Soc 7:1–25
28. Freitas E De Júnior M, Rüther R (2020) The influence of the solar radiation database and the photovoltaic simulator on the sizing and economics of photovoltaic-diesel generators. Energy Convers Manage 210:112737
29. Bhayo BA, Al-Kayiem HH, Gilani SIU, Ismail FB (2020) Power management optimization of hybrid solar photovoltaic-battery integrated with pumped-hydro-storage system for standalone electricity generation. Energy Convers Manage 215:112942
30. He Y, Su G, Zhou J, Wu F, Huang J, Pei H (2021) The quantitative techno-economic comparisons and multi-objective capacity optimization of wind-photovoltaic hybrid power system considering different energy storage technologies. Energy Convers Manage 229:113779
31. Vamba K, Bode O, Gamil MM (2023) A scenario-based multi-attribute decision making approach for optimal design of a hybrid off-grid system. Energy 265:125663
32. In E, Windt HJ Van Der, Olubayo B (2023) Energy for sustainable development scaling up the electricity access and addressing best strategies for a sustainable operation of an existing solar PV mini-grid : a case study of Mavumira village in Mozambique. Energy Sustain Dev 72:58–82

Optimal Placement and Sizing of Active Power Filters in RDS Using TLBO for Harmonic Distortion Reduction

Ashokkumar Lakum, Deepak Bhonsle, and Mahesh Pandya

Abstract This paper focuses the optimal placement and sizing (OPAS) of active power filters (APFs) in a radial distribution system (RDS) for harmonic reduction. The optimization problem is constrained and nonlinear. The placement of APFs is achieved through a recently developed technique called nonlinear load (NL) position-based current injection (NLPCI). The optimal size of APFs to reduce the total harmonic distortion in voltage (THD_V) to meet IEEE standards, a teaching-learning-based optimization (TLBO) algorithm is used, and its performance is compared to that of flower pollination (FPA) and bird swarm (BSA) algorithms. The results indicate that the TLBO outperforms the other algorithms in computational performance, as demonstrated using an IEEE 69-bus RDS with NL and nonlinear distributed generation (NLDG).

Keywords Active power filter · Power quality · Radial distribution system · Teaching-learning-based optimization

1 Introduction

Distributed generation (DG) is a key element of the smart grid. It has garnered significant attention due to its numerous benefits. One of the essential tasks is integrating DG into the radial distribution system (RDS). Power quality issues come from improper DG integration due to the harmonics generated by the converter [1]. A converter-based DG is referred to as a nonlinear DG (NLDG) when it introduces harmonics into the RDS [2]. Total harmonic distortion in voltage (THD_V) and individual harmonic distortion in voltage (IHD_V) should be less than 5% and 3%, respectively, by IEEE standard 519 [3]. Harmonics need to be controlled to meet this

A. Lakum (✉) · M. Pandya
Lukhdhirji Engineering College, Morbi, Gujarat, India
e-mail: aclakum@lecollege.ac.in

D. Bhonsle
C. K. Pithawla College of Engineering and Technology, Surat, Gujarat, India

A. Kumar et al. (eds.), *Decarbonisation and Digitization of the Energy System*, Lecture Notes in Electrical Engineering 1099, https://doi.org/10.1007/978-981-99-7630-0_22

criterion. The straightforward solution to mitigate harmonic distortion in RDS is to install active power filters (APFs) at all NL buses with current ratings equal to those of the respective NLs. But the price is rather exorbitant. As a result, an optimization approach is needed for APF location, rating, and cost [4].

Numerous algorithms, including genetic algorithm, variants of particle swarm optimization [5], harmony search [4], firefly algorithm, and grey wolf optimizer [6] algorithms have been utilized for optimal placement and sizing (OPAS) of APF. Most of the used algorithms have algorithmic-specific parameters. It is too difficult to tune these parameters for the different problems. Improper tuning of these parameters may give false results.

The theory of "No Free Lunch" states that no single optimization algorithm can be considered the best for all issues and that several algorithms can tackle a particular optimization problem [7]. For the same issue, a different optimization procedure produces a different outcome.

Considering this aspect, Prof. Rao et al. have developed the teaching–learning-based optimization (TLBO) algorithm, and it operates without algorithm-specific parameters [8]. It is used for OPAS of APF and compared with nature-inspired flower pollination algorithm (FPA) [9] and bio-inspired bird swarm algorithm (BSA) [10].

Below are the main contributions of this paper in relation to the OPAS of APF, taking into account the integration of NLDG and nonlinear load:

- The TLBO is coupled with harmonic load flow and used to find optimal size of APF.
- The three algorithms (TLBO, FPA, and BSA) that have different bases are compared and analyzed for two cases: NL and NL plus NLDG.
- To determine the size of APF in the presence of NL and NLDGs, the TLBO, FPA, and BSA algorithms were utilized. The computational tests were conducted in terms of the best value. The results indicated that the APF requirement is higher when the NLDG is integrated into the RDS as compared to without NLDG.
- The computational tests revealed that TLBO outperformed FPA and BSA by producing the minimum value of APF current in both cases.

In the next section TLBO is explained.

2 TLBO

A teacher's influence on students' performance in a class is the basis for the algorithm known as TLBO. It is based on the ideas of the teaching-learning process. The two essential elements of the algorithm are the teacher and the students, who characterize two fundamental modes of learning: learning from the teacher (known as the teacher phase) and engaging with other students (known as the learner phase). It is well explained in [8].

2.1 Teacher Phase

In this stage of the algorithm, the focus is on replicating the learning process of students with the aid of a teacher. The teacher imparts knowledge to the learners and strives to improve the overall performance of the class. If there are "*m*" subjects (or design variables) available to a population of "*n*" learners (with $k = 1, 2, \ldots, n$), then at each sequential teaching-learning cycle *i*, *Mj*, *i* represents the average performance of the learners in a particular subject "*j*" (where $j = 1, 2, \ldots, m$).

The algorithm selects the most knowledgeable and experienced individual in the population as the teacher, based on their performance across all subjects. This is denoted by $X_{total}k_{best,i}$, which represents the best learner's result across all subjects, and who will serve as the teacher for the current cycle. The teacher is expected to make maximum efforts to improve the knowledge level of the entire class. However, the quality of learning gained by each individual learner is influenced by the quality of teaching provided by the teacher and the learners' abilities. To account for this, the difference between the teacher's result and the mean result of the learners in each subject is calculated as follows:

$$Difference\_Mean_{j,i} = r_i(X_{j,kbest,i} - T_f M_{j,i}) \quad (1)$$

The difference between the teacher's result (X_j, $k_{best,i}$) in a given subject "*j*" and the mean result of the learners in that subject is expressed in the equation. The value of the teaching factor, T_f, determines the extent of change to the mean result, and is determined by a random number, r_i, within the range of 0 to 1. T_f can take on a value of either 1 or 2, and its specific value is determined randomly.

$$T_f = round[1 + rand(0, 1)(2 - 1)] \quad (2)$$

The teaching factor (T_f) is not a fixed parameter in the TLBO algorithm and is not provided as an input. Instead, its value is randomly determined by the algorithm using Eq. (2).

$$X'_{j,k,i} = X_{j,k,i} + Difference\_Mean_{j,i} \quad (3)$$

During the teacher phase, based on $Difference\_Mean_{j,i}$ the existing solution is updated according to Eq. (3), where $X'_{j,k,i}$ represents the updated value of $X_{j,k,i}$. This updated value is accepted only if it leads to a better function value. All accepted function values at the end of the teacher phase are retained and used as input for the learner phase.

2.2 Learner Phase

In this stage of the algorithm, the focus is on simulating the learning process of students through interactions with one another. Learners have the opportunity to gain new knowledge by engaging in discussions and interactions with other learners. If one learner possesses more knowledge than another, the latter can learn from the former.

The next section deals with problem formulation. The simulated results are discussed later on and it is followed by the conclusion in the last.

3 Problem Formulation

RDS with NLs, NLDGs, and APFs are mathematically modeled as per [6]. The bus injection to branch current (BIBC) and branch current to the bus voltage (BCBV) based harmonic load flow is employed in the simulation [11]. Then it is coupled with the TLBO, FPA, and BSA.

3.1 Modeling of RDS

The RDS is modeled in terms of impedance, which includes the resistance and inductance of the RDS. It is calculated considering harmonics environment.

3.2 Modeling of Nonlinear Load

The representation of nonlinear load can be formulated by keeping the source of harmonic current injection as the base as shown in [5, 12]. The magnitude of rms nonlinear current is given below,

$$I_{nl}^{(h)} = I_{nl,r}^{(h)} + jI_{nl,im}^{(h)} \tag{4}$$

$$I_{\mathrm{nl}} = \sqrt{\sum_{h=2}^{H} (I_{nl,r}^{2(h)} + I_{nl,im}^{2(h)})} \tag{5}$$

Here, $I_{nl,r}^{(h)}$ is the real part of nonlinear current while $I_{nl,im}^{(h)}$ is the imaginary part of nonlinear current, I_{nl} indicates the rms value, the highest order of harmonics is H.

3.3 Modeling of APF

The APF is represented as a current source [5, 12]. Calculation of I_{apf} is done on the basis of net nonlinear current at PCC. Net nonlinear current (I_{nl}) is a vector addition of nonlinear currents of loads.

$$I_{apf} = k(I_{nl}) \tag{6}$$

Here, k is proportionate of net nonlinear current. It has the value such that *Iapf* satisfies the constraints.

APF is represented as a set of current sources. It injects different orders of harmonics at a PCC as per the value of k. The current of APF is expressed as

$$I_{apf}^{(h)} = I_{apf,r}^{(h)} + jI_{apf\,,im}^{(h)} \tag{7}$$

$$I_{apf} = \left[\sum_{h=2}^{H} (I_{apf,r}^{2(h)} + I_{apf,im}^{2(h)})\right]^{1/2} \tag{8}$$

Here, $I_{apf,r}^{(h)}$, $I_{apf,im}^{(h)}$, and I_{apf} are the real part, imaginary part, and the rms value of current of APF respectively.

3.4 Modeling of NLDG

The DG which is a current source [13, 14], supplies the nonlinear current into the system. The fundamental current is estimated [13] by power rating of the DG and therefore, NLDG harmonic current is calculated from spectrum as

$$I_{hdg}^{(h)} = K_{dg} I_{dg} \tag{9}$$

Here, $I_{hdg}^{(h)}$, K_{dg}, and I_{dg} are the harmonic current of DG, part of harmonic current as per the spectrum of DG, and the fundamental current of DG respectively.

3.5 Harmonic Load Flow

The bus injection to branch current (BIBC) and branch current to the bus voltage (BCBV) based harmonic load flow is used in this paper [11]. Then it is coupled with the optimization algorithm. THDv is calculated at all buses using harmonic load flow and it is essential for identification of critical buses which violates the standard limits.

The THD_V is given as,

$$THDv = \frac{\sqrt{\sum_{h=2}^{H} (IHDv)^2}}{V_f^1} \quad (10)$$

Here V_f^1 is fundamental frequency voltage.

From the value of THDv at all buses, the vital buses where the IEEE standard limits are not satisfied are found. It helps in deciding the policy to alleviate the harmonics; i.e. the placement and sizing of APF.

4 Optimization Process

The optimization algorithms are employed to determine the optimal size. The objective function is based on the APF current. Here the main objective is to reduce the APF current such that the constraints are satisfied. It is depicted as:

$$OF_{IFT} = \min \sum_{m=1}^{M} \sqrt{\sum_{h=2}^{H} \left| I_{apf,m}^{h} \right|^2} + DP \quad (11)$$

where, "*m*" represents the bus number, "*M*" represents the total number of buses, "*h*" denotes the order of harmonics, "*H*" represents the highest order of harmonics, and "*DP*" stands for a dynamic penalty.

Subjected to

$$THD_V - 0.05 \leq 0, THD_{Vindi} - 0.03 \leq 0, I_{apf} \leq I_{apf,MAX} \quad (12)$$

By using a black box approach, as illustrated in Fig. 1, TLBO has the advantage of analyzing a system. It offers the system with input variables, such as the current of APFs, and observes the resulting output, which is the objective function's value. The TLBO continues to adjust the inputs of the system based on the feedback it receives (output) until the specified end criteria are met. Figure 2 shows the flowchart for optimization process.

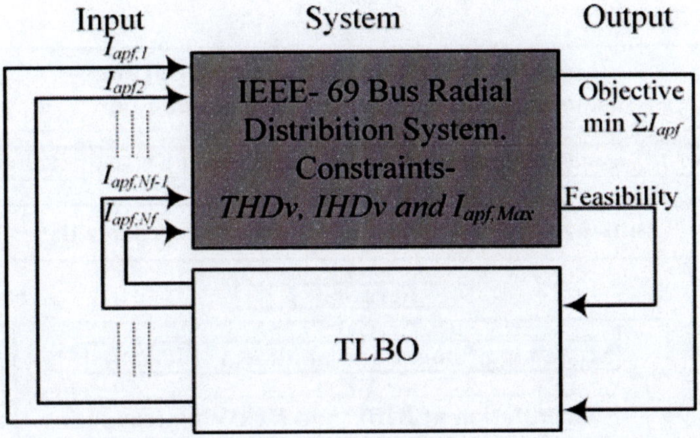

Fig. 1 Block diagram of TLBO for OPAS of APFs in RDS

5 Results Analysis

This section covers the result analysis.

5.1 Initial Data

The IEEE-69 bus [15] RDS with modifications is taken into account, for simulation (Fig. 3). Here, it is assumed that the NLDGs are connected to the NL buses. The NLs have a harmonic spectrum up to 49th order of harmonics as per the six-pulse converter [16]. The NLDG exhibits a harmonic spectrum, as illustrated in Fig. 4 [17]. The NLs are placed at buses 27, 46, and 65, i.e., at end nodes.

To find the location of APF, here recently developed NL position-based APF current injection (NLPCI) technique is used [6].

Table 1 represents the situation of buses regarding the availability of APF. Here, all the possible combinations of APF allocation are considered. The NLs are at three buses. So, the possible combinations of APF allocation are ($2^3 - 1 = 7$) seven, viz. 1 through 7. The investigations are carried out for seven states as furnished in Table 1, wherein "+" denotes the availability of APF and "*" denotes its non-availability. For finding the rating of APF, the three algorithms viz. TLBO, FPA, and BSA are implemented to determine APF current in a manner that meets the constraints.

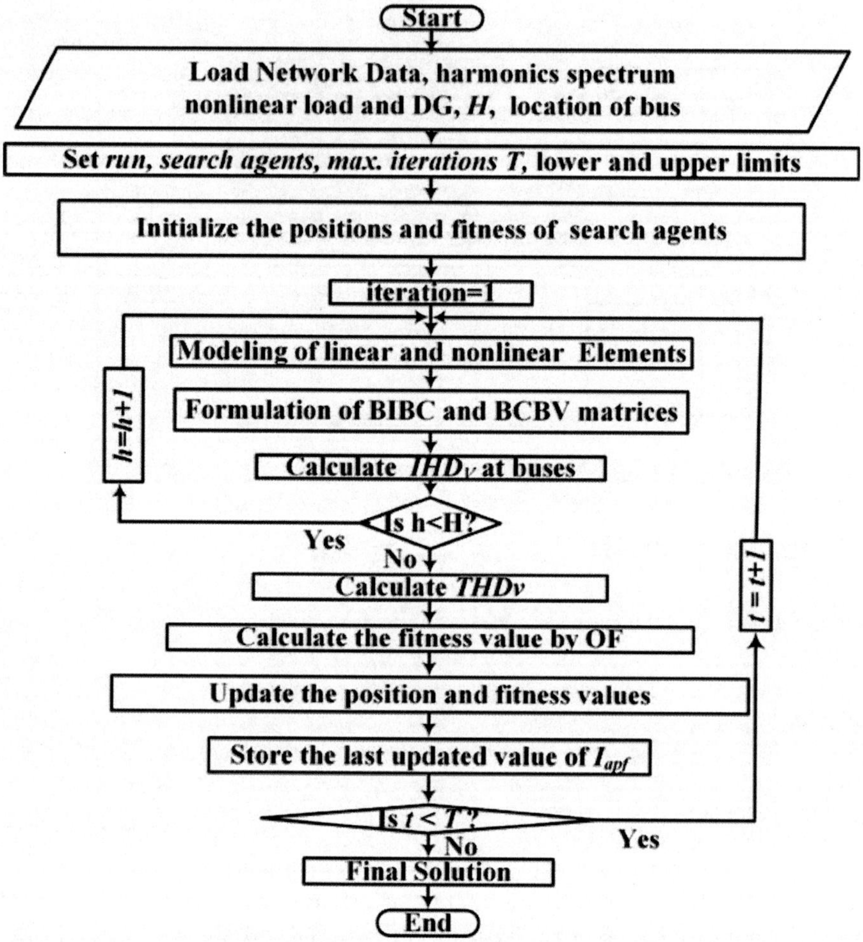

Fig. 2 Flowchart for OPAS of APF

5.2 *Result Analysis*

The NLs and NLDGs are connected at buses 27, 46, and 65. There are four different cases simulated; (i) Only NL (without APF), (ii) NL (With APF), (iii) NL plus NLDG (without APF), and (iv) NL plus NLDG (with APF).

Case 1 Only NL (without APF): In this case, only NLs are connected with RDS at buses 27, 46, and 65. The results obtained from THD_V without using APF are tabulated in Table 2. Bus 65 has a maximum THD_V (7.97%), compared to the other NL buses. Outs

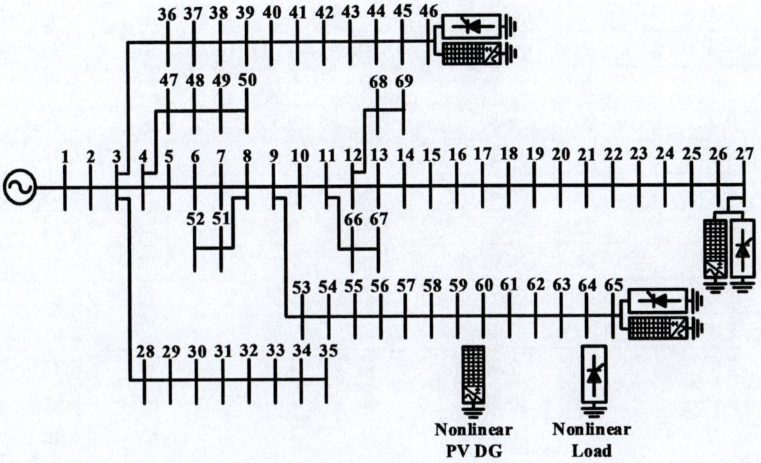

Fig. 3 IEEE-69 bus RDS with NLs and NLDGs

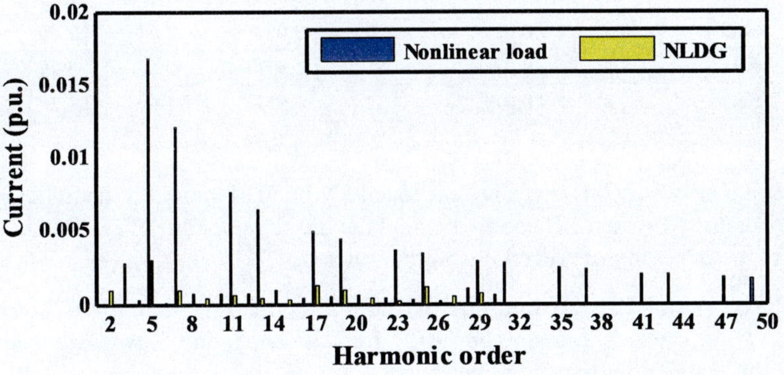

Fig. 4 Harmonics spectrum of NL and NLDG

Table 1 NLPCI matrix

State/APF bus	Number of APFs	27	46	65
1	One APF	+	*	*
2		*	+	*
3		*	*	+
4	Two APFs	+	+	*
5		+	*	+
6		*	+	+
7	Three APFs	+	+	+

"+" APF; "*" No APF

Table 2 THDv (%) at different buses (no APF)

Bus	THDv (%)	Bus	THDv (%)	Bus	THDv (%)	Bus	THDv (%)
2	0.01	19	**6.15**	36	0.04	53	2.26
3	0.02	20	**6.30**	37	0.37	54	2.48
4	0.03	21	**6.54**	38	0.62	55	2.79
5	0.15	22	**6.55**	39	0.70	56	3.09
6	0.94	23	**6.66**	40	0.70	57	4.24
7	1.77	24	**6.91**	41	2.49	58	4.80
8	1.97	25	**7.44**	42	3.25	59	**5.02**
9	2.07	26	**7.66**	43	3.35	60	**5.27**
10	2.65	27	**7.78**	44	3.38	61	**5.82**
11	2.79	28	0.02	45	3.67	62	**5.92**
12	3.29	29	0.02	46	3.67	63	**6.08**
13	4.02	30	0.02	47	0.03	64	**6.85**
14	4.75	31	0.02	48	0.03	65	**7.97**
15	**5.51**	32	0.02	49	0.03	66	2.79
16	**5.65**	33	0.02	50	0.03	67	2.79
17	**5.91**	34	0.02	51	1.97	68	3.29
18	**5.91**	35	0.02	52	1.97	69	3.29

of a total of twenty buses, THD_V exceeded 5% in all of them. It is an indication of a highly distorted system. These are beyond the limits as per the IEEE-519 standard. It shows the necessity of APF/APFs in this system.

Case 2 NL (With APF): To minimize the harmonics up to standard limits, according to Table 1, for states 1, 2, and 3, one APF; for states 4, 5, and 6, two APFs; and for state 7, three APFs are located as per NLPCI.

The size of the APFs is found using optimization methods at these placements. The TLBO, FPA, and BSA are employed in IEEE-69 bus RDS to test the computation performance. For considered algorithms, the common parameters—number of populations and total iterations are set to 40 and 100, respectively. Table 3 summarizes the best fitness value of an objective function for all the states. When the algorithms have not fulfilled the constraints, they exhibit a high value of the best fitness score as a result of the penalty applied. This effect of the penalty can also be observed in the early stages of the iterations (Fig. 7). After locating the APF/APFs, for states 1 to 3, only one APF, the algorithms are not converged as per Table 3. It shows that more than one APF is needed to achieve the goal. For states 4 to 6, two APFS are located as per location matrix. Here, for state 4 and state 6, algorithms are not converged; i.e., two APFs at buses 27 and 46 for state 4; and at 46 and 65 for state 6 are no proper solutions to solve the problem.

For state 5 the algorithms are converged. For state 5, two APFs are at buses 27 and 65 the algorithms are converged. Among the optimization algorithms tested, the

Table 3 Comparison of the value of objective functions for both cases

State	Number of APFs and bus	Algorithm	I_{apf} (p.u.) (only NL)	I_{apf} (p.u.) (NL + NLDG)
1	One APF at 27	TLBO	29.6727	39.0036
		FPA	29.7775	39.3697
		BSA	29.6727	39.0036
2	One APF at 46	TLBO	54.4317	61.8787
		FPA	54.4317	61.8787
		BSA	54.4612	61.8787
3	One APF at 65	TLBO	27.3056	37.0769
		FPA	27.3271	37.1788
		BSA	27.3058	37.1088
4	Two APFs at 27 and 46	TLBO	29.5767	38.9426
		FPA	29.8321	39.4435
		BSA	37.2652	47.7294
5	**Two APFs at 27 and 65**	**TLBO**	**0.0184**	**0.0247**
		FPA	0.0245	0.0649
		BSA	0.0300	0.0500
6	Two APFs at 46 and 65	TLBO	27.1974	37.0102
		FPA	27.4594	38.3496
		BSA	27.8629	37.0813

best value was obtained using TLBO with a result of 0.0184 p.u., outperforming FPA (0.0245 p.u.), and BSA (0.0300 p.u.). It shows the limitation of BSA; it tends to suffer from premature convergence and easily gets trapped in local minima [10, 18]. Similarly, premature convergence and poor exploitation are shortcoming of FPA [19].

As per Table 3, for states 1 to 4 and 6, one APF and two APFs have not fulfilled the constraints; therefore, all three algorithms failed to converge. The algorithms are converged for state 5. As shown in Fig. 5, two APFs for state 5 fulfill the constraints, and all buses have THD_V less than 5%. It is found that the TLBO outperforms other algorithms in terms of performance.

Case 3 NL plus NLDG (without APF): The considered RDS is simulated with NLs and NLDGs. Due to the inclusion of NLDGs, the system is more distorted compared to only NL. Previously twenty buses have exceeded the acceptable limit for THD_V; but after incorporating NLDGs total of twenty-two buses have THD_V exceeded the standard

limit. It proves that the penetration of NLDG in RDS increases the harmonic distortion. As per Fig. 6, the maximum THD_V is observed at bus 65 (8.84%).

Case 4 NL plus NLDG (With APF): The scenario for state 5 is shown in Fig. 8. Here, two APFs are placed at buses 27 and 65. These two APFs have fulfilled the constraints.

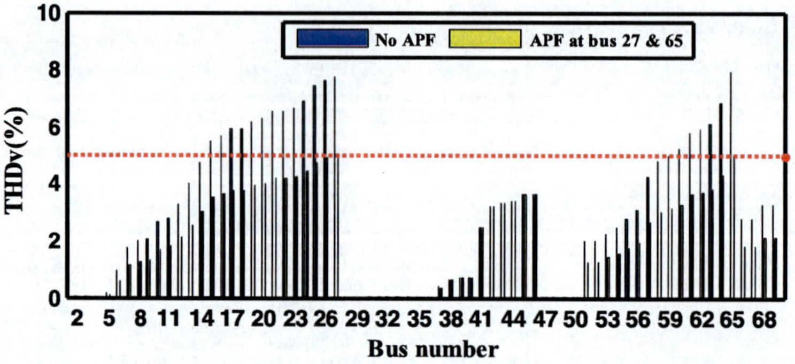

Fig. 5 THD$_V$ (%) at all buses without APF and placement of APF at buses 27 and 65 (only NL)

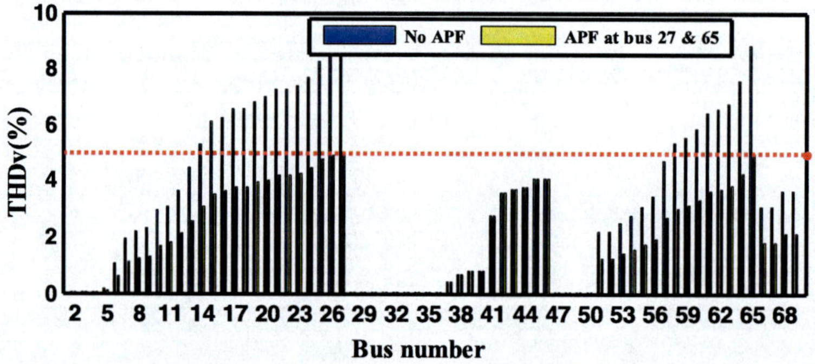

Fig. 6 THD$_V$ (%) at all buses without APF and placement of APF at buses 27 and 65 case 2 (NL plus NLDG)

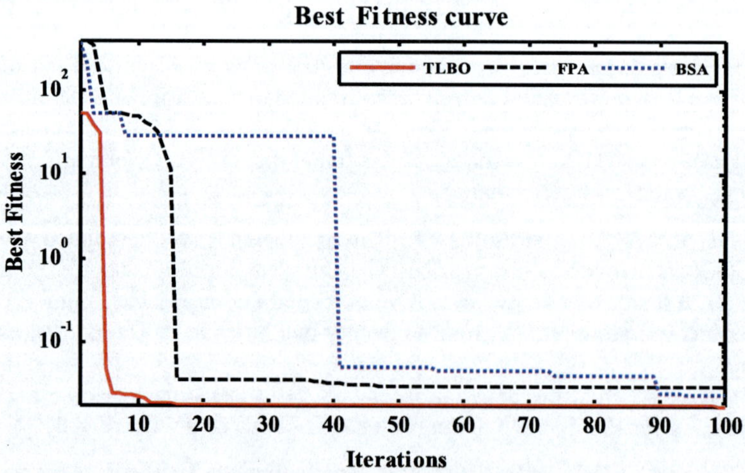

Fig. 7 Best fitness curve for State 5 (case 2)

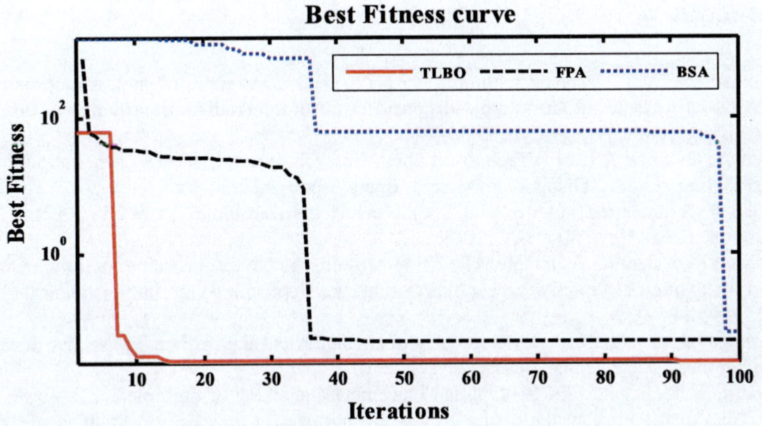

Fig. 8 Best fitness curve for state 5 (case 4)

All three algorithms have fulfilled the constraints for considered harmonic environment and converged. It is again proved that the TLBO has the lowest value compared to other considered algorithms. Out of the algorithms tested, TLBO produced the best value with a result of 0.0247 p.u., surpassing the results obtained with FPA (0.0649 p.u.) and BSA (0.0500 p.u.). It is also observed that the FPA and BSA are poor in the exploration phase compared to TLBO.

According to Table 3, it is clearly noticed that the inclusion of NLDG in RDS increases the harmonic distortion. The APFs required for this case have more ratings compared to case 2.

6 Conclusion

In this study, the OPAS of APF problem was effectively addressed using the TLBO algorithm, and its results were compared to those obtained from the FPA and BSA algorithms. The performance of the algorithms is demonstrated for NLs plus NLDGs in the IEEE-69 bus RDS. The TLBO algorithm implementation is straightforward and doesn't require adjusting any algorithm-specific parameters. The obtained numerical results have shown that TLBO has given the lowest APF current (0.0184 p.u. and 0.0247 p.u.). Also, it has a better convergence characteristic and the ability to converge near better optimal solution for the given problem. Moreover, it is observed that harmonic distortion is increased due to the addition of NLDGs in RDS. Therefore, the required APF current is also increased 1.34 times compared to NL only. The APF current has increased by a factor of 1.34 when compared to the scenario where only the NLs are present.

References

1. Hengsritawat V, Tayjasanant T, Nimpitiwan N (2012) Optimal sizing of photovoltaic distributed generators in a distribution system with consideration of solar radiation and harmonic distortion. Int J Electr Power Energy Syst 39:36–47
2. Kumar N, Kumar A (2017) Techno-economic analysis of non-linear DG penetration in radial distribution systems. Distr Gener Alternat Energy Journal 32:54–74
3. IEEE Recommended Practice and Requirements for Harmonic Control in Electric Power Systems. IEEE, New York, NY (2014)
4. Shivaie M, Salemnia A, Ameli MT (2014) A multi-objective approach to optimal placement and sizing of multiple active power filters using a music-inspired algorithm. Appl Soft Comput 22:189–204
5. Ziari I, Jalilian A (2010) A new approach for allocation and sizing of multiple active power-line conditioners. IEEE Trans Power Delivery 25:1026–1035
6. Lakum A, Mahajan V (2019) Optimal placement and sizing of multiple active power filters in radial distribution system using grey wolf optimizer in presence of nonlinear distributed generation. Electr Power Syst Res 173:281–290
7. Wolpert DH, Macready WG (1997) No free lunch theorems for optimization. IEEE Trans Evol Comput 1:67–82
8. Rao RV, Savsani VJ, Vakharia D (2011) Teaching–learning-based optimization: a novel method for constrained mechanical design optimization problems. Comput Aided Des 43:303–315
9. Yang X-S, Karamanoglu M, He X (2014) Flower pollination algorithm: a novel approach for multiobjective optimization. Eng Optim 46:1222–1237
10. Meng X-B, Gao XZ, Lu L, Liu Y, Zhang H (2016) A new bio-inspired optimisation algorithm: bird Swarm Algorithm. J Exp Theor Artif Intell 28:673–687
11. Teng J-H, Liao S-H, Leou R-C (2014) Three-phase harmonic analysis method for unbalanced distribution systems. Energies 7:365–384
12. Shivaie M, Salemnia A, Ameli MT (2013) Optimal multi-objective placement and sizing of passive and active power filters by a fuzzy-improved harmony search algorithm. Int Trans Electr Energy Syst 25:520–546
13. Sakar S, Balci ME, Aleem SHA, Zobaa AF (2017) Increasing PV hosting capacity in distorted distribution systems using passive harmonic filtering. Electr Power Syst Res 148:74–86
14. Sakar S, Balci ME, Aleem SHA, Zobaa AF (2018) Integration of large-scale PV plants in non-sinusoidal environments: considerations on hosting capacity and harmonic distortion limits. Renew Sustain Energy Rev 82:176–186
15. Savier J, Das D (2007) Impact of network reconfiguration on loss allocation of radial distribution systems. IEEE Trans Power Delivery 22:2473–2480
16. Fuchs E, Masoum MA (2011) Power quality in power systems and electrical machines. Academic press
17. Oliva AR, Balda JC (2003) A PV dispersed generator: a power quality analysis within the IEEE 519. IEEE Trans Power Delivery 18:525–530
18. Zhang D, Yang J, Yang P (2019) An improved chaos bird swarm optimization algorithm. J Phys Conf Ser 022016
19. Cui W, He Y (2018) Biological flower pollination algorithm with orthogonal learning strategy and catfish effect mechanism for global optimization problems. Math Prob Eng 2018

Comparison of LSTM and GRU for Predicting Market Clearing Price in Open Electricity Market

Mrinal Kanti Dey and Saurabh Chanana

Abstract The Indian power exchange market is rapidly evolving. As a result of which consistent forecasting of Market Clearance Price (MCP) has become a very crucial task for trading electricity to any part of the country. An analysis of the fifteen-minute MCP forecasting of the Indian Energy Exchange (IEX) is done. Firstly, only a single data point is predicted using Long-Short-Term Memory (LSTM) based Recurrent Neural Network (RNN) and Gated Recurrent Unit (GRU) based RNN. Furthermore, twenty data points which represent five hours ahead forecasting are done using both models. The comparative analysis between the experimental results and the real data from IEX proves the effectiveness of the developed models that GRU is better when it comes to speed and memory, but in terms of accuracy LSTM has got a slight edge over GRU.

Keywords Indian energy exchange · Market clearing price · Recurrent neural networks · Long-short-term memory · Gated recurrent unit

1 Introduction

With the rising power demand from different parts of India, sometimes the state distribution entities are unable to supply demand. To deal with this problem the Indian Government has adapted one nation one grid policy in 2013. The Indian power sector is mainly categorized into five regions: the northern region, the eastern region, the north-eastern region, the western region, and the southern region. Furthermore, in 2019 another initiative was launched by the Indian government as a part of the National Electricity policy 2021 which is the one nation one grid one price policy. This gave a boost to the power market in India. There are two major power exchange

M. K. Dey (✉) · S. Chanana
National Institute of Technology, Kurukshetra, India
e-mail: mrinal_32114310@nitkkr.ac.in

S. Chanana
e-mail: saurabh@nitkkr.ac.in

A. Kumar et al. (eds.), *Decarbonisation and Digitization of the Energy System*, Lecture Notes in Electrical Engineering 1099, https://doi.org/10.1007/978-981-99-7630-0_23

companies that are currently operating in India: Power Exchange India Limited (PXIL) and Indian Energy Exchange (IEX), out of which IEX has around 91% (Nov-22) of the market share [1]. Moreover, IEX has recently expanded its power beyond India to create an integrated South-Asian power market.

Electricity price forecasting is divided in three major categories, namely, short-term price forecasting, mid-term price forecasting, and long-term price forecasting. Short-term price forecasting deals with predicting electricity prices for a few hours to a few weeks, whereas mid-term price forecasting refers to predicting electricity prices for a week to a year, and long-term price forecasting deals with electricity price forecasting of a year ahead.

Electricity price forecasting has been done for several decades. Several researches have been conducted on electricity price forecasting. Short-term electricity price forecasting can be categorized into two sections: Traditional methods and artificial intelligence-based methods. Artificial intelligence-based methods can be further classified into two sections namely: Regression-based models and statistical-based models. The statistical-based models can further be categorized into several categories: Fuzzy logic, Evolutionary algorithms, knowledge-based expert systems, and neural networks.

An optimization model is presented for integrated portfolio management in wholesale and retail power markets with the help of GAMS [2]. In IEX both the customers and the suppliers need accurate price forecasts in order to maximize their investments [3]. Statistical model GARCH is implemented in [4, [5]. Another statistical model SARIMA is compared with GARCH in which the SARIMA easily outperformed the GARCH model [6]. A study of electricity price forecasting is done with the help of wavelet transformation along with ARIMA [7]. Some research on the basis of the time-series-based methods are done in [8, 9]. A comparison between ANN, CNN, CatBoost, AdaBoost, and XGBoost is done in [10]. A detailed comparison of various machine learning-based techniques is done in [11], among which LSTM (Long-Short-Term Memory) outperforms all of them followed by SVM (Support Vector Machine). In [11], hourly, daily, weekly, and monthly predictions are done with the help of Simple Linear Regression, Support Vector Machines, K nearest neighbors, and LSTM, in which the LSTM model clearly surpasses the rest with a good margin and the K nearest neighbors were trailing having predictions which achieved an RMS value of 523.65 in the yearly predictions.

Electricity price forecasting deals with several parameters such as time of the day, time of the week, month and year, historical prices, demand, and generation outages. It establishes a cyclic pattern around the year, so for this kind of data, RNN (Recurrent Neural Network) is the perfect candidate as it can save the data of the previous output and implement it as an activation function to the next input.

In this paper a thorough comparison is made between LSTM (Long-Short-Term Memory) based RNN and GRU (Gated Recurrent Unit) based RNN and the appropriate scenarios are mentioned for the proper use of these models.

2 Indian Energy Exchange

The electricity market of IEX is executed through the Day-Ahead Market (DAM) and Term-Ahead Market (TAM). DAM is a physical electricity trading market for deliveries for any/all/some fifteen-minute time blocks in 24 h of the next day starting from midnight. The prices and quantum of electricity to be traded are determined through a double-sided close auction bidding process. Some characteristics of this market are: this is active since June 2008, the deliveries in this market are done for the next day, it includes cross-border trade which was launched on April 21, 2021, and the price discovery is determined through a closed and double-sided auction. Whereas, TAM provides a range of products, allowing participants to buy or sell electricity on a term basis for a duration of up to eleven days ahead. The price at which the trading takes place in DAM is termed as Market Clearing Price (MCP). So, MCP forecasting plays a significant role for market participants to make accurate investment decisions.

3 Recurrent Neural Networks

It is a type of Neural Network in which the output of the previous dataset is saved by the hidden layers which is utilized for the next dataset as an activation function in the hidden layers. So, in this neural network sequence is prioritized and is very important for the next dataset. As mentioned above electricity price forecasting follows a cyclic pattern around time. RNN is currently used for applications like speech recognition, video recognition, and image recognition.

In spite of RNN being a proper candidate for electricity price forecasting, it has its limitations while dealing with larger datasets. While dealing with larger datasets or while having many hidden layers, the traditional RNN encounters a problem of vanishing gradient and exploding gradient which results in short-term memory problems. To counter this problem there are two methods namely: LSTM and GRU. Both of these abovementioned methods can be used to solve problems like exploding gradient and vanishing gradient which occur in the traditional RNN.

In LSTM, there are three gates: input gate, output gate, and forget gate. Here in this forget is responsible for deciding the amount of relevant data that is to be stored and the amount of irrelevant that is to be forgotten. This relevant data is then used in the next steps for the activation function. The basic block diagram of a LSTM-based RNN model is shown in Fig. 1.

The mathematical formulation of LSTM is described from Eq. (1a)–(1f), where x_t is the input vector, and h_t is the output of the current block. C_t is referred to as the memory from the current block. The corresponding weights of the input, output, and memory blocks are mentioned as W_{xi}, W_{xo}, W_{xf}, W_{xc}, W_{hi}, W_{ho}, W_{hf}, W_{hc}, and W_{ci}, W_{co}, W_{cf}, respectively. Here, b_1, b_2, b_3, b_4 are termed as the bias vectors of the respective blocks.

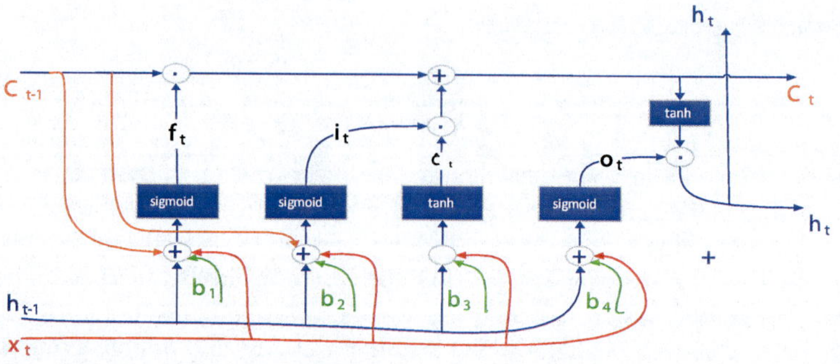

Fig. 1 Block diagram of LSTM-based RNN

$$i_t = \sigma(W_{hi} * h_{t-1} + W_{xi} * x_t + W_{ci} * C_{t-1} + b_2) \tag{1a}$$

$$o_t = \sigma(W_{ho} * h_{t-1} + W_{xo} * x_t + W_{co} * C_t + b_4) \tag{1b}$$

$$f_t = \sigma\left(W_{hf} * h_{t-1} + W_{xf} * x_t + W_{cf} * C_{t-1} + b_1\right) \tag{1c}$$

$$\hat{C}_t = \tanh(W_{hc} * h_{t-1} + W_{xc} * x_t + b_3) \tag{1d}$$

$$C_t = i_t \odot \hat{C}_t + f_t \odot C_{t-1} \tag{1e}$$

$$h_t = o_t \odot \tanh(c_t) \tag{1f}$$

In the above equations, the activation functions taken are sigmoid and tan$h$, which can be altered according to the dataset.

GRU has 2 gates: the update gate and the reset gate as shown in Fig. 2. Here the amount of relevant past information that is to be stored for future states is decided by the update gate whereas the reset gate decides the amount of information that is to be neglected. GRU can be interpreted as the lighter version of LSTM.

The mathematical formulation of GRU-based RNN is described from Eq. (2a)–(2d). Here, x_t and h_t are referred to as the input and the output vector respectively. $\hat{h}_t$ is defined as the candidate activation vector. z_t and r_t are termed as the update gate vector and the reset gate vector, respectively.

$$r_t = \sigma(u_r * h_{t-1} + w_r * x_t + b_r) \tag{2a}$$

$$z_t = \sigma(u_z * h_{t-1} + w_z * x_t + b_z) \tag{2b}$$

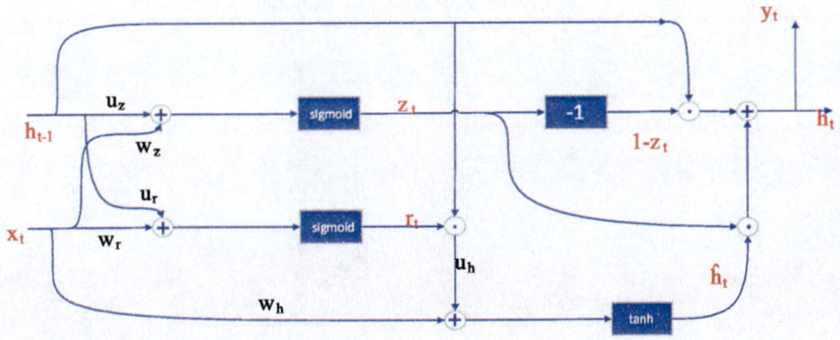

Fig. 2 Block diagram of GRU-based RNN

$$\hat{h}_t = \tan h\left[u_h * (r_t \odot h_{t-1}) + w_h * x_t + b_h\right] \tag{2c}$$

$$h_t = z_t \odot \hat{h}_t + (1 - z_t) \odot h_{t-1} \tag{2d}$$

4 Methodology

4.1 Preparing Data

The dataset used here is the 15 min MCP (Market Clearing Price) from 01.01.2014 to 31.12.2021 which has a total of around 2,80,512 entries which is obtained from IEX [12]. The electricity price data is plotted in Fig. 3.

This dataset is then normalized using Minmax scaler in the range 0 to +1, and the normalized dataset is plotted in Fig. 4. The purpose of this normalization is to replace the data of the numeric columns in the dataset to utilize a common scale, without deforming the differences in the range of values or losing any data.

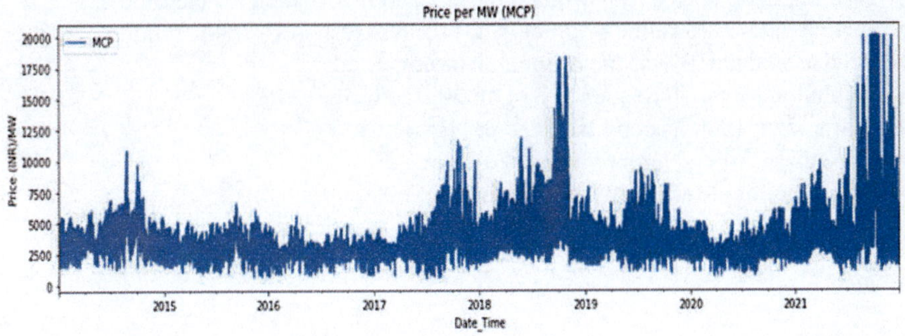

Fig. 3 MCP from 01.01.2014 to 31.12.2021

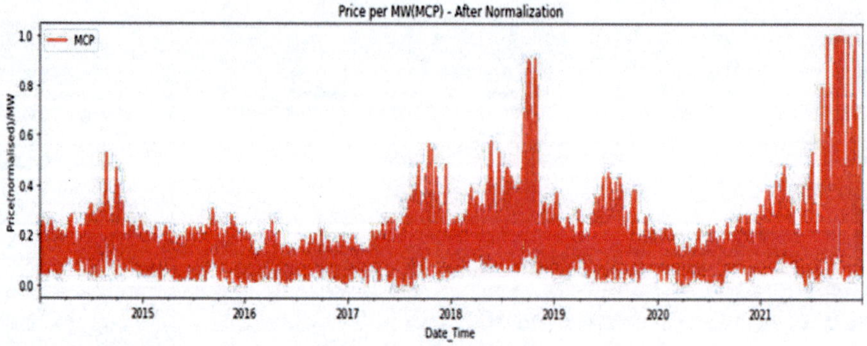

Fig. 4 Normalized MCP data from 01.01.2014 to 31.12.2021

$$X_i\prime = \frac{X_i - X_{min}}{X_{max} - X_{min}} \tag{3}$$

The formulation of the scaling method used which is Minmax scaler is presented in Eq. (3). Where X_i represents the data point i that is to be scaled, X_{min} and X_{max} are the minimum and maximum values of the dataset used. X_i' is the scaled value of the ith data point.

The formula used by the min–max scaler for scaling our entire dataset in the range from 0 to +1 is represented in Eq. (3). This formula is further used to de-normalize our data points.

4.2 RNN Model Building

The dataset is thoroughly studied and two appropriate models are built according to the dataset and the parameters of these models are described. Both the models are trained using the same parameters and then they are evaluated.

- Training data: 80% of the original dataset and is reshaped for the models.
- Testing data: 10% of the original dataset and is reshaped for the models.
- Validation data:10% of the original dataset is used.
- Validation steps: 50 Sequence length: 20 Batch size: 256
- Buffer size: 1000 Epochs:10 Steps per Epoch: 300
- Optimizer: Adam Number of hidden layers: 2
- Loss function: MAE Activation function: Selu

Table 1 Model metrics

	R2 score	MAPE
LSTM	0.9554	0.0631
GRU	0.9558	0.0573

4.3 Evaluation

For evaluating the LSTM model and GRU model, we call their predicted method on the test dataset and evaluate their performance utilizing the r2_score function and the MAPE (Mean Absolute Percentage Error) function of the Sklearn metrics.

$$R2 = 1 - \frac{\sum(e_i - \tilde{e}_i)^2}{\sum(e_i -)^2} \tag{4a}$$

$$MAPE = \frac{1}{v} * \sum_{i=1}^{v} \frac{(e_i - \tilde{e}_i)}{e_i} \tag{4b}$$

The formulas used for calculating the R2 and MAPE scores are represented in Eq. (4a) and (4b), respectively. Here, e_i is the actual value, $\bar{e}$ is the mean value, and $\tilde{e}_i$ is the predicted value from our models. Here, v is the size of the sample. Simulations are done multiple times in these 2 models and the worst-performing R2 scores of both models are shown in Table 1.

4.4 Prediction

Based on the results of Sect. 5, a comparative study between LSTM-based RNN and GRU-based RNN is done in Table 2. In both the scenarios first, only one data point is predicted and later on, a total of 20 data points are predicted, which means a prediction of Market Clearing Price is made 5 h prior.

Table 2 Comparison between LSTM and GRU in real-life scenarios

	Most accurate prediction (₹)	Real data (₹)	Accuracy (%)	Time taken	Time stamp
LSTM	1992.01	1999.06	99.647	1 min 35 s	01-01-2022, 4:30
GRU	1993.5	1999.32	99.708	1 min 21 s	01-01-2022, 4:45

5 Results

The appropriate parameters are set and the following results are obtained and discussed in this section. Here, we have used the LSTM-based RNN model and the GRU-based RNN model. The important parameters such as the dataset, number of hidden layers, loss function, and optimization function are kept the same for both cases. The predicted 20 data points are matched with the real-time MCP values of 01.01.2022 and are noted in Table 2.

Figure 5 shows the prediction of a single data point utilizing the LSTM-based RNN model. Thereafter, 19 more data points are predicted using the same model as shown in Fig. 6 and afterward these 20 data points are de-normalized using Eq. (3) and these data points are matched with the real MCP data of 01.01.2022. The analysis of these 20 data points is shown in Table 2.

Figure 7 shows the single datapoint prediction utilizing the GRU-based RNN model. Furthermore, a total of 20 data points is forecasted using this model and the prediction of these data points is presented in Fig. 8. Similarly, as done before these

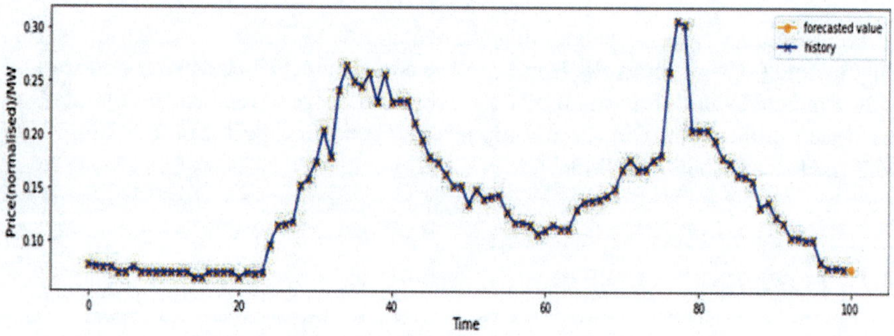

Fig. 5 Single point predicted using LSTM-based RNN

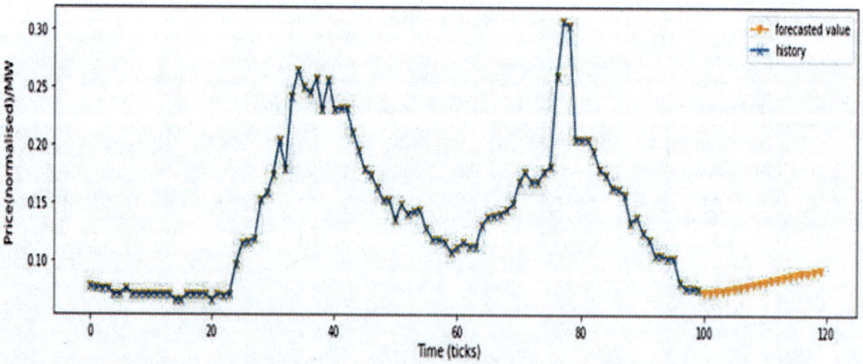

Fig. 6 Twenty data points predicted using LSTM-based RNN

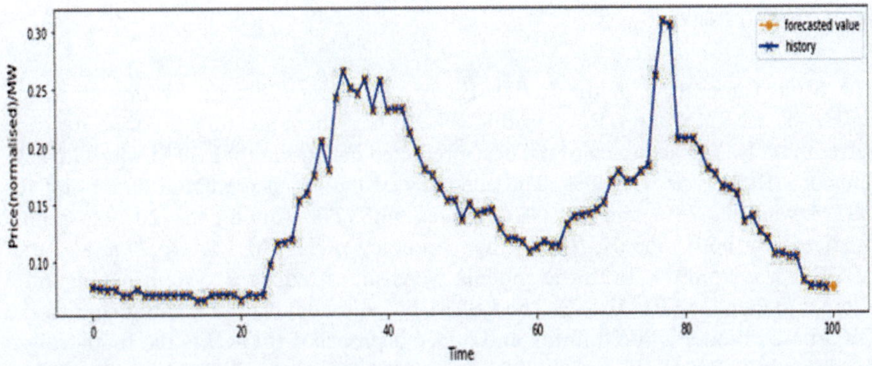

Fig. 7 Single data point predicted using GRU-based RNN

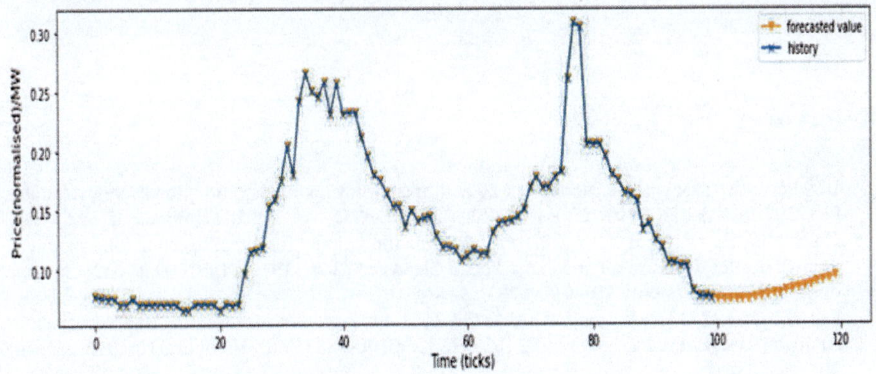

Fig. 8 Twenty data points predicted using GRU-based RNN

20 data points are de-normalized using Eq. (3) and these 20 data points are compared with the real MCP data of 01.01.2022. The analysis of this comparison between the de-normalized forecasted data points and the MCP data of 01.01.2022 is shown in Table 2.

Table 1 represent the R2 score and the MAPE score of both models. Among the predicted 20 data points by both models, the most accurate predicted data points along with the real MCP values of 01.01.2022 are presented in Table 2. For analyzing the time taken by both the models 10 simulations were performed for each one. The LSTM-based model had an average time duration of 1 min 34 s whereas the GRU-based model had an average time duration of 1 min 19 s. Among the 20 predicted data points by both the models, 11 data points of the LSTM-based model had an accuracy above 98% and 10 data points of the GRU-based model had an accuracy above 98% when compared with the respective original MCP data of 01.01.2022.

6 Conclusion

The MCP forecasting of the Indian Energy Exchange is done by implementing LSTM-based RNN and GRU-based RNN. Both of them are very accurate for time series models. The accuracy of the best-predicted data point for LSTM was 99.647% and for GRU it was 99.708%. The accuracy of the worst predicted data point for LSTM was 95.678% and for GRU it was 94.832%. Among the 20 data points predicted by both models, the average accuracy of LSTM was 96.74% and that of GRU was 95.89%. Both the models were simulated on a system having Intel iCore, 5th Gen, 16 GB DDR-3. The LSTM-based model was a bit sluggish than the GRU-based model while training and hence it proves that GRU is the light-weight version of LSTM. However while analyzing the accuracy of all the predicted data points, the LSTM had better accuracy than the GRU. Hence, from this analysis, it can be concluded that GRU is better when it comes to speed and memory, but in terms of accuracy, LSTM has a slight edge over GRU.

References

1. IEX Power Market Update, November 2022. https://www.iexindia.com/Uploads/NewsUpdate/06122022IEX%20POWER%20MARKET%20UPDATE_Nov%202022_Final.pdf
2. Koltsaklis NE, Dagoumas AS (2020) An optimization model for integrated portfolio management in wholesale and retail power markets. J Clean Prod 248:119198. ISSN 0959-6526. https://doi.org/10.1016/j.jclepro.2019.119198
3. Liu H, Shi J (2013) Applying ARMA–GARCH approaches to forecasting short-term electricity prices. Energy Econ 37:152–166. ISSN 0140-9883. https://doi.org/10.1016/j.eneco.2013.02.006
4. Garcia RC, Contreras J, van Akkeren M, Garcia JBC (2005) A GARCH forecasting model to predict day-ahead electricity prices. IEEE Trans Power Syst 20(2):867–874. https://doi.org/10.1109/TPWRS.2005.846044
5. Rajan P, Chandrakala KRMV (2021) Statistical model approach of electricity price forecasting for Indian electricity market. In: 2021 IEEE Madras section conference (MASCON), Chennai, India, pp 1–5. https://doi.org/10.1109/MASCON51689.2021.9563474
6. Conejo AJ, Plazas MA, Espinola R, Molina AB (2005) Day-ahead electricity price forecasting using the wavelet transform and ARIMA models. IEEE Trans Power Syst 20(2):1035–1042. https://doi.org/10.1109/TPWRS.2005.846054
7. Li G, Liu C-C, Lawarree J, Gallanti M, Venturini A (2005) State-of-the-art of electricity price forecasting. In: International symposium CIGRE/IEEE PES, 2005, New Orleans, LA, USA, pp 110–119. https://doi.org/10.1109/CIGRE.2005.1532733
8. Zhang L, Luh PB (2005) Neural network-based market clearing price prediction and confidence interval estimation with an improved extended Kalman filter method. IEEE Trans Power Syst 20(1):59–66. https://doi.org/10.1109/TPWRS.2004.840416
9. Anamika, Kumar N (2016) Market clearing price prediction using ANN in Indian electricity markets. In: 2016 international conference on energy efficient technologies for sustainability (ICEETS), Nagercoil, India, pp 454–458. https://doi.org/10.1109/ICEETS.2016.7583797
10. Boru İpek A (2021) Prediction of market-clearing price using neural networks based methods and boosting algorithms. Int Adv Res Eng J 5(2):240–246. https://doi.org/10.35860/iarej.824168

11. Chaudhury P, Tyagi A, Shanmugam PK (2020) Comparison of various machine learning algorithms for predicting energy price in open electricity market. In: 2020 international conference and utility exhibition on energy, environment and climate change (ICUE), Pattaya, Thailand, pp 1–7. https://doi.org/10.1109/ICUE49301.2020.9307100
12. https://www.iexindia.com/marketdata/areaprice.aspx

Small-Signal Stability Analysis of Synchronverter-Based AC Microgrid in Islanded Mode

Siddhant Singh Maurya, Trapti Jain, and Amod C. Umarikar

Abstract Synchronverter control of the inverter is a promising technology that can be used for controlling grid forming converters in an islanded AC microgrid. Synchronverter provides virtual inertia by emulating the characteristics of synchronous generators with simple control structure and better voltage and frequency regulations. This paper presents a small signal model of the synchronverter based microgrid with resistive(R), inductive(R-L) and Constant Power load (CPL) operating in islanded mode. The developed small signal model is validated by comparing it with the simulated model in MATLAB Simulink software. Further the microgrid is simulated on the Real Time Digital Simulator (RTDS) and the impact of load and PV array insolation changes are observed. The eigen value analysis is done to show the stability of the microgrid for R, R-L and CPL.

Keywords Small signal model · Synchronverter · Microgrid

1 Introduction

THE advancement of power electronics and their effective control over the years have led to the deployment of distributed energy resources (DER) based microgrid with more scalability and effectiveness. However, there are several challenges poised due to the integration of microgrid in the existing power systems [1]. A microgrid in the islanded mode of operation requires better controllability in order to keep the frequency and voltage within the permissible range.

The virtual synchronous machine [2] or virtual synchronous generator (VSG) has been one of the prime areas for research in control of grid forming converters. VSGs incorporate synchronous generator like characteristic in an inverter. Synchronverter i.e., the inverter that replicates the characteristics of the synchronous generators is proposed in [3], where virtual inertia is incorporated in the control to

S. S. Maurya (✉) · T. Jain · A. C. Umarikar
IIT Indore, Indore, India
e-mail: phd2101102008@iiti.ac.in

A. Kumar et al. (eds.), *Decarbonisation and Digitization of the Energy System*, Lecture Notes in Electrical Engineering 1099, https://doi.org/10.1007/978-981-99-7630-0_24

improve the dynamics of the system. Some improvements of the synchronverter control like self-synchronization or improving dynamic response have been presented in [4, 5].

It is inevitable for any system to remain unaffected due to the disturbance caused either by changes in load or generation. Hence, small signal stability analysis of the microgrid becomes important to examine the dynamic response of the system. Modelling and small signal stability analysis of the Inverter Interfaced Distributed Generations (IIDGs), filters and the network of an AC microgrid based on the droop control are proposed in [6]. However, the small signal model and the analysis presented assumed that the Distributed Generation (DG) provides constant voltage at the inverter input. Small signal model of the synchronverter has been proposed in [7] to obtain small signal transfer function, whereas the eigen value analysis for grid connected synchronverter based microgrid has been presented in [8]. An attempt has been also made to include a PV source at the inverter inputs and develop the complete microgrid small signal model based on synchronverter [9]. In [10] a geometrical approach is taken to study the small signal stability of grid forming converter, but utilising droop control and considering single converter connected to grid.

In most of the research papers, the small signal model of a DG with synchronverter is attempted for grid connected mode. Moreover, the parallel operation of the DGs with synchronverter is considered in [11] with grid connected mode but the inverter input is assumed to be constant irrespective of the load changes. However, in practical situations, the inverter input may not be constant due to the dynamics of PV converters, load condition and under certain conditions it may operate in islanded mode. The approach to include the dc-dynamics of the system is presented in [12] but renewable source such as PV array is not used, instead a dc-dc converter is chosen at the input for the impact on small signal stability of droop controlled inverter. This paper presents a small signal model of an islanded microgrid with synchronverter control of its IIDGs considering PV array as well as battery dynamics. The PV array with boost converter is connected in parallel with Li-ion battery as energy storage through a bidirectional dc-dc converter to provide a constant dc source [13]. Several different approaches for control of PV-Boost converter and Bidirectional converter have been proposed for such configurations. Simple approach for control of converters are used in [14] using MPPT algorithm for PV-Boost converter and Battery charge, discharge technique for controlling the Bidirectional dc-dc converter. Similar control is also utilised in the [15] with certain modifications. An optimised PI tuning is used for the control of Bidirectional dc-dc converter and incremental conductance method computes the duty cycle of PV-Boost converter [16]. Most of the synchronverter model uses time domain approach for simulation whereas the small signal model is developed in either synchronous or stationary reference frame. The adopted synchronverter control is developed in synchronous reference frame (SRF), which provides a simpler small signal model that reduces the complexity of modelling and simulation. The microgrid is simulated in RSCAD in order to get the operating points and the eigenvalue analysis is carried out for the entire microgrid including interconnecting lines and loads.

2 System Description and Control

2.1 PV/Battery System and Modelling of Power Conversion Chain

The PV array and Li-Ion battery is connected to a boost converter and a bidirectional converter, respectively and the converter outputs are connected in parallel as shown in Fig. 1. This configuration provides more flexibility and smooth control either in charging or discharging mode of the battery [13]. The simulation model is developed in RSCAD software of RTDS. The software provides two type of configurations for the PV array model: (1) Single Diode, Five Parameter Model and (2) Approximated Single Diode, Four Parameter Model. The RTDS PV array model has four and five parameter model of which four parameter model is used in the simulation. The current-voltage relationship of the single diode, four parameter model is given by: $I = I_{ph} - I_o(e^{\frac{V+R_sI}{N_caV_t}} - 1)$ where, I_{ph} is the current induced by the solar cell. The diode ideality factor a is a measure of how closely the diode matches the ideal diode equation. The series resistance R_s represents series resistance in the solar cell. The other symbols hold the usual meaning as is used in the diode equation such as V_t, diode thermal voltage, temperature T (in K) and number of series connected cells of a module that forms a PV array N_c. The small signal linearized equation for the PV-Boost converter and Bidirectional converter are given below:

$$\begin{bmatrix} \Delta \dot{i}_{LPV} \\ \Delta v_{PV} \end{bmatrix} = \begin{bmatrix} \frac{-r_{LPV}}{L_{PV}} & \frac{1}{L_{PV}} \\ \frac{-1}{C_{PV}} & \left(\frac{\partial i_{LPV}}{\partial v_{PV}}\right)\frac{1}{C_{PV}} \end{bmatrix} \begin{bmatrix} \Delta i_{LPV} \\ \Delta v_{PV} \end{bmatrix} + \begin{bmatrix} \frac{V_{dc}}{L_{PV}} & \frac{-(1-D_1)}{L_{PV}} \\ 0 & 0 \end{bmatrix} \begin{bmatrix} \Delta d_1 \\ \Delta V_{dc} \end{bmatrix} \tag{1}$$

$$\begin{bmatrix} \Delta \dot{i}_{LB} \\ \Delta v_{DC} \end{bmatrix} = \begin{bmatrix} \frac{-r_{LB}}{L_B} & \frac{-(1-D_2)}{L_B} \\ \frac{1-D_2}{C_{DC}} & 0 \end{bmatrix} \begin{bmatrix} \Delta i_{LB} \\ \Delta v_{DC} \end{bmatrix} + \begin{bmatrix} \frac{1}{L_B} & \frac{v_{DC}}{L_B} \\ 0 & \frac{-I_{LB}}{C_{DC}} \end{bmatrix} \begin{bmatrix} \Delta v_B \\ \Delta d_2 \end{bmatrix} + \begin{bmatrix} 0 \\ \frac{-1}{C_{DC}} \end{bmatrix} \Delta i_O \tag{2}$$

Equation 1 reveals that matrix depends on the derivative of the PV current with respect to the PV voltage $\partial i_{PV}/\partial v_{PV}$. Hence the non-linear model of the PV module which is $\partial i_{PV}/\partial v_{PV} < 0$, it is assumed to be $1/r_{PV}$ [17].

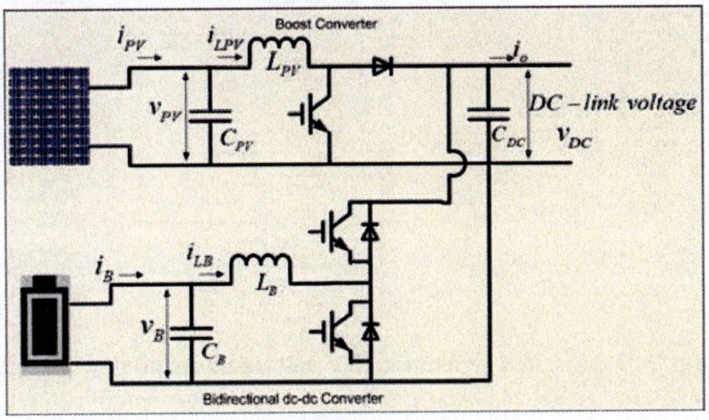

Fig. 1 Distributed generation (PV and Battery with power electronic interface)

2.2 PV-BESS Management and Controller Design for Power Conversion Chain

During the time of sufficient insolation, the PV array is capable to provide the power to the inverter connected through the dc-link capacitor and charge the battery through the bidirectional converter. During low insolation level, the battery discharges through the bidirectional converter providing the balance power as required by the load. For maximum power point tracking (MPPT), incremental conductance method is used to get the maximum voltage, V_{MPPT}. Also, the power across the PV array is measured and reference current, i_{LPVref} is calculated. The control is shown in Fig. 2 to obtain the duty ratio of the boost converter. The bidirectional dc-dc converter operates in boost mode during discharging and in buck mode during charging. It should be noted that the bidirectional converter is responsible for maintaining constant dc voltage across the capacitor, whereas the boost converter extracts the maximum power from the PV array.

As shown in Fig. 3, during the normal operation of the microgrid, the PV-boost converter will operate in the MPPT mode and the bidirectional converter may charge

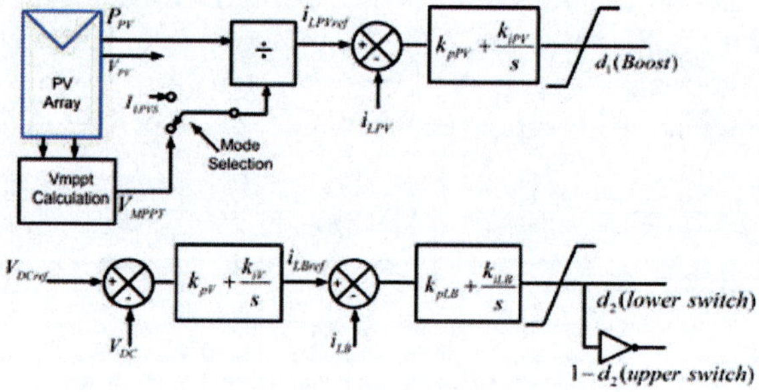

Fig. 2 Control of PV-boost converter and bidirectional dc-dc converter

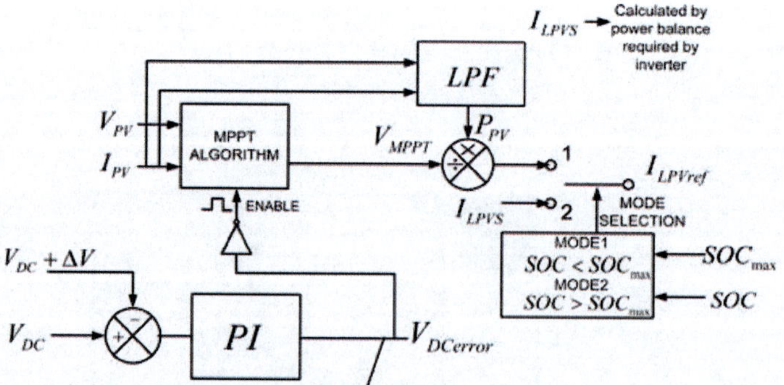

Fig. 3 Distributed generation (PV and Battery with power electronic interface)

or discharge depending upon the power required by the inverter and power generated by the PV array. The PI controller shown in Fig. 3 has no significance value as it is used to enable and disable the MPPT algorithm to obtain V_{MPPT}. During low insolation the battery will provide the complete power to the load until the State of charge(SOC) of the Battery goes below certain threshold. Below the certain threshold the microgrid in islanded can't provide a reliable supply and hence the microgrid needs to be connected to grid. In the condition when insolation is sufficient enough and also the battery is charged to a maximum level of SOC the Bidirectional converter needs to be disconnected and the MPPT mode is shifted to Non-MPPT mode. The PV inductor current reference may be generated using a look table or by calculating power balance equation. The constant dc voltage in that case must be held constant and must be incorporated through voltage control in the inverter which is not covered in the paper and is part of the further research. The small signal stability analysis is done around an operating point of MPPT. Cascade control is used to control the switches of the bidirectional converter. As shown in Fig. 2, bidirectional converter has an outer voltage control and an inner current control to get its duty ratio. The controller design for the bidirectional converter is achieved by using the transfer function mentioned in [13]. The values of controller are mentioned in Table 1.

Table 1 Steady state condition and test system parameters

Parameters	Values
$[r_{lPV}\ \ r_{lB}\ \ L_{PV}\ \ L_B]$	[4 mΩ 4 mΩ 3.74 mH 10 mH]
$[C_{dc}\ \ C_{PV}\ \ C_B]$	[5000 µF 1500 µF 1500 µF]
$[r_{lf}\ \ r_{lc}\ \ L_f\ \ L_c]$	[0.1 Ω 0.03 mΩ 2.35 mH 0.35 mH]
$[C_f\ \ C_B\ \ C_{PV}]$	[50 µF 1.5 mF 1.5 mF]
$[D_1\ \ D_2\ \ I_{lPV}\ \ I_{lB1}]$	[0.525 0.5Ω 46 A − 18.2 A]
$[I_{lB2}\ \ I_{lB3}\ \ V_{dc}]$	[−10.8A − 2.9A 800 V]
$[k_{pB}\ \ k_{iB}\ \ k_{pV}]$	[0.08153 140.57 0.21613]
$[k_{iV}\ \ k_{pPV}\ \ k_{iPV}]$	[43.225 12.474 180]
$[D_{p1}\ \ D_{p2}\ \ D_{p3}]$	[50.686 76 100]Ws/rad
$[D_{q1}\ \ D_{q2}\ \ D_{q3}]$	[0.6 0.9 1.5]kVAR/V
$[V_{oQ1}\ \ V_{oQ2}\ \ V_{oQ3}]$	[16.4 21.28 27]V
$[V_{oD1}\ \ V_{oD2}\ \ V_{oD3}]$	[339 344 342]
$[I_{oQ1}\ \ I_{oQ2}\ \ I_{oQ3}]$	[0.91 9.66 12.9]A
$[I_{oD1}\ \ I_{oD2}\ \ I_{oD3}]$	[11.8 16.8 22.35]
$[I_{lQ1}\ \ I_{lQ2}\ \ I_{lQ3}]$	[−5.22 4.26 7.6]A
$[I_{lD1}\ \ I_{lD2}\ \ I_{lD3}]$	[12.14 17.26 22.8]
$[I_{LineD1}\ \ I_{LineD2}]\ [I_{LineQ1}\ \ I_{LineQ2}]$	[−18.08 − 9.7] [−2.89 − 4.89]A
$[\delta_0\ \ \delta_1\ \ \delta_2]$	[0 3.6 − 0.9]°
$[r_{Line1}\ \ r_{Line2}\ \ L_{Line1}\ \ L_{Line2}]$	[0.23 Ω 0.23 Ω 11.1 mH 15.7 mH]
Load parameters	$R = 20\,\Omega R - L = 20\,\Omega$ 100 mH, $CPL = 15\,kW, 5\,kVAR$

$$\frac{i_{LPV}(s)}{d_1(s)} = \frac{\frac{V_{DC}}{L_{PV}}\left(s - \frac{1}{C_{PV}}\frac{\partial i_{PV}}{\partial v_{PV}}\right)}{s^2 + \left(\frac{r_{LPV}}{L_{PV}} - \frac{1}{C_{PV}}\frac{\partial i_{PV}}{\partial v_{PV}}\right)s + \left(\frac{1 - r_{LPV}\frac{\partial i_{PV}}{\partial v_{PV}}}{L_{PV}C_{PV}}\right)} \tag{3}$$

2.3 *IIDGs and Network Under Study*

The DG source i.e. solar PV and battery in our case is connected to the inverter inputs through a dc-link capacitor. The microgrid network under consideration is shown in Fig. 4. The filter parameters, currents and voltages are also shown, where r_f and L_f are the resistance and inductance of the filter inductor of the inverter. r_c and L_c are the resistance and inductance of coupling inductor connected to the 3 phase buses. C_f is the filter capacitor value of the LCL filter. i_l, i_o, and v_o denote the inverter inductor current, output current and the capacitor voltage. The 3 buses are connected to 3 different loads, i.e. R load, RL load and CPL for the purpose of analysis. IIDGs are controlled using the synchronverter control.

2.4 *Synchronverter Control*

Microgrid utilises fast acting inverters which are unable to provide rotational inertia as in the case of a synchronous generator. Synchronverter, a mathematical model of synchronous generators, is used to implement virtual inertia through its control for controlling VSCs. The implementation of the synchronverter control shown in Fig. 5 has been done in SRF in this paper. Synchronverter uses swing equation of the synchronous machine with a damping term in the equation. Since the microgrid is autonomous, the P_{set} and Q_{set} are set to zero in order to operate inverters as grid forming. P and Q represent the measured active and reactive power at the output terminals of the inverter. The synchronverter droop coefficients D_p and D_q are defined as below:

Fig. 4 Microgrid network

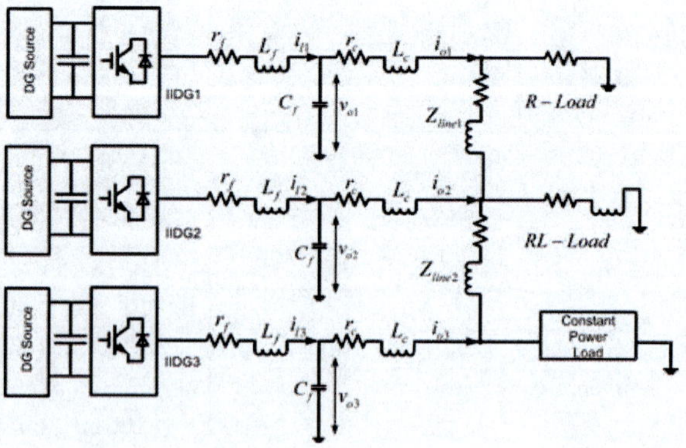

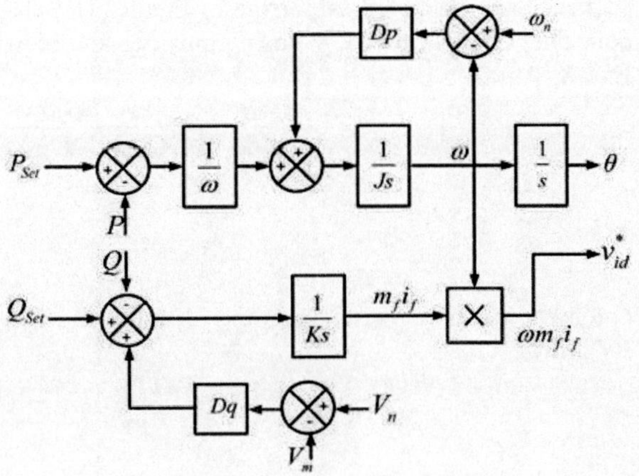

Fig. 5 Synchronverter control

$$D_p = \frac{\Delta T}{\Delta \omega} \tag{4}$$

$$D_q = \frac{\Delta Q}{\Delta V} \tag{5}$$

ΔT, $\Delta \omega$, ΔQ and ΔV represent the change in torque, frequency, reactive power and voltage respectively. The torque is calculated as the ratio of power to actual speed but for the analysis purpose, ratio of power to nominal speed is used as in steady state $\omega = \omega_n$ where, ω and ω_n is the actual and nominal frequency in rad/s. The inertia J, and design parameter K are selected using the equation below where τ_f andτ_v are the time constant of active and reactive power loop respectively.

$$J = D_p \tau_f \tag{6}$$

$$K = D_q \tau_v \omega \tag{7}$$

D_p and D_q can be said to be a part of active and reactive power loop, where the active power sets the frequency and reactive power loop establishes the voltage of the microgrid.

3 Small-Signal Model

3.1 Mathematical Equations and Small-Signal Model of DG

For simplicity, the small signal model is presented for a single DG and hence the subscripts are dropped in the variables. For multiple DGs, simply a subscript of a sequence can be added. Referring to Figures 1, 2 and 4, the averaged equation for controller design of the PV-boost converter and bidirectional converter is given in Eqs. 8, 10–13.

$$\frac{dv_{PV}}{dt} = \frac{i_{PV}}{C_{PV}} - \frac{i_{LPV}}{C_{PV}} \tag{8}$$

$$r_{pv} = -\frac{\Delta v_{PV}}{\Delta i_{PV}} \tag{9}$$

r_{PV} is the dynamic resistance of the PV array and C_{PV} is the capacitance across the PV array.

$$\frac{di_{LPV}}{dt} = \frac{1}{L_{PV}} v_{PV} - \frac{r_{LPV}}{L_{PV}} i_{LPV} - (1 - d_1)\frac{v_{DC}}{L_{PV}} \tag{10}$$

$$\frac{dv_B}{dt} = \frac{i_B}{C_B} v_{PV} - \frac{i_{LB}}{C_B} \tag{11}$$

$$\frac{di_{LB}}{dt} = \frac{1}{L_B} v_B - \frac{r_{LB}}{L_B} i_{LB} - (1 - d_2)\frac{v_{DC}}{L_B} \tag{12}$$

$$\frac{dv_{DC}}{dt} = \frac{(1 - d_1)i_{LPV}}{C_{DC}} + \frac{(1 - d_2)i_{LB}}{C_{DC}} - \frac{i_O}{C_{DC}} \tag{13}$$

r_{LPV}, L_{PV} and r_{LB}, L_B are inductor resistances and inductances of PV-Boost and Bidirectional converter respectively. d_1 and d_2 are the duty ratio of the PV-Boost and Bidirectional converter. C_{DC} and i_o are the capacitance of dc-link voltage and input current to the inverter. The small signal model after linearization are:

$$\Delta \dot{v}_{PV} = -\frac{\Delta v_{PV}}{C_{PV} r_{PV}} - \frac{\Delta i_{LPV}}{C_{PV}} \tag{14}$$

$$\Delta \dot{i}_{LPV} = \frac{1}{L_{PV}} \Delta v_{PV} - \frac{r_{LPV}}{L_{PV}} \Delta i_{LPV} - \frac{(1 - D_1)}{L_{PV}} \Delta v_{DC} + \frac{V_{DC}}{L_{PV}} \Delta d_1 \tag{15}$$

$$\Delta \dot{v}_B = -\frac{1}{C_B} \Delta i_{LB} + \frac{1}{C_B} \Delta i_B \tag{16}$$

$$\Delta \dot{i}_{LB} = \frac{1}{L_B} \Delta v_B - \frac{r_{LB}}{L_B} \Delta i_{LB} - \frac{(1 - D_2)}{L_B} \Delta v_{DC} + \frac{V_{DC}}{L_B} \Delta d_2 \tag{17}$$

$$\Delta \dot{v}_{DC} = \frac{(1 - D_1)}{C_{DC}} \Delta i_{LPV} + \frac{(1 - D_2)}{C_{DC}} \Delta i_{LB} - \Delta i_O - \frac{I_{LPV}}{C_{DC}} \Delta d_1 - \frac{I_{LB}}{C_{DC}} \Delta d_2 \tag{18}$$

From PV-Boost controller:

$$\frac{d\psi_{PV}}{dt} = i_{LPVref} - i_{LPV} \tag{19}$$

$$d_1 = k_{pPV}(i_{LPVref} - i_{LPV}) + k_{iPV} \int (i_{LPVref} - i_{LPV})dt \tag{20}$$

$$\Delta\dot{\psi}_{PV} = -\Delta i_{LPV} \tag{21}$$

$$\Delta d_1 = -k_{pPV}\Delta i_{LPV} + k_{iPV}\Delta\psi_{PV} \tag{22}$$

Here ψ_{PV} is assumed to be a PV electric flux and k_{pPV} and k_{iPV} are proportional and integral gains of the PI controller. From Eqs. 15 and 22:

$$\begin{aligned}\Delta\dot{i}_{LPV} = \frac{1}{L_{PV}}\Delta v_{PV} + \frac{v_{DC}k_{iPV}}{L_{PV}}\Delta\psi_{PV} - \frac{(r_{LPV} + V_{DC}k_{pPV})}{L_{PV}}\Delta i_{LPV} \\ - \frac{(1 - D_1)}{L_{PV}}\Delta v_{DC}\end{aligned} \tag{23}$$

From bidirectional dc-dc cascade controller:

$$\frac{\mathrm{d}\varphi_B}{\mathrm{d}t} = v_{DCref} - v_{DC} \tag{24}$$

$$\dot{\varphi_B} = -\Delta v_{DC} \tag{25}$$

$$i_{LBref} = k_{pV}(v_{DCref} - v_{DC}) + k_{iV} \int (v_{DCref} - v_{DC})dt \tag{26}$$

$$\Delta i_{LBref} = -k_{pV}\Delta v_{DC} + k_{iV}\Delta\varphi_B \tag{27}$$

$$\frac{\mathrm{d}\psi_B}{\mathrm{d}t} = i_{LBref} - i_{LB} \tag{28}$$

$$\Delta\dot{\psi}_B = \Delta i_{LBref} - \Delta i_{LB} \tag{29}$$

$$d_2 = k_{pLB}(i_{LBref} - i_{LB}) + k_{iLB} \int (i_{LBref} - i_{LB})dt \tag{30}$$

$$\Delta d_2 = k_{pLB}(\Delta i_{LBref} - \Delta i_{LB}) + k_{iLB}\Delta\psi_B \tag{31}$$

Here φ_B and ψ_B are assumed to be magnetic flux and battery electric flux respectively and k_{pLB} and k_{iLB} are proportional and integral gains of the inner current control PI controller and k_{pV} and k_{iV} are proportional and integral gains of the outer voltage control PI controller. From Eqs. 27 and 29

$$\Delta\dot{\psi}_B = -\Delta i_{LB} + k_{iV}\Delta\varphi_B - k_{pV}\Delta v_{DC} \tag{32}$$

From Eqs. 17, 24–31

$$\begin{aligned}\Delta\dot{i}_{LB} =& \frac{1}{L_B}\Delta v_B + \frac{V_{DC}k_{iLB}}{L_B}\Delta\psi_B - \frac{r_{LB}+V_{DC}k_{pLB}}{L_B}\Delta i_{LB} + \\ & \frac{V_{DC}k_{pLB}k_{iV}}{L_B}\Delta\varphi_B - \frac{(1-D_2)+v_{DC}k_{pLB}k_{pV}}{L_B}\Delta v_{DC}\end{aligned} \tag{33}$$

Similarly, the small signal model equation for the output capacitor can be given by:

$$\begin{aligned}\Delta\dot{v}_{DC} =& \frac{(1-D_1)+I_{LPV}k_{pPV}}{C_{DC}}\Delta i_{LPV} + \frac{(1-D_2)+I_{LB}k_{pLB}}{C_{DC}}\Delta i_{LB} \\ & - \frac{I_{LPV}k_{iPV}}{C_{DC}}\Delta\psi_{PV} - \frac{I_{LB}k_{iLB}}{C_{DC}}\Delta\psi_B - \frac{I_{ILB}k_{pLB}k_{iV}}{C_{DC}}\Delta\varphi_B \\ & + \frac{I_{ILB}k_{pLB}k_{iV}}{C_{DC}}\Delta v_{DC} - \Delta i_O\end{aligned} \tag{34}$$

The state space matrix ADG are formed using the above equation. The small signal model equations of the DG are:

$$\begin{aligned}[\Delta\dot{\mathbf{X}}_{\mathbf{DG}}] &= \mathbf{A}_{\mathbf{DG}}[\Delta\mathbf{X}_{\mathbf{DG}}] + \mathbf{B}_{\mathbf{X}}\Delta i_B + \mathbf{B}_{\mathbf{Y}}\Delta i_O \\ \Delta\mathbf{X}_{\mathbf{DG}} &= [\Delta v_{PV} \;\; \Delta\psi_{PV} \;\; \Delta i_{PV} \;\; \Delta v_B \;\; \Delta\psi_B \;\; \Delta i_{LB} \;\; \Delta\varphi_B \;\; \Delta v_{DC}]^T \\ \mathbf{B}_{\mathbf{X}} &= [0 \;\; 0 \;\; C_B{}^{-1} \;\; 0 \;\; 0 \;\; 0 \;\; 0 \;\; 0]^T \\ \mathbf{B}_{\mathbf{Y}} &= [0 \;\; 0 \;\; 0 \;\; 0 \;\; 0 \;\; 0 \;\; 0 \;\; -1]^T\end{aligned} \tag{35}$$

3.2 *Small-Signal Modelling of Synchronverter Control*

From Fig. 6, the small signal model equations for the synchronverter are developed in SRF to reduce the complexity of the modelling and present a simpler model to implement to control. The modelling is divided into subsections such as synchronverter control, IIDG, LC filter and coupling inductor for simplicity.
Active and Reactive Power Controller

$$\Delta\dot{\omega} = -\frac{D_p}{J}\Delta\omega - \frac{\Delta P}{J\omega_n} \tag{36}$$

$$\Delta\dot{m}_{fi_f} = -\frac{D_q}{K}\Delta V_m - \frac{\Delta Q}{K} \tag{37}$$

$$\frac{\mathrm{d}\theta}{\mathrm{d}t} = \omega,\; \Delta\dot{\theta} = \Delta\omega \tag{38}$$

$$\Delta P = \frac{3}{2}(V_{od}\Delta i_{od} + I_{od}\Delta v_{od} + V_{oq}\Delta i_{oq} + I_{oq}\Delta v_{oq}) \tag{39}$$

$$\Delta Q = \frac{3}{2}(V_{oq}\Delta i_{od} + I_{od}\Delta v_{oq} - V_{od}\Delta i_{oq} - I_{od}\Delta v_{od}) \tag{40}$$

$$V_m = \sqrt{v_{od}^2 + v_{oq}^2}, \Delta V_m = \frac{V_{od}\Delta v_{od}}{\sqrt{v_{od}^2 + v_{oq}^2}} + \frac{V_{oq}\Delta v_{oq}}{\sqrt{v_{od}^2 + v_{oq}^2}} \tag{41}$$

$$\begin{bmatrix} \dot{\Delta\theta} \\ \Delta\omega \\ \Delta m_f i_f \end{bmatrix} = \mathbf{A_1} \begin{bmatrix} \Delta\theta \\ \Delta\omega \\ \Delta m_f i_f \end{bmatrix} + \mathbf{B_{11}} \begin{bmatrix} \Delta P \\ \Delta Q \\ \Delta V_m \end{bmatrix}$$
$$\left\{ \mathbf{A_1} = \begin{bmatrix} 0 & 1 & 0 \\ 0 & -\frac{D_p}{J} & 0 \\ 0 & 0 & 0 \end{bmatrix}, \mathbf{B_{11}} = \begin{bmatrix} 0 & 0 & 0 \\ -\frac{1}{J\omega_n} & 0 & 0 \\ 0 & -\frac{1}{K} & -\frac{D_q}{K} \end{bmatrix} \right. \tag{42}$$

where V_m is the peak value of the phase voltage across the filter capacitor and $m_f i_f$ is the output of the reactive power loop. Equation 42 is derived from Eqs. 36–38where,

$$\begin{bmatrix} \Delta P \\ \Delta Q \\ \Delta V_m \end{bmatrix} = \mathbf{B_{21}} \begin{bmatrix} \Delta i_{od} \\ \Delta i_{oq} \end{bmatrix} + \mathbf{B_{22}} \begin{bmatrix} \Delta v_{od} \\ \Delta v_{oq} \end{bmatrix} \tag{43}$$

Equation 43 is derived from Eqs. 39–41 where,

$$\mathbf{B_{21}} = \begin{bmatrix} \frac{3}{2}V_{od} & \frac{3}{2}V_{oq} \\ \frac{3}{2}V_{oq} & -\frac{3}{2}V_{od} \\ 0 & 0 \end{bmatrix}, \mathbf{B_{22}} = \begin{bmatrix} \frac{3}{2}I_{od} & \frac{3}{2}I_{oq} \\ -\frac{3}{2}I_{oq} & \frac{3}{2}I_{od} \\ \frac{V_{od}}{\sqrt{V_{od}{}^2+V_{od}{}^2}} & \frac{V_{oq}}{\sqrt{V_{od}{}^2+V_{od}{}^2}} \end{bmatrix} \tag{44}$$

Voltage Equations and Interfacing Matrices

The VSC is assumed to be lossless and converter is able to reproduce the reference voltage as actual output voltage of the inverter. Therefore $v_{id}{}^* = v_{id}$, $v_{iq}{}^* = v_{iq}$. As $v_{iq} = 0$ and $v_{id} = (m_f i_f)\omega$,

$$\Delta v_{id} = m_f i_f \Delta\omega + \omega \Delta m_f i_f \tag{45}$$

$$\begin{bmatrix} \Delta v_{id} \\ \Delta v_{iq} \\ \Delta\omega \\ \Delta\delta \end{bmatrix} = \begin{bmatrix} 0 & m_f i_f & \omega \\ 0 & 0 & 0 \\ 0 & 1 & 0 \\ 1 & 0 & 0 \end{bmatrix} \begin{bmatrix} \Delta\delta \\ \Delta\omega \\ \Delta m_f i_f \end{bmatrix} \tag{46}$$

where,

$$\mathbf{C_{cs}} = \begin{bmatrix} 0 & m_f i_f & \omega \\ 0 & 0 & 0 \end{bmatrix}, \mathbf{C_{c\omega}} = \begin{bmatrix} 0 & 1 & 0 \end{bmatrix}, \mathbf{C_{c\theta}} = \begin{bmatrix} 1 & 0 & 0 \end{bmatrix} \tag{47}$$

$$\begin{bmatrix}\Delta i_{od}\\ \Delta i_{oq}\end{bmatrix} = \mathbf{C_2}\begin{bmatrix}\Delta \mathbf{i_{ldq}}\\ \Delta \mathbf{v_{odq}}\\ \Delta \mathbf{i_{odq}}\end{bmatrix}, \begin{bmatrix}\Delta v_{od}\\ \Delta v_{oq}\end{bmatrix} = \mathbf{C_3}\begin{bmatrix}\Delta \mathbf{i_{ldq}}\\ \Delta \mathbf{v_{odq}}\\ \Delta \mathbf{i_{odq}}\end{bmatrix} \tag{48}$$

where C_2 and C_3 are interfacing matrices given by:

$$\mathbf{C_2} = \begin{bmatrix}0\,0\,0\,0\,1\,0\\ 0\,0\,0\,0\,0\,1\end{bmatrix}, \mathbf{C_3} = \begin{bmatrix}0\,0\,1\,0\,0\,0\\ 0\,0\,0\,1\,0\,0\end{bmatrix} \tag{49}$$

LC Filter and Coupling inductance

The small signal equations for LC filter, coupling inductance and the transformation matrices required to interface with the whole system are given as in [6]:

$$\begin{bmatrix}\Delta \dot{\mathbf{i}}_{\mathbf{ldq}}\\ \Delta \mathbf{v_{odq}}\\ \Delta \mathbf{i_{odq}}\end{bmatrix} = \mathbf{A_{LCL}}\begin{bmatrix}\Delta \mathbf{i_{ldq}}\\ \Delta \mathbf{v_{odq}}\\ \Delta \mathbf{i_{odq}}\end{bmatrix} + \mathbf{B_{LCL1}}[\Delta \mathbf{v_{idq}}] + \mathbf{B_{LCL2}}[\Delta \mathbf{v_{bdq}}] + \mathbf{B_{LCL3}}[\Delta\omega] \tag{50}$$

$$\begin{bmatrix}\Delta \dot{\mathbf{i}}_{\mathbf{ldq}}\\ \Delta \mathbf{v_{odq}}\\ \Delta \mathbf{i_{odq}}\end{bmatrix} = \mathbf{A_{LCL}}\begin{bmatrix}\Delta \mathbf{i_{ldq}}\\ \Delta \mathbf{v_{odq}}\\ \Delta \mathbf{i_{odq}}\end{bmatrix} + \mathbf{B_{LCL1}}[\Delta \mathbf{v_{idq}}] + \mathbf{B_{LCL2}}[\Delta \mathbf{v_{bdq}}] + \mathbf{B_{LCL3}}[\Delta\omega] \tag{51}$$

$$\left[\Delta \mathbf{i_{oDQ}}\right] = \mathbf{T_S}\left[\Delta \mathbf{i_{odq}}\right] + \mathbf{T_C}[\Delta\delta], \left[\Delta \mathbf{v_{bdq}}\right] = \mathbf{T_S}^{-1}\left[\Delta \mathbf{v_{bDQ}}\right] + \mathbf{T_V}^{-1}[\Delta\delta] \tag{52}$$

The complete small signal model of the inverter is given below:

$$[\Delta \dot{\mathbf{X}}_{\mathbf{syn}}] = \mathbf{A_{inv}}[\Delta \mathbf{X_{syn}}] + \mathbf{B_{inv}}\begin{bmatrix}\Delta v_{bD}\\ \Delta v_{bQ}\end{bmatrix} \tag{53}$$

$$[\Delta \mathbf{Y_{syn}}] = \mathbf{C_{inv}}[\Delta \mathbf{X_{syn}}] \tag{54}$$

where $[\mathbf{X_{syn}}] = [\Delta\delta \quad \Delta\omega \quad \Delta m_f i_f \quad \Delta i_{ld} \quad \Delta v_{od} \quad \Delta v_{oq} \quad \Delta i_{od} \quad \Delta i_{oq}]^T$, $\mathbf{Y_{Syn}} = [\Delta\omega \quad \Delta i_{oD} \quad \Delta i_{oQ}]^T$ and $\mathbf{A_{inv}}$, $\mathbf{B_{inv}}$, $\mathbf{C_{inv}}$ are given in Eqs. 55 and 56.The complete state space of 1DG microgrid without load or network can be obtained as: $[\Delta \dot{\mathbf{X}}_{\mathbf{MGi}}] = \mathbf{A_{MGi}}[\Delta \mathbf{X_{MGi}}] + \mathbf{B_{MGi}}\left[\Delta \mathbf{v_{bDQi}}\right]$ and $[\Delta \mathbf{Y_{MGi}}] = \mathbf{C_{MGi}}[\Delta \mathbf{X_{MGi}}]$ where $[\Delta \mathbf{X_{MGi}}] = [\Delta \mathbf{X_{DGi}} \quad \Delta \mathbf{X_{Syni}}]$ and $\mathbf{A_{MGi}}$,$\mathbf{B_{MGi}}$,$\mathbf{C_{MGi}}$ are given in Eqs. 57 and 58. The complete small signal model of the entire system including lines and loads is obtained as given in the Eqs. 59–61. The final state space equation is:

$$[\Delta \dot{\mathbf{X}}_{\mathbf{MGi}} \quad \Delta \dot{\mathbf{i}}_{\mathbf{lineDQ}} \quad \Delta \dot{\mathbf{i}}_{\mathbf{loadDQ}}]^T = \mathbf{A_{MGNET}}[\Delta \mathbf{X_{MGi}} \quad \Delta \mathbf{i_{lineDQ}} \quad \Delta \mathbf{i_{loadDQ}}]^T$$

$$\mathbf{A_{inv}} = \begin{bmatrix}\mathbf{A_1} & \mathbf{B_{11}B_{21}C_2} + \mathbf{B_{11}B_{22}C_3}\\ \mathbf{B_{LCL1}C_{cs}} + \mathbf{B_{LCL2}T_V}^{-1}\mathbf{C_{c\theta}} + \mathbf{B_{LCL3}C_{c\omega}} & \mathbf{A_{LCL}}\end{bmatrix} \tag{55}$$

$$\mathbf{B_{inv}} = \begin{bmatrix} [\mathbf{0}]_{3\times 2} \\ \mathbf{B_{LCL2}T_S^{-1}} \end{bmatrix}, \mathbf{C_{inv}} = \begin{bmatrix} \mathbf{C_{pw}} \\ \mathbf{C_{pinv}} \end{bmatrix},$$

$$\begin{cases} \mathbf{C_{pw}} = [0\ 1\ 0\ 0\ 0\ 0\ 0\ 0\ 0] \\ \mathbf{C_{pinv}} = \left[\begin{bmatrix} -I_{od}\sin\delta_o - I_{oq}\cos\delta\ 0\ 0 \\ I_{od}\cos\delta_o - I_{oq}\sin\delta_o\ \ 0\ 0 \end{bmatrix} [\mathbf{T_{SC_2}}] \right] \end{cases} \tag{56}$$

$$\mathbf{A_{MGi}} = \begin{bmatrix} [\mathbf{A_{DGi}}] & \left[0\ 0\ 0\ \frac{I_{ld}}{\sqrt{I_{ld}^2+I_{lq}^2}}\ \frac{I_{ld}}{\sqrt{I_{ld}^2+I_{lq}^2}}\ 0\ 0\ 0\ 0\right] \\ [\mathbf{0}] & [\mathbf{A_{invi}}] \end{bmatrix}, \mathbf{B_{MGi}} = \begin{bmatrix} [\mathbf{0}] \\ [\mathbf{B_{invi}}] \end{bmatrix} \tag{57}$$

$$\mathbf{C_{MGi}} = \left[[\mathbf{0}]\ [\mathbf{C_{invi}}]\right], \mathbf{C_{pwMGi}} = \left[[\mathbf{0}]\ [\mathbf{C_{pwi}}]\right], \mathbf{C_{pcMGi}} = \left[[\mathbf{0}]\ [\mathbf{C_{pinvi}}]\right] \tag{58}$$

$$\mathbf{A_{MG}} = \begin{bmatrix} [\mathbf{A_{MG1}}]_{17\times 17} & [\mathbf{0}] & [\mathbf{0}] \\ [\mathbf{0}] & [\mathbf{A_{MG2}}]_{17\times 17} & [\mathbf{0}] \\ [\mathbf{0}] & [\mathbf{0}] & [\mathbf{A_{MG3}}]_{17\times 17} \end{bmatrix} \tag{59}$$

$$\mathbf{B_{MG}} = \begin{bmatrix} [\mathbf{B_{MG1}}]_{17\times 2} & [\mathbf{0}] & [\mathbf{0}] \\ [\mathbf{0}] & [\mathbf{B_{MG2}}]_{17\times 2} & [\mathbf{0}] \\ [\mathbf{0}] & [\mathbf{0}] & [\mathbf{B_{MG3}}]_{17\times 2} \end{bmatrix},$$
$$\mathbf{C_{pcMG}} = \begin{bmatrix} [\mathbf{C_{pcMG1}}]_{2\times 17} & [\mathbf{0}] & [\mathbf{0}] \\ [\mathbf{0}] & [\mathbf{C_{pcMG2}}]_{2\times 17} & [\mathbf{0}] \\ [\mathbf{0}] & [\mathbf{0}] & [\mathbf{C_{pcMG3}}]_{2\times 17} \end{bmatrix} \tag{60}$$

$$\mathbf{A_{MGNET}} = [\mathbf{A_{MGNET1}}\ \ \mathbf{A_{MGNET2}}]$$
$$\mathbf{A_{MGNET1}} = \begin{bmatrix} \mathbf{A_{MG}} + \mathbf{B_{MG}R_NM_{INV}C_{pcMG}} & \mathbf{B_{INV}R_NM_{NET}} \\ \mathbf{B_{1NET}R_NM_{INV}C_{pcMG}} + \mathbf{B_{2NET}C_{pwMG}} & \mathbf{A_{NET}} + \mathbf{B_{1NET}R_NM_{NET}} \\ \mathbf{B_{1LOAD}R_NM_{INV}C_{pcMG}} + \mathbf{B_{2LOAD}C_{pwMG}} & \mathbf{B_{1LOAD}R_NM_{NET}} \end{bmatrix} \tag{61}$$
$$\mathbf{A_{MGNET2}} = \begin{bmatrix} \mathbf{B_{INV}R_NM_{LOAD}} \\ \mathbf{B_{1NET}R_NM_{LOAD}} \\ \mathbf{A_{LOAD}} + \mathbf{B_{1LOAD}R_{NM_{LOAD}}} \end{bmatrix}$$

4 Results and Analysis

4.1 Eigenvalue Analysis

For the eigenvalue analysis, an operating point is chosen from the real time simulation. The insolation of PV array for the 3-DGs are assumed to be the same and the parameters of PV boost and bidirectional converter are kept similar. The parameters and the values required for the eigenvalue analysis are mentioned in Table I. Out of 61 eigenvalues, only the low frequency modes are shown in Fig. 6. Since the CPL loads are unstable, hence the 2 modes appear on the right side of the complex plane. The movement of low frequency modes with respect to change in droop coefficient is shown in Fig. 7. The poles can be seen moving towards the imaginary axis as the droop coefficient D_p decreases.

4.2 Simulation Waveforms

The AC microgrid model is prepared in RSCAD software and simulated in RTDS. The impact on active power, reactive power sharing and frequency due to change in CPL-load of 10 kW at 2.5 s, 4.4 kW of R-load at 5.5 s and 2.49 kW of RLload 8.5 s at bus 3,1 and 2 respectively is shown in Fig. 8. Even though the load change occurs on one bus but IIDGs share active power smoothly and get stabilises after some time. The frequency regulation due to load change is well within the acceptable range. Figure 9 shows the impact of insolation change from 1000 W/m^2 to 200 W/m^2. It can be seen that the DC link voltage gets affected due to change in insolation but gets

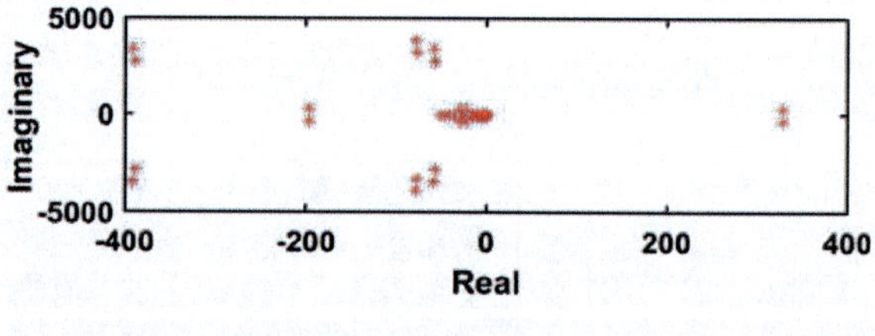

Fig. 6 Low frequency modes of the matrix A_{MGNET}

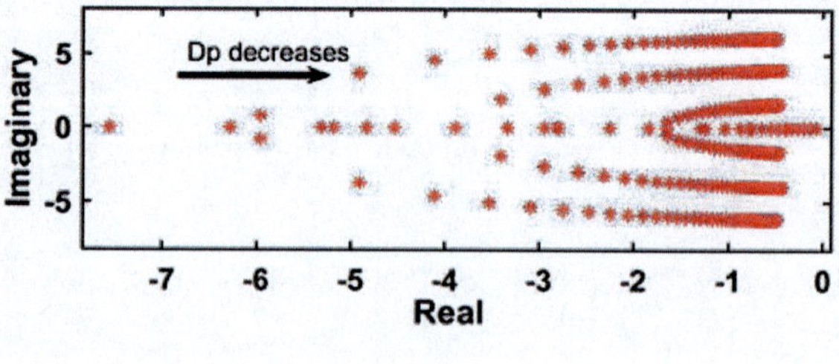

Fig. 7 Movement of poles for change in D_p

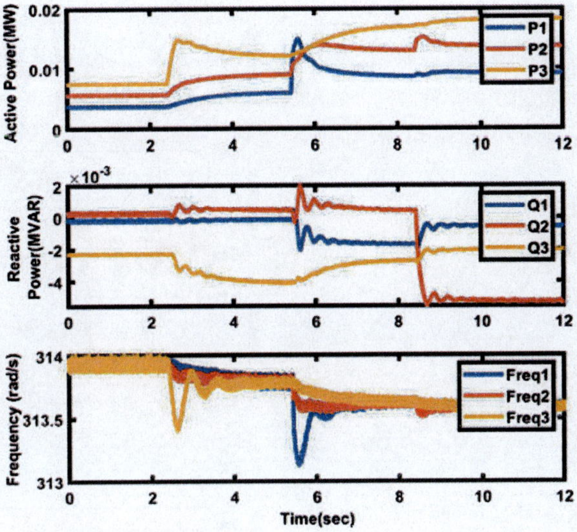

Fig. 8 P and Q response (change in load)

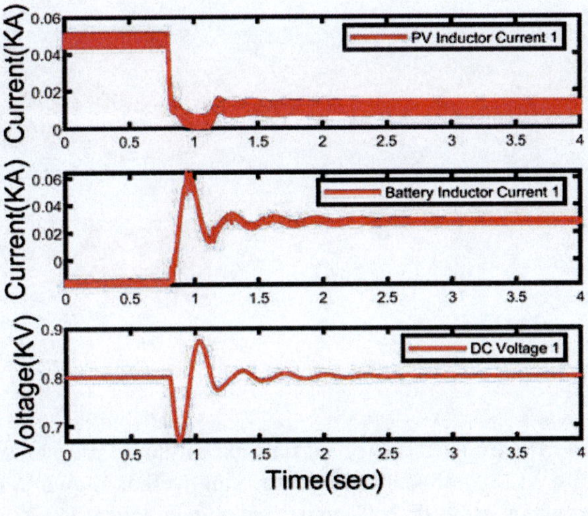

Fig. 9 Change in i_{LPV}, i_{LB} and V_{DC}

stabilised after around 2 s. As the insolation changes battery starts delivering power and bidirectional converter operates in discharging mode.

Two test cases are presented below subjected to disturbance:

Case 1: Bus 3 is subjected to increase of 20 kW CPL-load

Subjected to change of 20 kW, Bidirectional converter 2 and 3 changes mode form charging to discharging which is shown in Figs. 10 and 13. There is no change in

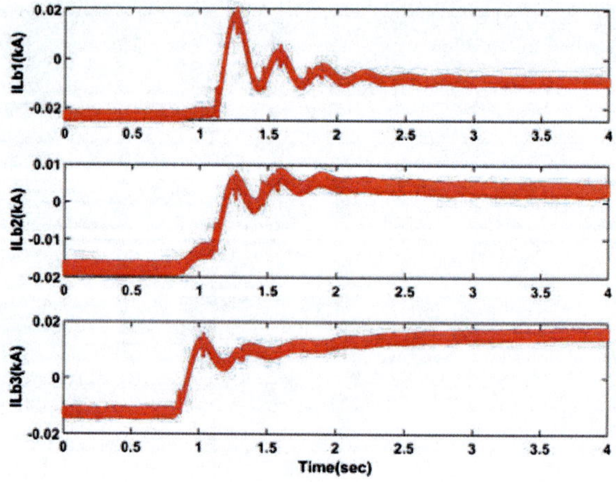

Fig. 10 Change in i_{LB}

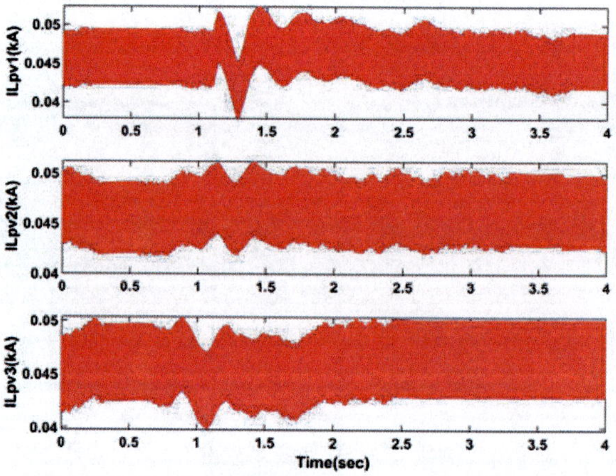

Fig. 11 Change in i_{LPV}

the PV-Boost inductor current except some momentarily disturbance as shown in Fig. 11. Power sharing is performing well as shown in Fig. 10 and v_{DC} is maintained constant at $800V$ following load change (Figs. 12, 13 and 14).

Case 2: PV array of Inverter 2 is subjected to insolation change of 800 to 200 W/m^2. Subjected to change of insolation of Inverter 2 the Bidirectional converter changes mode from charging to discharging as shown in Fig. 15 and the SOC starts decreasing thereafter which can be seen in Fig. 18. The PV-Boost inductor current of converter 1 and 3 remains unaffected as can be seen in the Fig. 16. The active power, reactive power and frequency remains unaffected but get disturbed momentarily during the change as shown in Figs. 17 and 18. In Fig. 19, it can be seen that the dc voltage of Inverter 2 is more affected as compared to the Inverter 1 and 3 as insolation change had ocurred on the Inverter 2.

Fig. 12 Change in active power, reactive power and frequency

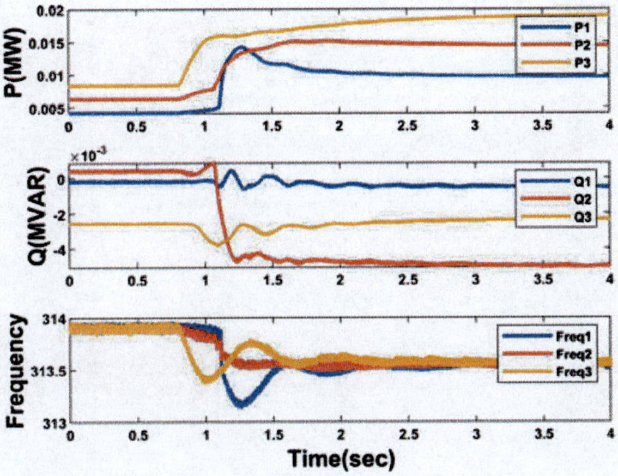

Fig. 13 Change in SOC

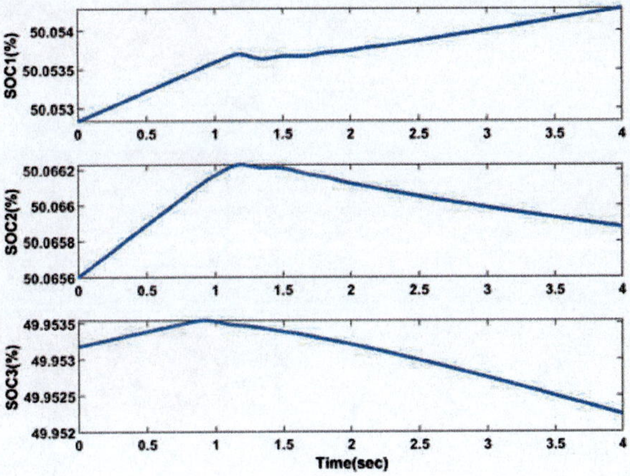

Fig. 14 Change in v_{DC}

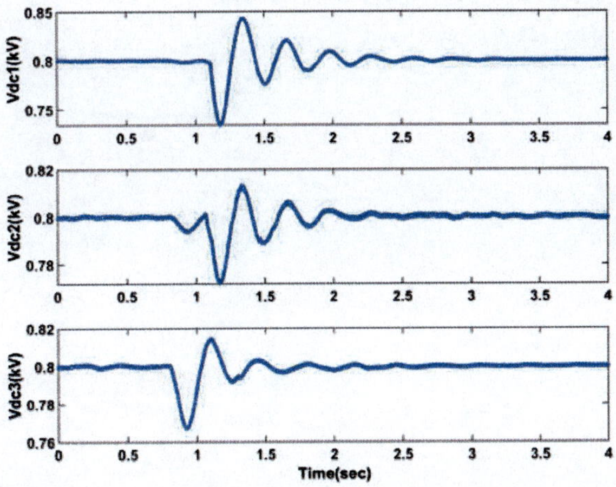

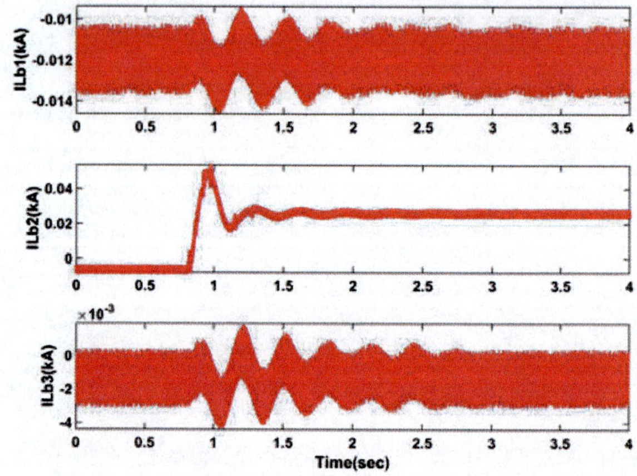

Fig. 15 Change in i_{LB}

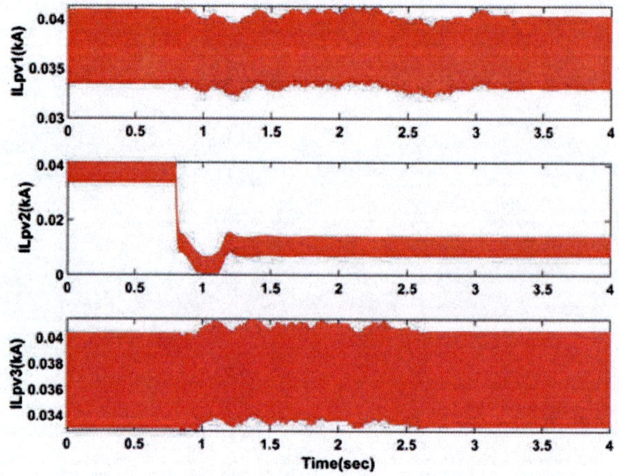

Fig. 16 Change in i_{LPV}

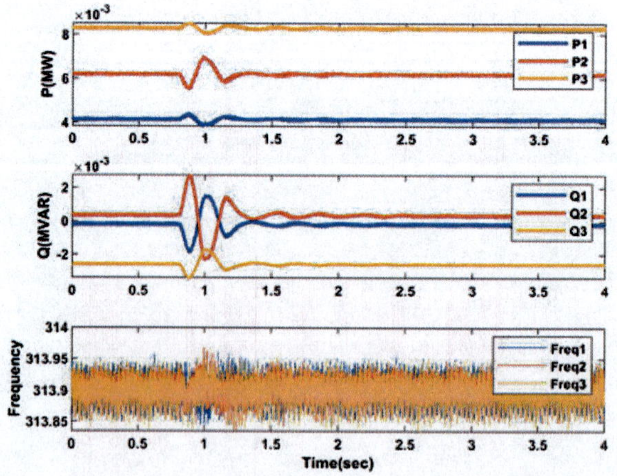

Fig. 17 Change in active power, reactive power and frequency

Fig. 18 Change in SOC

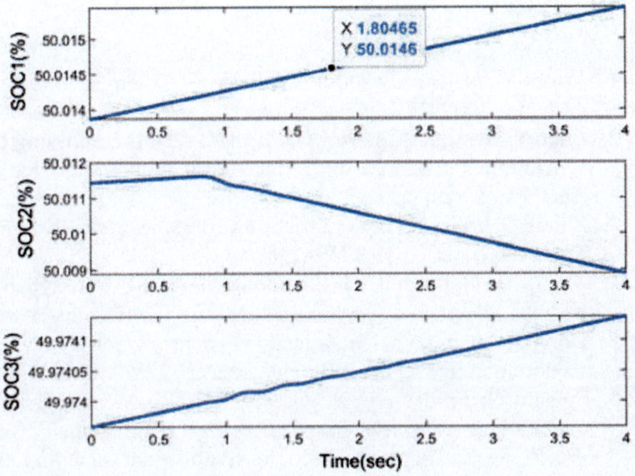

Fig. 19 Change in v_{DC}

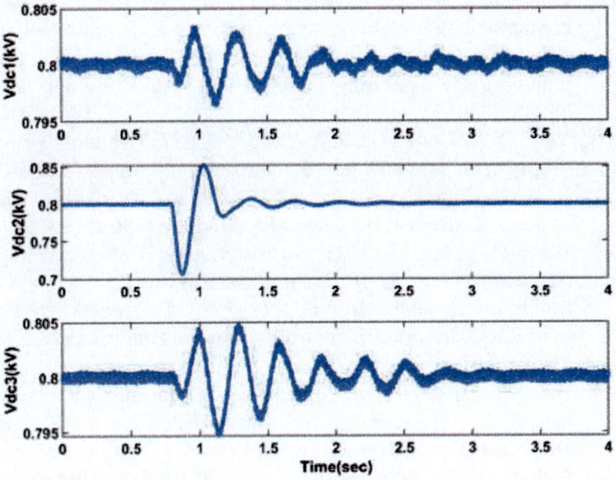

5 Conclusion

The PV array/Li-ion battery supported microgrid with synchronverter control operating in islanded mode and feeding various types of load is considered for small signal stability analysis. For this, the small signal model is developed and eigenvalue analysis is carried out to identify the unstable modes and study the impact of controller parameters on low frequency modes. The synchronverter is found to perform well with respect to power sharing, frequency regulation and voltage regulation under different loading conditions.

Acknowledgements The authors sincerely acknowledge the financial support provided by Science and Engineering Research Board (SERB), New Delhi, India under research grant SPG/2020/000358 dated 5th April 2021.

References

1. Colson C, Nehrir M (2009) A review of challenges to real-time power management of microgrids. In: 2009 IEEE power & energy society general meeting, pp 1–8
2. Chen Y, Hesse R, Turschner D, Beck H (2011) Improving the grid power quality using virtual synchronous machines. In: 2011 international conference on power engineering, energy and electrical drives, pp 1–6
3. Zhong Q, Weiss G (2011) Synchronverters: inverters that mimic synchronous generators. IEEE Trans Ind Electron 58:1259–1267
4. Zhong Q, Nguyen P, Ma Z, Sheng W (2014) Self-synchronized synchronverters: inverters without a dedicated synchronization unit. IEEE Trans Power Electron 29:617–630
5. Dong S, Chen Y (2017) Adjusting synchronverter dynamic response speed via damping correction loop. IEEE Trans Energy Convers 32:608–619
6. Pogaku N, Prodanovic M, Green T (2007) Modeling, analysis and testing of autonomous operation of an inverter-based microgrid. IEEE Trans Power Electron 22:613–625
7. Wei Z, Jie C, Chunying G (2015) Small signal modeling and analysis of synchronverters. In: 2015 IEEE 2nd international future energy electronics conference (IFEEC), pp 1–5
8. Rodríguez-Cabero A, Roldan-Perez J, Prodanovic M (2017) Synchronverter small-signal modelling and eigenvalue analysis for battery systems integration. In: 2017 IEEE 6th international conference on renewable energy research and applications (ICRERA), pp 780–784
9. Sonawane A, Umarikar A (2022) Small-signal stability analysis of PV-based synchronverter including PV operating modes and DC-link voltage controller. IEEE Trans Ind Electron 69:8028–8039
10. Yu J, Qi Y, Deng H, Liu X, Tang Y (2022) Evaluating small-signal synchronization stability of grid-forming converter: a geometrical approach. IEEE Trans Ind Electron 69:9087–9098
11. Rosso R, Engelken S, Liserre M (2019) Robust stability analysis of synchronverters operating in parallel. IEEE Trans Power Electron 34:11309–11319
12. Wang B, Verbic G (2021) Stability analysis of low-voltage distribution feeders operated as islanded microgrids. IEEE Trans Smart Grid 12:4681–4689
13. Mahmood H, Michaelson D, Jiang J (2015) Decentralized power management of a PV/battery hybrid unit in a droop-controlled Islanded microgrid. IEEE Trans Power Electron 30:7215–7229
14. Hasabelrasul H, Cai Z, Sun L, Suo X, Matraji I (2022) Two-stage converter standalone PV-battery system based on VSG control. IEEE Access 10:39825–39832
15. Jha S, Singh B, Mishra S (2023) Control of ILC in an autonomous AC-DC hybrid microgrid with unbalanced nonlinear AC loads. IEEE Trans Ind Electron 70:544–554
16. Chankaya M, Ahmad A, Hussain I, Singh B, Naqvi S (2022) Grid-interfaced photovoltaic-battery energy storage system with slime mold optimized adaptive seamless control. IEEE Trans Ind Appl 58:7728–7738
17. Petrone G, Ramos-Paja C, Spagnuolo G (2016) Modeling the PV power conversion chain. In: Photovoltaic sources modeling, pp 127–164

Feasibility Study of PV/Wind Hybrid System with Recycled Retired Electric Vehicle Batteries

Ambati Bhimaraju, Shomi Kumari, and Aeidapu Mahesh

Abstract The island mode Hybrid Renewable Energy Systems (HRES) are gaining more attention for rural electrification in remote areas. However, the island mode of HRES should be equipped with bulk storage backup due to the intermittent nature of renewable sources. So the adapted storage technology has a considerable effect on the overall cost of the HRES. In this paper, the recycling of Retired Electric Vehicles Batteries (REVBs) is used as an alternate solution for conventional battery storage technology in the HRES application. In this proposed case study, the feasibility analysis of the island mode of solar PV/Wind/REVB has been carried out with the Teaching Learning Based (TLBO) Algorithm. The Levelized Cost of Energy (LCE) and Net Present Cost (NPC) are used as the economic indicator to evaluate the economic benefits of the system while to Loss of Power Supply (LPSP) is a constraint. Further, the obtained results have been compared with PSO algorithms and included comparative results in the results section.

Keywords Hybrid renewable energy systems · Loss of power supply probability · Levelized cost of energy · Retired electric vehicles batteries

1 Introduction

The abrupt changes in environmental pollution are evidence of over-dependency on fossil fuels to meet the required energy demand. In that required demand, the majority portion is utilized in the form of electrical Energy. From the above sentence, the electricity sector is one of the main reasons for global warming. That's why it is time to search for alternative resources that can be most suitable for eco-friendly power generation. Among the available resources, solar energy and wind energy are the most attractive due to their abundant availability worldwide. These resources can solve future energy crises, but the task is strenuous because of their highly stochastic

A. Bhimaraju (✉) · S. Kumari · A. Mahesh
Sardar Vallabhbhai National Institute of Technology Surat, Surat, Gujarat, India
e-mail: bhimaraju.ambati@gmail.com

A. Kumar et al. (eds.), *Decarbonisation and Digitization of the Energy System*, Lecture Notes in Electrical Engineering 1099, https://doi.org/10.1007/978-981-99-7630-0_25

nature. Its results have less reliability and incur colossal costs. With these obstacles, hybrid renewable energy systems (HRES) came into the picture.

On the other hand, the economic feasibility of HRES is a challenging task; in this concern, numerous studies are conducted on optimal sizing of HRES on both grid-connected and stand-alone. The grid-connected HRES has advantages over a stand-alone HRES, like more reliability and required less storage capacity. Nevertheless, the PV and wind power outs are highly fluctuating in nature due to the unpredictable nature of solar insolation and wind speed. This waver nature of power output is a serious issue while HRES is integrated with the grid. From the available literature, the authors found optimal sizing of grid-connected HRES by taking various constraints to maintain the power balance and stability of the system. The optimal sizing of PV/WT/BES-based grid-tied HRES was designed by smoothing the BES (Battery energy storage) power fluctuations with power fluctuation as a constraint in [1]. The author intended PV/WT/hydro/BES-based HRES using energyPLAN software with an objective minimization of the system's annual cost and CO2 emission [2]. By adapting the DR (Demand Response) program, the share of renewable energy is increased; however, the author has not imposed any constraints related to surplus renewable power. In [3], the grid-connected smart energy hub has been investigated with GAMS software. In this framework, tri-objectives were considered to minimize operation costs and pollution emissions, and deviation of peak load demand. The grid-integrated model was developed based on FC/BES/PV- HRES by adapting the E-constrained method [4]. A similar model with information gap decision theory has been proposed in [5].

The primary contribution of this paper is conducting feasibility analysis of an islanded mode solar PV/Wind/REVBs configured system. The modeling of the residual capacity of the REVBs and its capacity degradation has been presented. Finally, the results obtained with the SSR and PSO algorithms have been discussed.

2 Detail Modeling of Components

2.1 *Mathematical Modeling of Wind Turbine*

To determine the average wind power, begin by measuring the wind speed at the specified location over a one-hour interval. Then, compare this measured wind speed with the following values: cut-out speed (w_{cout}), cut-in speed (w_{cin}), wind speed at turbine altitude (w), and rated speed (w_r). These selected values are used to calculate the average wind power output provided by the wind turbine.

$$f(x) = \begin{cases} 0\,When,\ w < w_{in}\ and\ w \geq w_{cout} \\ P_{wtr}\left(\frac{w^3 - w_{cin}^3}{w_{cr}^3 - w_{cin}^3}\right) When\, w_{cin} \leq w \leq w_r \\ P_r\, When\, w_r < w < w_{cout} \end{cases} \quad (1)$$

2.2 Mathematical Modeling of PV

The power generated by the PV module depends on solar irradiation and temperature. The power output by the panel is given by "$P_{pv}(t)$"

$$P_{pv}(t) = \frac{I(t)}{1000} \times P_{pvr} \times f_{dr} \times \eta_{pv}[1 - \alpha_T(T_c(t) - T_{c,STC})] \tag{2}$$

where "$I(t)(w/m^2)$" is the solar irradiation,"P_{pvr}" is the rated power at STC, "η_{pv}" is the efficiency, "$T_{c,STC}$" is the cell temperature under STC, "α_T" is the temperature coefficient of the PV panel.

2.3 Retired Electric Vehicle Battery Modeling

The estimation of the state of charge (SOC) for a battery pack is an essential parameter for the reliable and safer operation of the storage system as well as the efficient operation of the battery management system (BMS). The SOC of the battery being estimated for both the charging and discharging processes. The accurate SOC modeling of REVB is an essential part of BMS to regulate the power flow from the storage system. The SOC of REVB can be expressed as

$$SOC(t + \delta 1) = SOC(t)(1 - \sigma) - \frac{\eta_{revb}(t).I_m(t).\delta t}{C_m(t)} \tag{3}$$

where $SOC(t + \delta 1)$ and $SOC(t)$ are the current and previous state of charges in battery, respectively.

The ampere flow of the battery module is expressed:

$$I_{mod} = \frac{P_{revb}}{N_{revb} V_m(t)} \tag{4}$$

where P_{revb} is the total capacity of REVB module pack in W; $V_m(t)$ is the module voltage in V; N_{revb} is the number of REVB modules connected in parallel

the module voltage can be estimated by Eq. (5)

$$V_m(t) = OCV - I_{mod}(t) \times r \tag{5}$$

where OCV represents the module open circuit voltage; r is the dynamic internal resistance of the battery, which has a significant effect on the SOC and SOH of the battery.

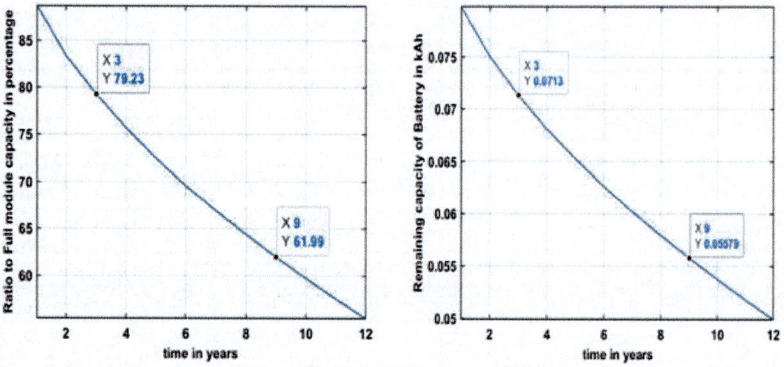

a) Remaining-percentage-of-capacity (year) b) Remaining-maximum-capacity (year)

Fig. 1 REVB capacity degradation

Since the SOC of recycled REVB pack is constrained in intervals of 20–80% of its nominal capacity. The plot between SOC and OCV is a flat line in the considered range. Therefore, the internal impedance also varies in the small range.

As mentioned earlier, a relatively new concept of capacity fading of REVB is being considered for its modeling. Equation (6) used to evaluate the capacity loss of Li-ion cells

$$C_{loss} = B_{exp}\left(\frac{-3100 + 370.3C_{rate}}{RT}\right)(Ah)^{0.55} \tag{6}$$

where C_{loss} is the capacity loss in (%); B represents the pre-exponential factor; R is gases constant for REVB; T is the absolute temperature; C_{rate} is the charge rate; Ah is Ah-throughput (Fig. 1).

3 Methodology

3.1 Total Cost Estimation

The overall expenses associated with the system comprise different costs, including investment cost (C_{inv}), operation and maintenance cost (C_{om}), and replacement cost (C_{rep}), which have been sourced from [6].

The total cost of the system can be expressed as

$$T_{cost} = C_{inv} + C_{om} + C_{rep} \tag{7}$$

Levelized Cost of Energy (LCE). The LCE represents the relationship between the total cost and energy generated by the HRES, including power sold to the grid. Essentially, it quantifies the cost of energy produced by the HRES per kilowatt-hour. This parameter plays a crucial role in determining the most cost-effective and optimal combination of the hybrid system.

$$LCE = \frac{T_{cost}}{E_{Total}} \tag{8}$$

In this context, the "E_{Total}" refers to the total energy supplied, and "T_{cost}" represents the total annual cost of the HRES. The optimal choice is determined by identifying the combination that yields the lowest LCE, provided that criteria and conditions are satisfied.

3.2 Objective Function

The objective function of the study is to minimizing the unit cost of power generation of a configured hybrid system while satisfying the imposed constraints and operational parameters. The objective function can be termed as a levelized cost of the energy (LCE).

Objective Function:

$$\min f = \min(LCE) \tag{9}$$

Subjected to Constraints:

$$0 < LPSP \leq LPSP_{max}, LPSP_{max} = 5\% \tag{10}$$

Sizing variables: The sizing variables pertain to the system components that were fine-tuned to achieve the optimal configuration of the system.

$$N_{pv}^{min} \leq N_{pv} \leq N_{pv}^{max} \tag{11}$$

$$N_{wt}^{min} \leq N_{wt} \leq N_{wt}^{max} \tag{12}$$

$$N_{revb}^{min} \leq N_{revb} \leq N_{revb}^{max} \tag{13}$$

where "N_{wt}", "N_{revb}", and "N_{pv}", are the number of Wind turbines, REVB, and PV panels, respectively.

4 Optimization Techniques

4.1 Teaching–Learning-Based Optimization

TLBO, a highly efficient algorithm inspired by the teaching–learning process, was invented by Prof. R.V. Rao [7]. This optimization technique begins by generating an initial population, which represents students, in a random manner. This optimization aims to enhance the individual's performance by closer to the best student, by taking the current mean of the students. Initially, the highest graded student is considered as the teacher, and the algorithm strives to improve the performance of individual students during both the teacher and learner phases. A student's grade is replaced only if the new solution is superior to the previous one. This iterative process continues until convergence is achieved or the maximum number of iterations is reached. The highest graded student represents the solution to the optimization problem. The grades of the learners are updated by the following equation.

$$X_{new} = X_i + r(X_{best} - T_f \times X_{mean}) \tag{14}$$

Here, "r" represents a random number ranging between 0 and 1, while "T_f" signifies the teaching factor, which can take values of either 1 or 2.

During the learner phase, the student enhances their grade by learning from another student, denoted as "X_p." The grade updates are carried out according to the following equation:

If grade of i th student is better than partner (p) grade, then

$$X_{new} = X_i + r(X_p - X_i) \tag{15}$$

Otherwise,

$$X_{new} = X_i - r(X_p - X_i) \tag{16}$$

5 Details of Case Study

The data for the case study is taken from the EU SCIENCE HUB of The European Commission's science and knowledge service. As the data varies with location, one suitable location, i.e., rich in solar irradiance and has an average optimum wind speed, is chosen and the study is performed. Using the same data, output power for solar panel and wind turbine is calculated. The load pattern is selected in a suitable way for the chosen location, with a peak load of 779 kW and an annual average load of 466.787 kW taken from [8]. The technical specifications of various components such as wind turbines, photovoltaic panels, battery energy storage systems, and inverters are provided in the given information [9].

6 Results and Discussion

The optimal sizing of the HRES depends on the location of meteorological data and load pattern. However, the capacity of PV, wind, and battery is selected in such a way that minimizes power purchased from the grid and LCE. In this proposed case study, the optimization has been carried out with the TLBO algorithm, and the various source capacities are obtained as follows Npv = 5827, Nwt = 1674, and Nrevb = 1897. With this optimal configuration, the LCE achieved was 0.5025 $/kWh while satisfying the imposed constraint LPSP with around 0.5. Further, the sizing of the designed system is also carried out with well-known algorithm PSO to validate the results obtained with TLBO. Each algorithm runs with 100 iterations, and the population size is 50. All utilized algorithms converged well toward an optimal solution. However, the TLBO algorithm converges quickly and consistently. From Table 1, it can be observed that TLBO achieved a better optimal solution, minimizing the LCE to 0.5025 $/kWh. The MATLAB environment is used for the proposed system implementation (Fig. 2).

Table 1 Tabulation of optimal results obtained with various algorithms

Optimization algorithm	N_{pv}	N_{wt}	N_{revb}	LCE ($/kWh)	Annual cost ($ $* 10^5$)
PSO	7168	1598	1493	0.5213	8.5889
TLBO	5827	1674	1897	0.5025	8.5477

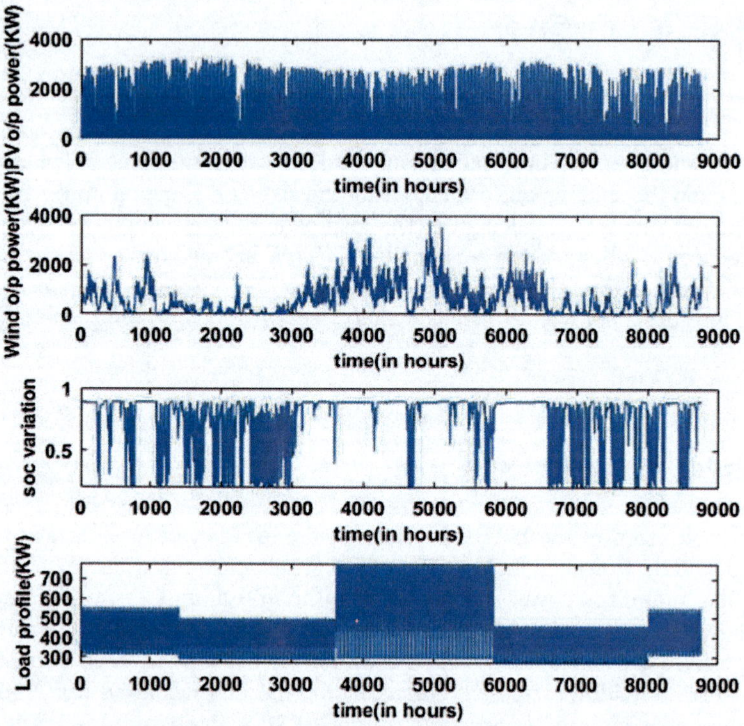

Fig. 2 Power output of various sources

7 Conclusion

In this study, the TLBO algorithm was proposed to determine the optimal sizing of PV/Wind/REVB hybrid renewable energy system. All the system components are detailed and the mathematical modeling of each system component is done. The objective of the optimization problem is to determine the optimal sizing of the hybrid renewable energy system by minimizing the LCE while maintaining the system constraints and satisfying the load reliably. For the configured system LPSP is the reliability index. A PMS (Power management strategy) has been implemented to manage the power flow among the system components due to variable renewable generation and load demand. The number of PV panels, wind turbines, and batteries is considered as the decision variables which are optimally determined by the proposed approach subject to constraints. Further, the Melapidavoor district in the state of Tamil Nadu, India is considered as the study area for the case study. TLBO as the optimization algorithm provides better solutions in solving an optimization problem that is economical and reliable in the case of stand-alone HRES compared to PSO. With the analysis of capacity fading we can find out the number of years the battery can be used as an energy storage system. In this case, it is found to be 5–6 years and this number of extra years will vary with the specification of the battery and its total number of cycles needed to reduce to 60% of its maximum capacity. With capacity fading the REVB would discharge or charge to its allowed SOC more easily and frequently in the last years of its life span, but we can easily use them for the second life of REVB as charging or discharging of the battery is working perfectly fine with the standard data taken. Then, this capacity fading effect over 9 years of battery lifespan shows that when it is used with a perception of having an LPSP of 0.0499, the actual LPSP would have increased to 0.0656 by the end of its lifecycle.

References

1. Sandhu KS, Mahesh A (2018) Optimal sizing of PV/wind/battery hybrid renewable energy system considering demand side management. Int J Electr Eng Inf 10(1):79–93
2. Tahir MF, Haoyong C, Khan A, Javed MS, Laraik NA, Mehmood K (2019) Optimizing size of variable renewable energy sources by incorporating energy storage and demand response. IEEE Access 7:103 115–103 126
3. Chamandoust H, Derakhshan G, Hakimi SM, Bahramara S (2019) Tri-objective optimal scheduling of smart energy hub system with schedulable loads. J Clean Prod 236:117584
4. Nojavan S, Majidi M, Najafi-Ghalelou A, Ghahramani M, Zare K (2017) A cost-emission model for fuel cell/PV/battery hybrid energy system in the presence of demand response program: _ -constraint method and fuzzy satisfying approach. Energy Convers Manag 138:383–392
5. Nojavan S, Majidi M, Zare K (2017) Risk-based optimal performance of a PV/fuel cell/battery/ grid hybrid energy system using information gap decision theory in the presence of demand response program. Int J Hydrogen Energy 42(16):11 857–11 867
6. Murty VV, Kumar A (2020) Optimal energy management and techno-economic analysis in microgrid with hybrid renewable energy sources. J Modern Power Syst Clean Energy 8(5):929–940
7. Rao RV, Savsani VJ, Balic J (2012) Teaching–learning-based optimization algorithm for unconstrained and constrained real-parameter optimization problems. Eng Optimiz 44(12):1447–1462
8. Bhimaraju A, Mahesh A, Nirbheram Joshi S (2022) Techno-economic optimization of grid-connected solar-wind-pumped storage hybrid energy system using improved search space reduction algorithm. J Energy Storage 52:104778
9. Mohammed AQ (2020) Optimal combination and sizing of a stand—Alone hybrid energy system using a nomadic people optimizer 8

Recurrent Neural Network for the Identification of Nonlinear Dynamical Systems: A Comparative Study

Kartik Saini, Narendra Kumar, Rajesh Kumar, and Bharat Bhushan

Abstract This study compares three different recurrent neural network topologies in order to assess their estimated capabilities. The three neural networks under detailed examination are the Elman recurrent neural network (ERNN), the diagonal recurrent neural network (DRNN), and the Jordan recurrent neural network (JRNN). A dynamical backpropagation algorithm is applied for developing and updating the parameters linked to each of these neural nets. The comparative analysis is performed by considering one nonlinear dynamical system with varying degrees of complexity.

Keywords Identification · JRNN · ERNN · DRNN

1 Introduction

The importance of nonlinear dynamical systems in control systems significantly increased during the past several decades [1]. The main issue with system theory is identifying the nature of the system's dynamics. The operation of several control systems depends on a basic mathematical theory of the given system [2]. The system identification procedure involves choosing an intelligent system or tool and modifying its parameters to make sure that its output matches the desired outcome. A number of approaches are available for system identification [3, 4]. Various intelligence approaches have been developed to characterize nonlinear systems, including

K. Saini (✉) · N. Kumar · B. Bhushan
Department of Electrical Engineering, Delhi Technological University, New Delhi 110042, India
e-mail: kartiksaini17@gmail.com

N. Kumar
e-mail: narendrakumar@dtu.ac.in

B. Bhushan
e-mail: bharat@dce.ac.in

R. Kumar
National Institute of Technology, Kurukshetra, Haryana 136119, India
e-mail: rajeshmahindru23@nitkkr.ac.in

A. Kumar et al. (eds.), *Decarbonisation and Digitization of the Energy System*, Lecture Notes in Electrical Engineering 1099, https://doi.org/10.1007/978-981-99-7630-0_26

fuzzy systems, ANNs, and genetic algorithms [5]. With the use of input–output data and neural networks, any unknown function may be roughly predicted. A black-box estimate is a kind of identification method that employs neural networks. The black-box has been estimated to contain all possible fundamental uncertainty [6]. A neural network's primary benefit is its capacity to learn from input–output relations [4]. The quantity of units in one hidden layer determines the size of a neural network. Recurrent neural networks have drawn more attention in identifying dynamic systems because feedforward networks lose the dynamic memory observed in RNNs and because recurrent neural networks are especially adept at modeling systems without external feedback [7]. There are several methods available for changing a parameter or the weights in an equation. The simplest one is founded on the back propagation method and gradient descent. For modifying the synaptic weights of an artificial neural network (ANN), a variety of optimization techniques have been proposed in the literature. The power of each of these algorithms has been established [8].

Recurrent neural networks are powerful tools that have been effectively used in a variety of technological fields, including time series forecasting, system identification, and controlling [9]. Most current attempts at neural network-based system identification are based on MLFNNs with learning through the backpropagation method or other more successful iterations of this method [10]. These techniques have demonstrated satisfactory behavior in actual processes [1]. Since RNN models feature internal feedback links, they are better able to handle dynamic systems than FFNN models. Many other RNN model types, such as the ERNN model, the JRNN model, etc., are available in the literature as a result of the various methods in which these recurrent connections may be constructed. The ERNN model, which has four layers overall, comprises the input layer, a hidden layer, an output layer, and a fourth layer called the context layer. The JRNN model's unit-delayed output is coupled to each hidden layer neuron through an adjustable feedback loop [3].

This paper is broken up into six pieces. Section 1 of the article provides a general outline. The nonlinear dynamical systems that were employed in this work are thoroughly detailed in Sect. 2. In Sect. 3, an ANN is used for nonlinear dynamic system identification, and Sect. 4 uses a backpropagation method for parameter training. Section 5 provides simulation results and modeling for system identification. In Sect. 6, the conclusions are provided. A list of references is then given.

2 Structures of ERNN, DRNN, and JRNN Models

Figures 1, 2, and 3 demonstrate the relevant structural diagrams for the ERNN, JRNN, and DRNN systems for all three of these neural network structures, the hidden layer count is assumed in this study as being equal to 1. The Elman recurrent neural network's topology is illustrated in Fig. 1, and DRNN and JRNN structures are illustrated similarly in Figs. 2 and 3. There are three hidden neurons in each of these neural networks' hidden layers. The induced field vector V(k) stands for the output of the hidden neurons. In these neural network architectures, the induced field vector (sum of synaptic) of hidden neurons is expressed by $V(k) = \{v_1(k), v_2(k)....v_r(k)\}$ where r is the number of hidden neurons. Within each of these neural networks, the linear activation function is used for output neurons, whereas the tangent hyperbolic function is used for hidden neurons. since the output of each can have a wide range of values. The output weight vector (which links the output of recurrent neurons to the output neuron) is represented by $W^O(k) = \{w_1^0(k), w_2^0(k) \ldots w_r^0(k)$. All of these neural networks take into account r = 3 hidden neurons. The following formula may be used to calculate the induced field of any rth recurrent neuron. Every rth hidden neuron's output at any kth moment is:

$$S_r(k) = f[V_r(k)] = f\left[\sum_p W^I(\text{k})X_p(k)\right] \quad (1)$$

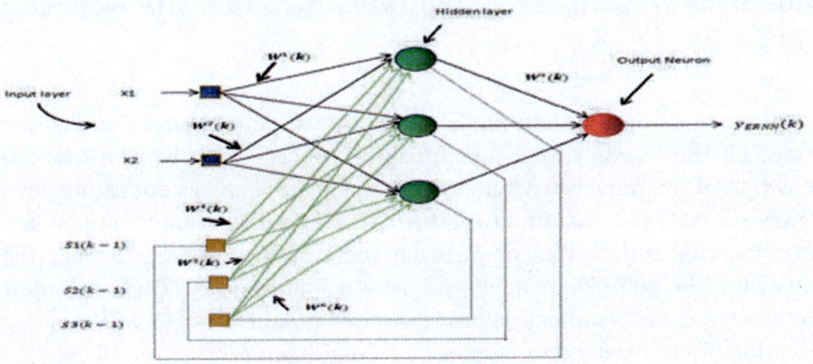

Fig. 1 Architecture of ERNN model

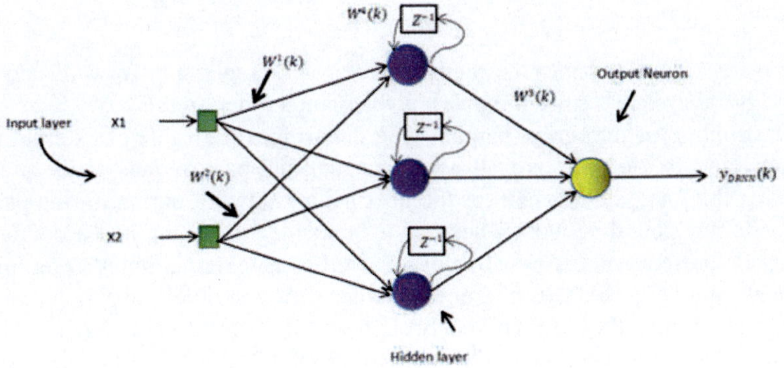

Fig. 2 Architecture of DRNN model

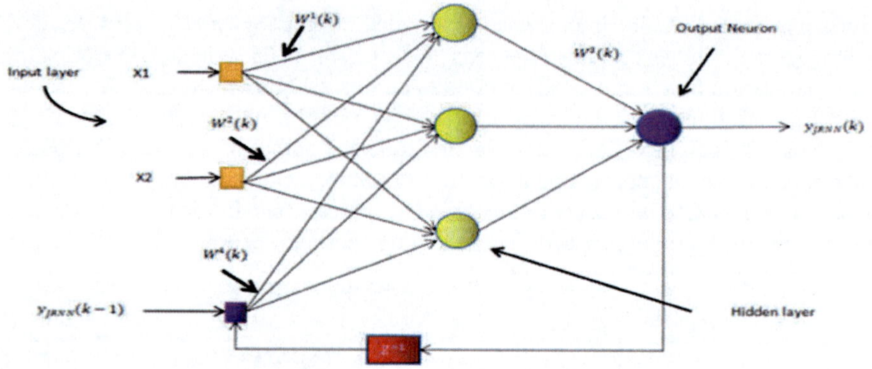

Fig. 3 Architecture of JRNN model

3 Nonlinear Dynamical System Identification with Neural Networks

A particular standard of system model quality must be met after the identification operation. Therefore, the system identification process must be properly selected. The selection of the model structure to which the supplied system belongs is a step in the identification process. In the mathematical representation of a plant, an input x is given to a function f, which produces an output y. To complete the identification challenge, use the following function f approximation [11]. There are two ways for nonlinear system identification process series-parallel identification and parallel identification. To understand the process of identification,

$$y_p(k+1) = f[y_p(k), y_p(k-1) \ldots y_p(k-n+1)] + \sum_{i=0}^{m-1} b_i x(k-i) \tag{2}$$

where $y_p(k+1)$ represents the plant's value for a single step forward and $x(k)$ represents the system's current input. n is the plant's order, in this case.

The structure of the series–parallel type determines the value of external input both now and in the past, as well as the current and past outputs of the plant to determine the immediate benefit of the next neural network output. In the parallel type mode, the neural network's next value is decided by taking into account both its output's present and historical values as well as its external input's present and historical values [12,13]. The difference between series–parallel and parallel training is that in the former, the output is sent back into the network's input, by utilizing the previous values of the series–parallel configuration, it is able to predict the system's upcoming output.

4 Parameter Training Using Backpropagation Algorithm

The weight update equations are obtained using a learning approach. We employed the dynamic back propagation approach to obtain the weight modification equations. This section just briefly discusses this strategy because it is well known and often used in the literature. It minimizes an objective function using the gradient descent technique. The aim of the training is to modify the model weights to decrease the identification error e(k). The difference between the predicted and actual answers is what is referred to as the error. Error = Expected output−Actual output. It may be claimed that the mistake is the crucial element that defines how the weights of the perceptron should be changed [4]. The weight update equation is constructed by first selecting an objective function to be used. Instantaneous mean square error serves as the study's main parameter (MSE) [14]. The description is as follows:

$$\mathrm{E(k)}= \frac{1}{2}e^2(k) \tag{3}$$

$$\mathrm{e}(k) = y(k) - y_n(k) \tag{4}$$

5 Simulation Results and Discussion

The performance of the ERNN, JRNN, and DRNN models are compared using one nonlinear system. In the simulated example, the initial weights are randomly selected for better convergence, the learning rate is set to 0.0046, and only 3 hidden neurons are taken into consideration within every sample.

Example:- Let's assume that the difference equation provides the plant's dynamics as given in [15]:

$$y(k) = f\{y(k-1), y(k-2), y(k-3), r(k-1), r(k-2), r(k-3)\} \tag{5}$$

$$f = \frac{y(k-1)y(k-2)y(k-3)r(k-2)\{r(k-3)-1\} + r(k-1)}{1 + [y(k-2)]^2 + [y(k-3)]^2} \tag{6}$$

The plant's current value is determined by six inputs, including delayed values of its own and three earlier values of the input, as can be seen from the given difference equation. For all identifiers, the two signals are treated as inputs.

Where the external input considered is

$$\mathrm{r(k)} = \begin{cases} sin(pi * k/45)\ if\ 0 < k \le 650 \\ 1.05 * sin(pi * k/45)\ if\ 650 < k \le 900 \end{cases} \tag{7}$$

A comparison of the mean square error (MSE) in Fig. 4 and the mean average error (MAE) in Fig. 5 obtained during the training procedure shows that the plot response of the DRNN identification model is superior to that of the JRNN and ERNN identification models in the scenario where the learning rate is set at 0.0046 and the learning is ongoing for 500 epochs.

In this illustration, the comparative analysis is carried out using simulation results and discussion, and the outcomes are presented in Table 1 in terms of mean square error (MSE) and mean average error (MAE). Figures 6, 7 and 8 illustrate the validation results from the ERNN, JRNN, and DRNN models, respectively, and in Fig. 8, DRNN identification model is also validated with some random inputs.

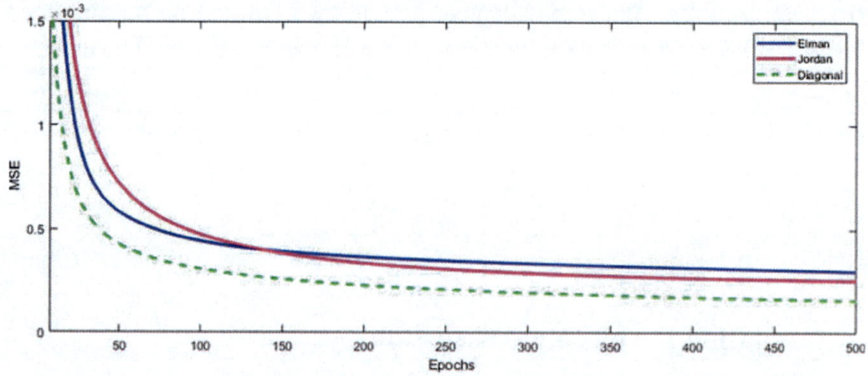

Fig. 4 MSE achieved during training with multiple neural networks

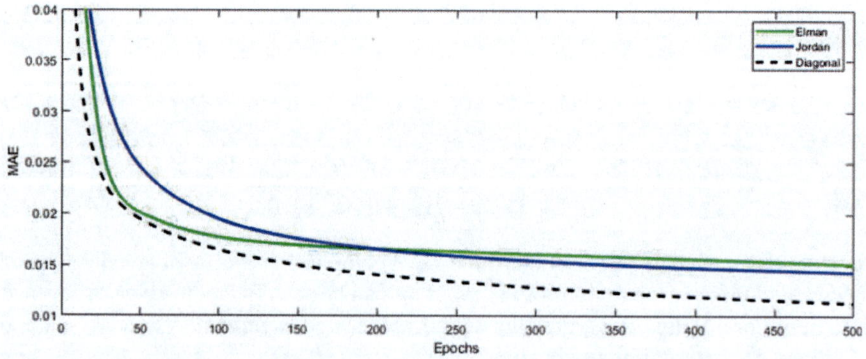

Fig. 5 MAE achieved during training with multiple neural networks

Table 1 Comparison of statistical error obtained at the end of training

Sr. no.	Model	MSE	MAE
1	ERNN	0.00029	0.0151
2	JRNN	0.00025	0.0143
3	DRNN	**0.00015**	**0.0113**

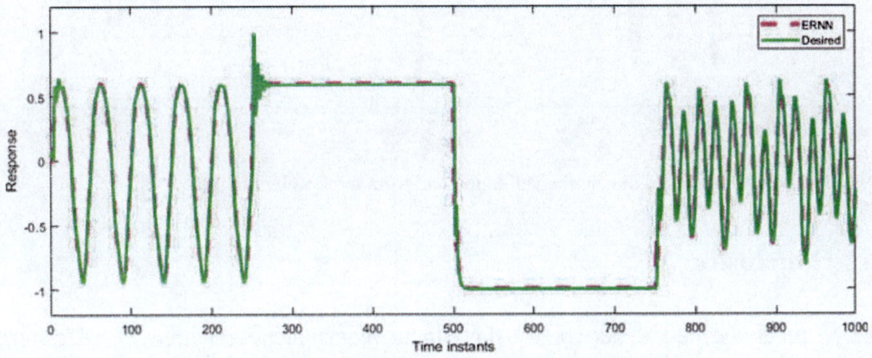

Fig. 6 Response received during the validation test from the ERNN model

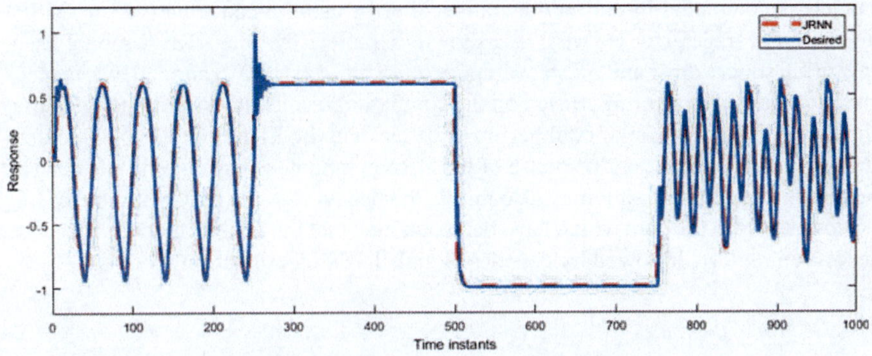

Fig. 7 Response received during the validation test from the JRNN model

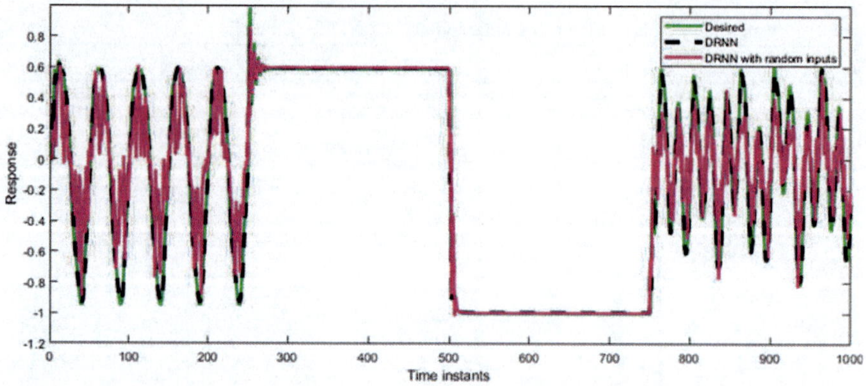

Fig. 8 Response received during the validation test from the DRNN model

6 Conclusion

Three recurrent neural networks—the Elman recurrent neural network (ERNN), Jordan recurrent neural network (JRNN), and diagonal recurrent neural network (DRNN)—are compared in this article. Backpropagation learning approach is employed to adjust the weights and for tuning. A performance evaluation of the models is conducted based on the statistical error using mean square error (MSE) and mean average error (MAE); It is shown in Table 1 that the DRNN model has a low mean square error and mean average error, which is 0.00015 and 0.0113, respectively at the end of training. Based on the simulation results, it can be concluded that the DRNN identification model performs better than the ERNN and JRNN identification models. The output response of the DRNN model is very close to the desired response, suggesting that it may also mimic the behavior of the physical system. It is discovered from the data which have been obtained that the DRNN model's response recovered rapidly, followed by the ERNN and JRNN identification models.

References

1. Grino R, Cembrano G, Torras C (2000) Nonlinear system identification using additive dynamic neural networks-two on-line approaches. IEEE Trans Circuits Syst I: Fundam Theory Appl 47(2):150–165
2. Li C, Yan H (2018)Identification of nonlinear time-delay system using multi-dimensional taylor network model. In: 2018 IEEE international conference on manipulation, manufacturing and measurement on the nanoscale (3M-NANO), Hangzhou, China, pp 87–90
3. Kumar R (2022) A Lyapunov-stability-based context-layered recurrent pi-sigma neural network for the identification of nonlinear systems. Appl Soft Comput 122:108836. ISSN 1568-4946
4. Nidhil Wilfred KJ, Sreeraj S, Vijay B, Bagyaveereswaran V (2015) System identification using artificial neural network. In: 2015 international conference on circuits, power and computing technologies [ICCPCT-2015], pp 1–4

5. Gautam P (2016) System identification of nonlinear Inverted Pendulum using artificial neural network. In: 2016 international conference on recent advances and innovations in engineering (ICRAIE), Jaipur, India, pp 1–5
6. Kumar R, Srivastava S, Gupta JRP et al (2019) Comparative study of neural networks for dynamic nonlinear systems identification. Soft Comput 23:101–114
7. Ren X, Fei S (2000) Recurrent neural networks for identification of nonlinear systems. In: Proceedings of the 39th IEEE conference on decision and control (Cat. No. 00CH37187), vol 3, pp 2861–2866. IEEE
8. Kumar R, Srivastava S, Gupta JRP (2018) Comparative study of neural networks for control of nonlinear dynamical systems with Lyapunov stability-based adaptive learning rates. Arab J Sci Eng 43:2971–2993
9. Şen GD, Günel GÖ, Güzelkaya M (2020) Extended Kalman filter based modified Elman-Jordan neural network for control and identification of nonlinear systems. In: 2020 innovations in intelligent systems and applications conference (ASYU), Istanbul, Turkey, pp 1–6
10. Liu H, Song X (2015) Nonlinear system identification based on NARX network. In: 2015 10th Asian control conference (ASCC), Kota Kinabalu, Malaysia, pp 1–6
11. Ku C-C, Lee KY (1992) Nonlinear system identification using diagonal recurrent neural networks. In: [Proceedings 1992] IJCNN international joint conference on neural networks, Baltimore, MD, USA, vol 3, pp 839–844
12. Pandey K, Bhanacharjee S, Lau S, Tushir M (2018)A comparative study of fuzzy systems and neural networks for system modeling and identification. In: 2018 2nd IEEE international conference on power electronics, intelligent control and energy systems (ICPEICES), pp 876–880
13. Efe MO, Kaynak O (1999) A comparative study of neural network structures in identification of nonlinear systems. Mechatronics 9(3):287–300
14. Samir L, Said G, Nora K, Youcef S (2017)Improved Pi-Sigma neural network for nonlinear system identification. In: 2017 5th international conference on electrical engineering—Boumerdes (ICEE-B), pp 1–5
15. Kumpati SN, Kannan P (1990) Identification and control of dynamical systems using neural networks. IEEE Trans Neural Netw 1(1):4–27

A Crest Factor-Based Voltage Sag Quantification Method

Priyanka Yadav, Padmanabh Thakur, Ashutosh Dixit, and Parvesh Saini

Abstract A fast detection and an accurate estimation of the voltage sags duration are highly desirable in the design and development of power quality (PQ) indices & mitigating devices. Therefore, in this work, a new method, based on crest factor (*C.F*), has been presented for the accurate quantification of the voltage sag. The window-based technique has been used for the estimation of the '*C.F*'. Further, the voltage sag signal, from the IEEE database, has been considered to test the efficacy and correctness of the proposed method. It has been revealed that the estimated duration is exactly the same as the real voltage sag duration. Also, the proposed method shows efficacy by providing the exact information about the voltage sag initiation and recovery. The outcomes of the proposed method are compared with the other existing voltage sag quantification methods. It is shown that the proposed method provides 0% error in the estimation of sag duration while the R.M.S, peak, and hybrid voltage have 34%, 40%, and 3%, respectively. Additionally, the detection delay of the proposed method is found zero while the peak voltage method provides a maximum delay of 16.75 ms. Therefore, the method presented in this work is not only efficient in the estimation of the duration but also capable of the detection of voltage sag and waveform distortion.

Keywords Crest factor · Peak voltage · R.M.S voltage · Voltage sag · Detection · Point on wave

P. Yadav · P. Thakur · A. Dixit (✉) · P. Saini
Department of Electrical Engineering, Graphic Era Deemed to be University, Dehradun, Uttarakhand, India
e-mail: ashutoshdixit@geu.ac.in

P. Yadav
e-mail: rao.priyanka91@gmail.com

P. Thakur
e-mail: tonu_arth@rediffmail.com

P. Saini
e-mail: parvesh.saini.eee@geu.ac.in

A. Kumar et al. (eds.), *Decarbonisation and Digitization of the Energy System*, Lecture Notes in Electrical Engineering 1099, https://doi.org/10.1007/978-981-99-7630-0_27

1 Introduction

Voltage sags are regarded as one of the most serious issues for the utilities as well as for the end users because of their high frequency of occurrence and tremendous cost of downtime (COD) [1–7]. The CODs, existing due to the voltage sags, are estimated as $50 000 per event [6, 7] or may be as high as 10% of the yearly turnover of the industry [8]. Also, the financial impacts, associated with the voltage sags, are found more stronger than those of due to short interruption because of their high frequency of occurrence, i.e., 20–30 times per year [6, 7]. In addition, the various adverse impacts of the voltage sags, such as spurious tripping of ASDs [9], partial or complete extinguishing of LED Lamps [10], PLC malfunction [11], disconnection of a wind turbine from the grid [12], stalling of directly-fed motors [13], etc., were already highlighted in various research works. Therefore, characterization, classification, detection, and mitigation of voltage sags have now become the top business priority to reduce the huge economic impacts connected with it.

Generally, a reduction in the root mean square (R.M.S) voltage for a specific duration of time is considered as the voltage sags [14]. The magnitude and duration are widely adopted as the most important indices for the characterization and classification of voltage sags [1–5]. The voltage sag magnitude lies between 10 and 90% of the nominal voltage with a duration between 0.5 cycles and 30 cycles for the instantaneous, 30 cycles-3 s for the momentary, and 3 s^{-1} min for the temporary voltage sags [14, 15]. Also, the reduction in the R.M.S voltage magnitude, between 1 and 90% [16] or short interruption [17], has been considered as the special condition of voltage sags. The consistent definition for the duration of voltage sag also differs from one standard to another. Usually, the voltage sags duration, as defined in various standards, is given in the range from 0.5 cycles to 1 min [14, 15] or usually less than 1 s [18].

Basically, the majority of the industrial equipment trips due to the voltage sags existing due to power system faults [1–6]. However, the energization of a transformer or starting large 3-phase induction motor (3-Φ IM) has also the capability of producing voltage sags but the impacts of such sags on equipment are not very severe [1]. Therefore, in this work, voltage sags, existing due to the power system faults, are considered for the analysis. The voltage sag, existing due to the power system faults and starting of three phase induction motor are simulated in MATLAB® and are shown in Fig. 1a, b, respectively.

It is obvious from Fig. 1a, b that the voltage sag magnitude due to induction motor starting is approximately 0.81 p.u. whereas it is 0.36 p.u due to the fault existing in the power systems. In other words, it can be said that the voltage sags that arise due to power system fault is more severe than that due to induction motor starting. Therefore, most of the research works, related to the characterization, classification, and detection of voltage sags, address the voltage sag existing due to the power system faults [1–6].

Also, the various power acceptability curves, such as CBEMA, ITIC, SEMI, and IEC 61000-4-11, were developed by the industrial expert to know the acceptability

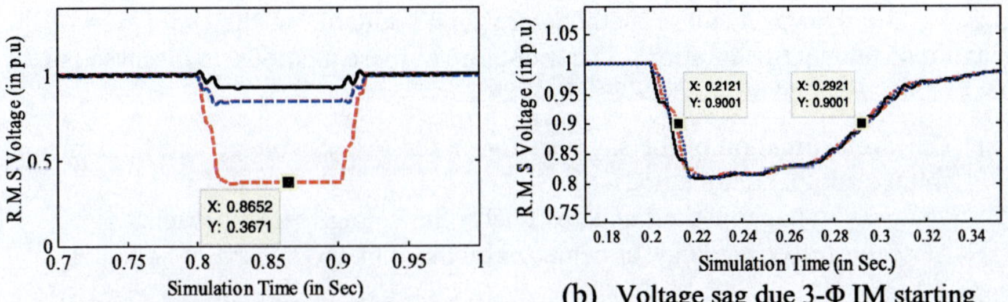

(a) Voltage sag due to faults

(b) Voltage sag due 3-Φ IM starting

Fig. 1 Voltage sags due to power system faults and 3-Φ IM starting

limits of the quality of power supply for the proper working of the equipment [4, 19]. In fact, these curves are useful to test the sensitivity of industrial processes to voltage sag. The ITIC and IEC curves are represented in Fig. 2 for better understanding [19]. In Fig. 2, twenty-three voltage sag data have been considered to check their acceptability. It is obvious that only twelve voltage sag data pass the limit, as defined by ITIC, whereas only one data fails to qualify the voltage acceptability as defined by IEC. The purpose of representing Fig. 2 is only to reveal the importance of the accurate estimation of voltage sag magnitude and duration. Any erroneous quantification of these characteristics may result in the wrong location of the point in acceptability curves. Consequently, the voltage quality that qualifies in the acceptability curve test may start malfunctioning in real-time environment. Further, using the magnitude and duration of the voltage sag various power quality (PQ) indices were also developed to get relevant information from the database. The PQ indices are found suitable for deciding the equipment and safety specifications.

Therefore, several research works were carried out for the quantification of the voltage sags. Three methods, namely R.M.S, peak, and, fundamental voltage component methods were presented for the characterization and detection of the voltage

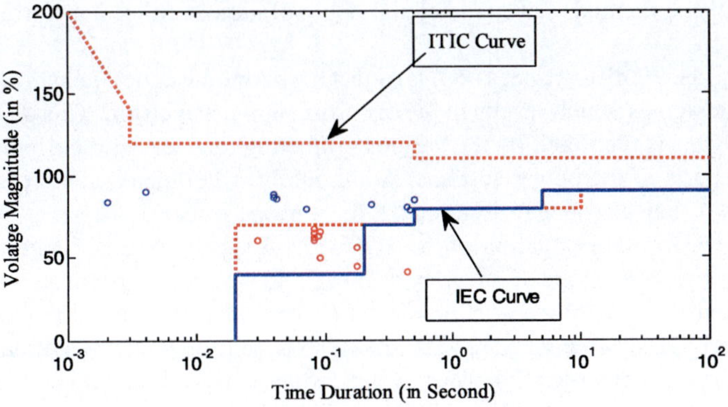

Fig. 2 The ITIC and IEC power acceptability curves

sags [1, 24]. However, these methods are found suitable in the estimation of the magnitude but not the durations. The limitation of these methods, as discussed in the next section, can be highlighted as [1, 4, 24]:

(1) The overestimation of the sag duration in the R.M.S voltage and fundamental voltage component methods.
(2) The peak voltage method underestimates the voltage sags duration.
(3) Also, the detection delay, as achieved by these methods, is very large.

Further, a hybrid method has also been suggested for the accurate estimation of the duration using the combination of peak and R.M.S voltages [4]. However, the error produced by the hybrid method is small, i.e., 3% (underestimation) but its implementation is found complex as it takes two-step for the estimation of sag duration. In [20], a wavelet transform (WT) based approach has been considered for the estimation of various characteristics of the voltage sags. In the study, as presented in [20], firstly fundamental voltage component is derived and then the WT is applied for the quantification of voltage sags, accurately. Further, the point of wave characteristics is estimated using the help of instantaneous signals. In the same line, as presented in [20], the magnitude and duration of voltage characteristics are estimated using the most suitable mother wavelet in [21]. However, the characteristics, as presented in [4, 20, 21], are more precise than the R.M.S, peak, and fundamental voltage component method but complex. Further, the hybridization of integrator and delay transform has been used in [22]. In this study, firstly integrator has been used to extract the phase angle and fundamental voltage, and then delay transform is implemented to construct a quadrature component. However, this method detects the sag in 1 ms but with the high computational burden. Further, the d-q transform and WT have been applied for the quantification of the voltage sags [23]. The d-q transform fails to detect the start and recovery of the voltage sags [23]. Therefore, a WT-based approach has been suggested for the identification of the start and end of the voltage sags [23]. The major issues with the WT-based methods are large the computational burden and the selection of an accurate mother wavelet [22]. In addition, there are several other methods of sag detection have also been presented but most of the methods have a detection delay of almost either 2 ms or more than 2 ms [24].

Usually, the PQ disturbances (PQDs) inject the harmonic component in the voltage and current signals which results in the distortion in the waveform. This distortion in the waveforms is identified by the various protecting devices for the tripping of the unhealthy parts of the power systems. Additionally, a PQ index, namely, the 'crest factor (*C.F*)' has also been discussed in the various research works as well as in the standards for the evaluation of the PQDs [25–28]. The normal range of *C.F* in proper working conditions of linear, non-linear, and computer systems has also been incorporated in [22]. Any deviations in '*C.F*' from the normal value indicate the injection of PQDs. Moreover, the '*C.F*' has also been used in various research work for analyzing the power quality in power substations from educational buildings [26], investigating the impacts of CFL and LED on PQ [27], the impacts of voltage sags on adjustable speed drive [28], etc. However, the importance of '*C.F*' in the assessment of various impacts was incorporated in several research works but has seldom been used in the quantification of the voltage sags. Nowadays, measuring instruments, with the capability of measurement of the '*C.F*', are commonly available. Therefore, this PQ index can also be used for the quantification of voltage sags.

Considering the importance of the '*C.F*', a new method for the quantification of the voltage sags has been discussed in this paper. It is shown that the proposed '*C.F*' based quantification method is not only accurate in the quantification of the various characteristics of the voltage sags but at the same time it also provides the sensitivity and level of the harmonic components available in voltage sag signals.

2 Crest Factor-Based Quantification Method

2.1 Fundamental of the Crest Factor

According to the IEEE Std, the ratio of peak to R.M.S voltage is termed as crest factor (*C.F*) [25]. Quantitively, it is given as [25]

$$C.F = \frac{Peak\ Voltage}{R.M.S\ Voltage} = \frac{V_{\max}}{V_{R.M.S}} \tag{1}$$

For a pure sinusoidal waveform, the R.M.S and peak voltage are related as

$$V_{R.M.S} = \frac{V_{\max}}{\sqrt{2}} \tag{2}$$

Therefore, from (1) and (2), the magnitude of '*C.F*' for a pure sinusoidal signal will always be a constant and equal to 1.414. Any deviation and oscillation in '*C.F*' from the normal value represent the injection of PQDs in the voltage or current signals. Therefore, through the deviation and oscillation of '*C.F*' the duration of the PQDs can easily be estimated. Indeed, the point of initiation and recovery of the voltage sags contains a large amount of the harmonic's component. Consequently, there will be a sharp change in '*C.F*' at these points. Hence, it would be easy to identify a point on wave (POW) of voltage sag recovery and initiation through the proposed method. Further, the time interval between these POWs provides the duration of the voltage sags and hence the proposed method is not only suitable for the estimation of the duration but also proficient in the detection of the voltage sag event. Accurate detection of voltage sags is highly desirable for the quick restoration of power systems and consequently in the reduction of the cost of downtime associated with it.

2.2 *Estimation of Crest Factor*

The foremost challenge of the proposed technique is to get '*C.F*' under the condition of voltage sag as neither the R.M.S voltage nor the peak voltage remains constant during the event due to the presence of high harmonic components [1]. There is an oscillation in both R.M.S and peak voltage during the voltage sag intervals [1]. Therefore, in the proposed method, the following steps have been considered for the estimation of '*C.F*' of recorded voltage signals:

(1) The R.M.S voltage of the voltage sag signal is evaluated by means of a window-based technique as suggested in [1, 24]. The R.M.S voltage is quantitively given as:

$$V_{R.M.S} = \sqrt{\frac{1}{N}\sum_{k=1}^{N} v_k^2} \tag{3}$$

In the method, as suggested in [1], the R.M.S voltage of any sample point is the R.M.S voltage of the samples, existing in one cycle, before the point.

(2) The peak value is estimated, using the quantitative definition, as given in (4) [1, 4, 24]

$$V_{\max} = peak(v(k - N + 1) : k)) \tag{4}$$

Here, the peak value on any sample point is the peak value of one cycle window behind this sample point. The same technique has also been used in [1, 4].

(3) Now using (1), (3), and (4), the '*C.F*' has been estimated for fast detection of voltage sag start and recovery as well as for the estimation of the duration of the event.

3 Validation of the Proposed Technique

The test voltage signal, as given in the IEEE database and in [1, 4], is considered in this study to check the effectiveness of the '*C.F*' based quantification method. The recorded voltage sag test signal is shown in Fig. 3 [1].

Using the concept, as discussed in [1], the R.M.S and peak voltage of the test signal are estimated and are shown in Fig. 4. The threshold value for the estimation of the voltage sags duration is considered as 90% of the nominal voltage [1, 14, 15].

The following observations can be highlighted in Fig. 4:

(1) The sag duration, as estimated with the R.M.S voltage method, is 3.2 cycles while the real duration of voltage sag in the test signal is 2.4 cycles [1, 4]. Therefore, the R.M.S voltage method provides the overestimation in the sag duration. The same result has also been discussed in [1] and [4].

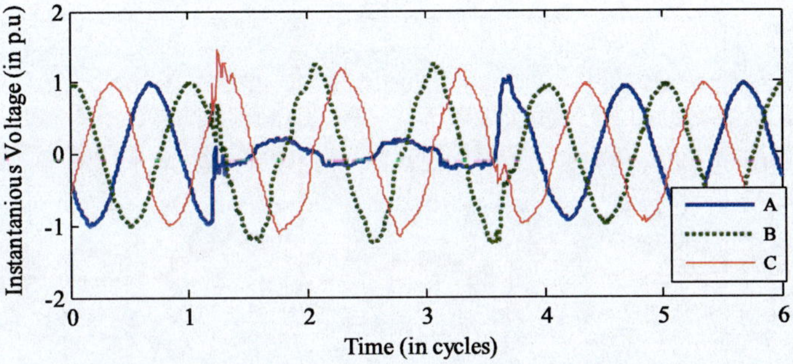

Fig. 3 Recorded voltage sag signal [1]

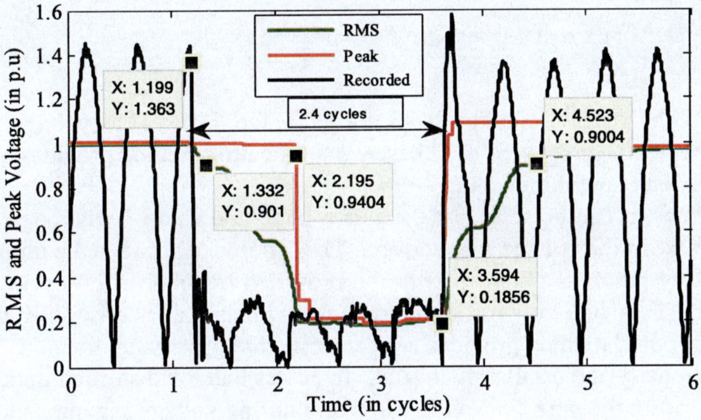

Fig. 4 R.M.S and peak voltage of the recorded signal [1, 4]

(2) The duration, as obtained with the peak voltage method, is 1.44 cycles and hence underestimation in the sag duration as the real duration is 2.4 cycles.
(3) Further, it is obvious from Fig. 4, that both peak and R.M.S voltage give delay in detection of POW of sag initiation. The delay in the detection of the voltage sags is not desirable when fast mitigation is required.

Considering the limitations, as discussed in points 1–3, the proposed method uses the PQ index, namely, '*C.F*' to get accurate duration and fast detection of the voltage sag. Using (1), (3), and (4), the '*C.F*' of the recorded signal has been evaluated and shown in Fig. 5.

The mean value of the '*C.F*' in the complete cycle, as estimated by the proposed method, is 1.60 while in the deeper portion of the waveform, it is found as 1.44. From Fig. 5, the attributes of the proposed method can be highlighted as:

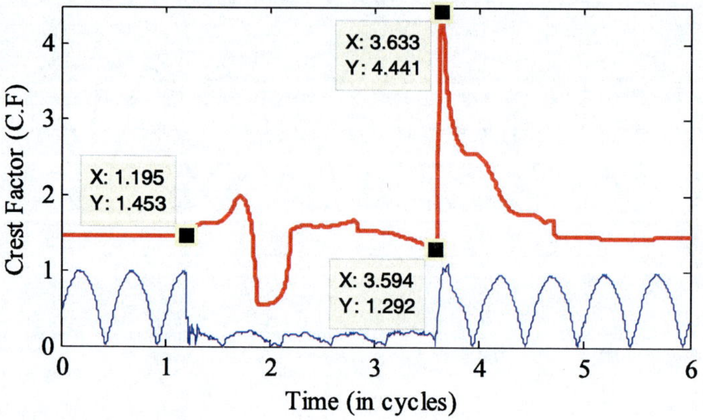

Fig. 5 The '*C.F*' of the recorded voltage signal

(1) The '*C.F*' starts oscillation from normal value (=1.414) as voltage sag start. Therefore, the proposed '*C.F*' based quantification method provides zero delay in the detection of the voltage sags.
(2) Similarly, it can be seen that there is a large and sharp deviation in '*C.F*' at the POW of the voltage sag recovery. The difference between the initiation and recovery point, as obtained with the proposed method, is 2.4 cycles (=3.59–1.19). The actual duration of the recorded signal is also 2.4 cycles and hence the proposed method provides zero error in voltage sag duration.
(3) Additionally, the oscillation in '*C.F*' indicates that the distortion in the voltage signal from the pure sinusoidal waveform during voltage sag interval.
(4) The proposed method detects the deviation in '*C.F*' due to the presence of the noise or harmonics contents existing in the signals. Hence, the presence of the noise makes the proposed method robust.

Further, the T.H.D of the recorded signal, as shown in Fig. 3, has also been estimated and is shown in Fig. 6. Firstly, three intervals, that include (i) point of sag initiation, (ii) point in during sag interval, and (iii) point of recovery, have been selected as the starting point and then one cycle window length has been considered for the estimation of the T.H.D. It is evident from Fig. 6 that the values of T.H.D are high, i.e., more than 40%, for the windows that include the point of initiation and recovery points of the voltage sags. The high values of T.H.D at these points indicate the existence of a large amount of the harmonic contents. Also, it can be seen from Fig. 5, that there large oscillation in the '*C.F*' at the point of initiation and recovery.

Based on the results, as shown in Figs. 5 and 6, it can be said that the oscillation in the '*C.F*' is more noticeable in the transition interval, i.e., when the amount of harmonics components are higher than the steady state conditions. Therefore, the proposed method of sag detection is also robust under the condition of transition.

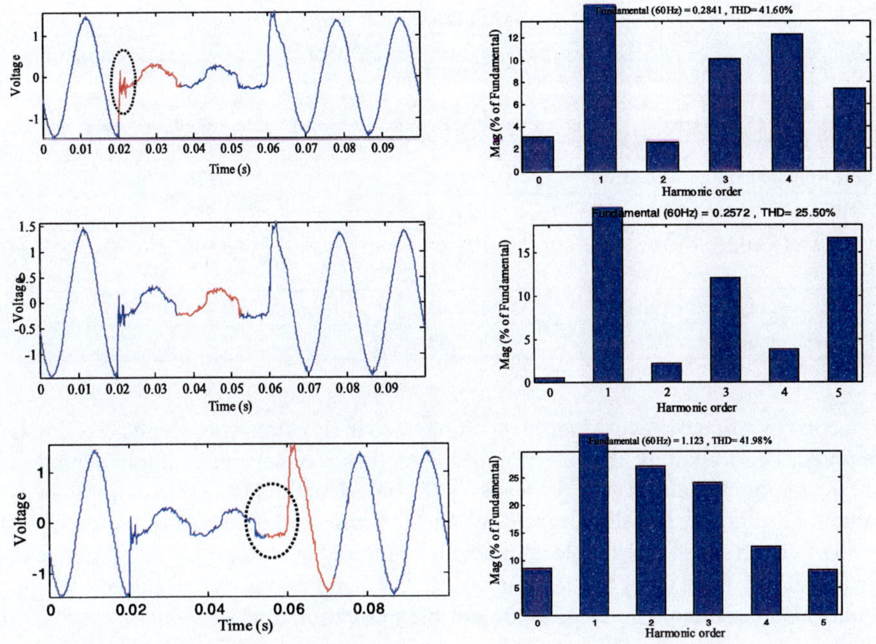

Fig. 6 Value of T.H.D in different intervals

Further, the results, as obtained with the proposed '*C.F*'-based method, are compared with R.M.S voltage, peak voltage, hybrid, and WT-based methods to check its accuracy and effectiveness. The comparative analysis is summarized in Tables 1 and 2.

It is evident from the comparative results of Tables 1 and 2, that the proposed method provides more accurate result than the R.M.S, peak voltage, hybrid, WT-based, and wavelet energy-based methods. However, the methods, as presented in [20, 21], provide almost the same result as obtained with the proposed method but this method is complex. However, the WT-based methods for the quantification and

Table 1 The comparative analysis of the estimated duration

Methods	Actual duration (In cycles)	Estimated duration (In cycles)	% Error
R.M.S voltage [1]	2.4	3.2	34
Peak voltage [1]	2.4	1.44	40
Hybrid method [4]	2.4	2.32	3.33
WT-based method [20]	2.4	2.41	0.41
Wavelet energy [21]	2.4	2.401	0.083
Proposed method	2.4	2.4	0

Table 2 The comparative analysis of delay in detection

Methods	Real POW of sag Initiation (in ms)	Estimated POW of sag initiation (in ms)	Delay in detection (in ms)
R.M.S voltage [1]	19.83	22.16	2.33
Peak voltage [1]	19.83	36.58	16.75
Hybrid method [4]	19.83	22.16	2.33
WT-based method [20]	19.83	20.41	0.583
Wavelet energy [21]	19.83	20.03	0.2
Proposed method	19.83	19.83	0

detection of voltage sags are highly recommended in research where high precision is required. Unfortunately, the selection of the mother wavelet and its implementation are the major challenges. The proposed '*C.F*' based method is simple in implementation and efficient in estimating POW and duration, simultaneously. Further, the level of distortion and harmonic components, required in voltage sag mitigation, can easily be identified with the magnitude of 'C.F' and hence the proposed method would also be suitable in the designing of the mitigation device.

4 Conclusions

In this work, a novel method for the quantification and detection of the voltage sag has been presented. The PQ index, namely, crest factor ('*C.F*') has been used for the quantification and detection of the voltage sags. Using the window-based technique, the '*C.F*' of the recorded voltage sag signal, is estimated for the quantification of the voltage sag. The results, as obtained with the proposed method, are compared with the existing methods, such as R.M.S voltage, peak voltage, hybrid, and WT-based methods. Through the comparative analysis, it has been revealed that the proposed method has only 0% error in the voltage sag duration while R.M.S and peak voltage methods have 34% and 40%, respectively. However, the WT-based technique also provides a very small % error in the value of sag duration, i.e., on 0.41%, quantification with WT is found complex. Further, it has been shown the proposed '*C.F*' based quantification method is not only accurate in the estimation of the duration but also identifies the POW precisely with zero delay. The voltage sag detection delay in the peak voltage method is found highest, i.e., 16.75 ms, and hence this method is usually not recommended for the detection of voltage sags. In a nutshell, the attributes of the proposed method can be highlighted as:

(1) The % error in the estimation of duration is found zero.
(2) The delay in the detection is found zero.
(3) Also, provides information about the level of waveform distortion.

Therefore, considering these attributes, the proposed '*C.F*'-based technique can be recommended as the fast, simple, and accurate voltage sag detection and quantification technique.

References

1. Bollen MHJ (2000) Understanding power quality problems: voltage sags and interruptions. IEEE Press, New York
2. Thakur P, Singh AK (2013) Unbalance voltage sag fault-type characterization algorithm for recorded waveform. IEEE Trans Power Deliv 28(2):1007–1014
3. Thakur P, Singh AK, Singh SB (2012) Type detection of voltage sags through voltage unbalance factor. In: 15th international conference on harmonics and quality of power, pp 892–896. IEEE, Hongkong
4. Thakur P, Singh AK (2013) A novel way to quantify magnitude of voltage sag. Electr Eng 95(4):331–340
5. Thakur P, Singh AK, Bansal RC (2013) Novel way for classification and type detection of voltage sag. IET Gener Transm Distrib 7(4):398–404
6. Thakur P, Singh AK (2017) A novel method for joint characterization of unbalanced voltage sags and swells. Int Trans Electr Energy Syst 27(9):1–10
7. Suma J, Mishra MK (2015) An improved direct AC–AC converter for voltage sag mitigation. IEEE Trans Ind Electron 62(1):21–29
8. Vegunta SC, Milanovic JV (2011) Estimation of cost of downtime of industrial process due to voltage sags. IEEE Trans Power Deliv 26(2):576–587
9. Bollen MHJ (1997) Characterisation of voltage sags experienced by three phase adjustable speed drives. IEEE Trans Power Deliv 12(4):1666–1671
10. Uddin S, Shareef H, Mohamed A, Hannan MA (2012) Analysis of voltage sag sensitivity of LED lamps. In: IEEE international conference on power and energy (PECon), 2012, pp 667–670
11. Maria de Carvalho Filho J, Policarpo G de Abreua J, Caminha Noronha JC, Arango H (2000) Analysis of power system performance under voltage sags. Electric Power Syst Res 55(3):211–218
12. Gómez-Lázaro E, Fuentes JA, Molina-Garciá A, Canas-Carreton M (2009) Characterization and visualization of voltage dips in wind power installations. IEEE Trans Power Deliv 24(4):2071–2078
13. Leiria A, Nunes P, Morcheda A, Teresa Correia de Barros M (2006) Induction motor response to voltage dips. Electric Power Syst Res76(8):676–680
14. IEEE recommended practice for monitoring electric power quality, IEEE Std. 1159-2009 (2009)
15. IEEE recommended practice for monitoring electric power quality, IEEE Std. 1159–1995 (1995)
16. Voltage characteristics of the electricity supplied by public distribution systems, European/British Standard EN (Euro Norms) BS/EN 50160, CLC, BTTF 68-6, November 1994.
17. Bollen MHJ, Sabin DD, Thallam RS (2003) Voltage sag indices—draft 5. Technical report, Working document for IEEE P1564, November
18. Becker C, Braun W, Jr., Carrick K, Diliberti T, Grigg C, Groesch J, Hazen B, Imel T, Koval D, Mueller D, St.John T, Conrad LE (1994) Proposed chapter 9 for predicting voltage sags (dips) in revision to IEEE Std 493, the gold book. Transmission and Distribution Conference, Chicago, IL, April, pp 7–14
19. Aggiag MMA, Tan RHG (2017) Evaluation of power quality experience in UCSI University North Wing Campus. In: 2nd international conference sustainable and renewable energy engineering (ICSREE), pp 129–133. IEEE Hiroshima

20. Upadhya M, Singh AK, Thakur P (2019) Wavelet based voltage sag characterization and detection method. In: 2019 international conference on electrical, electronics and computer engineering (UPCON), Aligarh, India, pp 1–6
21. Upadhya M, Singh AK, Thakur P, Nagata EA, Ferreira DD (2022) Mother wavelet selection method for voltage sag characterization and detection. Electric Power Syst Res 211:108246
22. Wang Z, Guo X, Li Z, Lin H, Chen G (2022) A fast voltage sag detection method based on second order generalized integrator and delay transform method. Energy Rep 8(13):549–556
23. Xiu J, Guangye X, Xiangping M, Guilin D (2021)Voltage sag detection method based on dq transform and complex wavelet transform. In: 2021 IEEE international conference on electrical engineering and mechatronics technology (ICEEMT), Qingdao, China, pp 429–434
24. Naidoo R, Pillay P (2007) A new method of voltage sag and swell detection. IEEE Trans Power Deliv 22(2):1056–1063
25. IEEE recommended practice for powering and grounding electronic equipment, IEEE Std. 1100-2005 (2005)
26. Popa GN, Iagăr A, Diniş CM (2017) Some power quality issues in power substation from residential and educational buildings. In: 2017 10th international symposium on advanced topics in electrical engineering (ATEE), Bucharest, Romania, pp 545–550
27. Verma P, Patel N, Nair N-KC (2016) Power quality impacts during CFL to LED transition. In: 2016 Australasian universities power engineering conference (AUPEC), Brisbane, QLD, Australia, pp 1–6
28. Lee K, Venkataramanan G, Jahns TM (2008) Modeling effects of voltage unbalances in industrial distribution systems with adjustable-speed drives. IEEE Trans Ind Appl 44(5):1322–1332